Animal Nutrition Conference of Canada (ANCC 2022)

Feed Production and Formulation Technologies: Considerations for a Sustainable Industry

Saskatchewan, Canada
10 – 12 May 2022

ISBN: 978-1-7138-5548-4

Printed from e-media with permission by:

Curran Associates, Inc.
57 Morehouse Lane
Red Hook, NY 12571

Some format issues inherent in the e-media version may also appear in this print version.

Animal Nutrition Society of Canada
150 Metcalfe Street, Suite 1301
Ottawa, Ontario K2P 1P1
Canada

Phone: (613) 241-6421
Fax: (613) 241-7970

info@anacan.org

Additional copies of this publication are available from:

Curran Associates, Inc.
57 Morehouse Lane
Red Hook, NY 12571 USA
Phone: 845-758-0400
Fax: 845-758-2633
Email: curran@proceedings.com
Web: www.proceedings.com

ANCC

2022 ANIMAL NUTRITION CONFERENCE OF CANADA

CNAC

2022 COLLOQUE DE NUTRITION ANIMALE DU CANADA

Proceedings
Cahier de conférences

10-12
May/mai 2022

Saskatoon
Saskatchewan

Feed production and formulation technologies: considerations for a sustainable industry

Technologies de production et de formulation des aliments du bétail : considérations pour une industrie durable

Pre-conference symposium sponsor
Commanditaire de symposium pré-colloque

ADM
Unlocking Nature. Enriching Life.

Animal Nutrition Association of Canada

ANAC

Association de nutrition animale du Canada

Welcome from ANAC

I am pleased to welcome you to the first ever hybrid edition of the Animal Nutrition Conference of Canada (ANCC). Whether you are able to join us in person or online, I am confident that this year's conference will provide you with exceptional learning and networking opportunities.

The Animal Nutrition Association of Canada is honoured to host the 6th edition of this world class event. Thanks to the dedication of the organizing committee, generosity of sponsors and amazing research shared from around the world, the ANCC has truly become recognized as "the" Canadian conference in livestock nutrition.

The livestock feed industry has faced many challenges over the last couple of years with the impacts of the global pandemic, droughts, floods, ongoing supply chain disruptions, war in the Ukraine and the highly pathogenic avian influenza outbreak. Despite these challenges, as nutritionists, you continued to adapt and innovate to ensure the best feed was available for Canadian livestock.

Moving into the future, increased pressures will be on the livestock industry to reduce methane emissions and production's impact on climate change. This year's conference will bring you some of the latest research and innovation in sustainable feed and nutrition practices to help our industry contribute to meeting the goals put forward by the federal government.

An event of this magnitude is not possible without the generous contributions of our many industry partners. Thank you to our sponsors who have shown incredible support for the conference despite the challenges faced by our industry over the past two years. We extend our appreciation as well to our dedicated speakers for sharing their expertise.

The Organizing Committee has once again succeeded in putting together an exceptional program and we hope that you will take the opportunity to enjoy all the opportunities the conference has to offer. Engage with colleagues and speakers in person or using the interactive digital tools and connect with the new generation of feed industry professionals during the student poster competition.

Enjoy learning and connecting at ANCC 2022 and I hope to see many of you again at next year's conference in Montreal, Quebec.

Melissa Dumont, agr.
Executive Director

Animal Nutrition Association of Canada

ANAC

Association de nutrition animale du Canada

Bienvenue de l'ANAC

Je suis heureuse de vous accueillir à la toute première édition hybride du Colloque de nutrition animale du Canada (CNAC). Que vous vous joigniez à nous en personne ou en ligne, je suis convaincue que le colloque de cette année vous offrira des possibilités exceptionnelles d'apprentissage et de réseautage.

L'Association de nutrition animale du Canada est honorée d'accueillir la 6e édition de cet événement de classe mondiale. Grâce au dévouement du comité organisateur, à la générosité des commanditaires et à l'incroyable partage de recherches provenant du monde entier, le CNAC est véritablement reconnu comme « le » colloque canadien en matière de nutrition animale.

L'industrie de l'alimentation animale a été confrontée à de nombreux défis au cours des deux dernières années avec les impacts de la pandémie mondiale, les sécheresses, les inondations, les perturbations continues de la chaîne d'approvisionnement, la guerre en Ukraine et l'épidémie de l'influenza aviaire hautement pathogène. Malgré ces défis, en tant que nutritionnistes, vous avez continué à vous adapter et à innover afin de garantir que les meilleurs aliments soient disponibles pour le bétail canadien.

À l'avenir, l'industrie des productions animales sera soumise à des pressions accrues pour réduire les émissions de méthane et l'impact de la production sur le changement climatique. L'événement de cette année vous fera découvrir les dernières recherches et innovations en matière de pratiques durables dans le domaine de l'alimentation animale et de la nutrition afin d'aider notre industrie à contribuer à l'atteinte des objectifs mis de l'avant par le gouvernement fédéral.

Un événement de cette ampleur n'est pas possible sans les généreuses contributions de nos nombreux partenaires industriels. Nous remercions nos commanditaires qui ont fait preuve d'un soutien incroyable pour la conférence malgré les défis auxquels notre industrie a été confrontée au cours des deux dernières années. Nous exprimons également notre reconnaissance à nos conférenciers dévoués qui ont accepté de partager leur expertise.

Le comité organisateur a, une fois de plus, réussi à mettre sur pied un programme exceptionnel et nous espérons que vous profiterez de toutes les opportunités que le colloque a à offrir. Engagez un dialogue avec vos collègues et les conférenciers en personne ou en utilisant les outils numériques interactifs et connectez-vous avec la nouvelle génération de professionnels de l'industrie de l'alimentation animale lors du concours d'affiches pour étudiants.

Profitez de l'apprentissage et de la connexion au CNAC 2022 et j'espère revoir beaucoup d'entre vous au colloque de l'année prochaine qui aura lieu à Montréal, au Québec.

Melissa Dumont, agr.
Directrice exécutive

Organizing Committee 2022 / *Comité organisateur 2022*

Welcome everyone!

On behalf of the Organizing Committee, we are honoured to welcome you to the 6th annual Animal Nutrition Conference of Canada.

The ANCC continues to build on its tradition of bringing together researchers, students and feed industry specialists to share expertise, ideas and camaraderie as well as the latest scientific and technological developments in the livestock industry. In order to continue the success of this conference, this year the Organizing Committee has selected the theme "Feed Production and Formulation Technologies: Considerations for a Sustainable Industry." In today's global economic climate, perhaps now more than ever, it is increasingly clear that we must work towards improving sustainability across all sectors of livestock production in order to better overcome both current and future challenges.

The Organizing Committee recognizes the importance that animal nutrition will continue to play in keeping Canadian agriculture sustainable, as well as the importance that technology will have in the advancement of our industry. On the same note, we see that diversity and sustainability mean advancements not only for our strongest commodities, but also for small ruminants, aquaculture and other smaller but still important sectors of the livestock industry. Therefore, the Organizing Committee has brought together a variety of presenters, both home-grown and international, established and early-career, to ensure that our attendees are provided with all the components necessary for networking and development within their specialties. In addition, we are proud to highlight novel research from Canadian students and the next generation of talent in the Canadian livestock industry.

The goal of this conference is to be inclusive and to provide a platform for exchange and interaction about the latest developments in feed formulation, feed production and feeding technologies. To do this, we will be offering hybrid-type attendance, and we have decided to create a roundtable session to facilitate an enriched and unique opportunity for our attendees to interact with a panel of experts on today's most noteworthy topics.

On a final note, the Organizing Committee would like to thank the Animal Nutrition Association of Canada, and the ANCC 2022 sponsors for their support and for giving us the opportunity to connect and share with our colleagues in these unparalleled times.

Thank you to all our attendees for joining us, have fun and stay safe.

Bienvenue à tous!

Au nom du Comité organisateur, nous avons l'honneur de vous souhaiter la bienvenue au sixième Colloque annuel sur la nutrition animale du Canada.

Le CNAC continue de croître en respectant sa tradition de réunir des chercheurs, des étudiants et des spécialistes de l'industrie de l'alimentation pour animaux pour partager des connaissances, des idées et de la camaraderie, ainsi que les dernières avancées scientifiques et technologiques dans l'industriedu bétail. Dans le but de prolonger le succès de ce colloque, le comité organisateur a choisi cette année le thème « Technologies de production et de formulation des aliments du bétail : considérations pour une industrie durable ». Dans le climat économique mondial actuel, peut-être maintenant plus que jamais, il est de plus en plus clair que nous devons travailler à l'amélioration de la durabilité dans tous les aspects de la production animale afin de mieux surmonter les difficultés actuelles et futures.

Le Comité organisateur reconnaît l'importance du rôle que la nutrition animale continuera de jouer dans le maintien de la durabilité de l'agriculture canadienne, ainsi que la place qu'occupera la technologie dans l'avancement de notre industrie. Dans le même ordre d'idées, on constate que la diversité et la durabilité sont synonymes d'avancées, et ce, non seulement en ce qui concerne nos produits les plus importants, mais aussi pour ce qui est des petits ruminants, de l'aquaculture et d'autres segments plus modestes, mais néanmoins importants, de l'industrie du bétail. C'est pourquoi le Comité organisateur a réuni une variété de présentateurs, tant nationaux qu'internationaux, qu'ils soient bien établis ou en début de carrière, afin de s'assurer que nos participants aient à leur disposition tous les éléments nécessaires au développement et au réseautage au sein de leurs domaines respectifs. De plus, nous sommes fiers de mettre en valeur les recherches novatrices des étudiants canadiens et de la prochaine génération de talents de l'industrie canadienne du bétail.

L'objectif du colloque est de favoriser l'inclusion et de fournir une plateforme d'échange et d'interaction sur les derniers développements en matière de formulation d'un régime pour animaux, de production d'aliments pour animaux et de technologies d'alimentation. Pour ce faire, nous proposerons une participation de type hybride et avons opté pour la création d'une table ronde afin d'offrir à nos participants une expérience enrichissante et une occasion unique d'interagir avec un groupe d'experts sur les sujets les plus pertinents du moment.

Pour conclure, le Comité organisateur tient à remercier l'Association de nutrition animale du Canada et les commanditaires du CNAC de 2022 de nous avoir offert leur soutien et donné l'occasion de nous réunir et de partager entre collègues en cette période particulière.

Merci à tous nos participants de se joindre à nous. N'oubliez pas de vous amuser, et restez en sécurité!

Renée Petri, Agriculture and Agri-Food Canada (Program Chair/*Présidente du programme*)

Holly McGill, Wallenstein Feed & Supply Ltd. (Sponsorship Chair/*Présidente des commandites*)

Rex Newkirk, University of Saskatchewan (Academic Chair/*Président de l'académie*)

Renée Benczik, Wallenstein Feed & Supply Ltd. (Student Events Chair/*Présidente des événements pour étudiants)*

Jeff Bond, Prairie Micro-tech Inc. (PMT)

Murray Feist, Saskatchewan Ministry of Agriculture

Sarah Hopkins, Animal Nutrition Association of Canada (ANAC)

Dinesh Kumar, Federated Cooperatives Ltd.

Marie-Pierre Létourneau-Montminy, Université Laval

Keshia Paddick, More Than Just Feed Inc.

Ilona Parenteau, Halchemix Canada Inc.

Sharon Robinson, Halchemix Canada Inc.

Nancy Stonos-Smith, Animal Nutrition Association of Canada (ANAC)

Kristin Thompson, New Life Mills

Table of Contents / *Table des matières*

Pre-conference Symposium / *Symposium pré-colloque*

Conference / *Colloque*

Graduate Student Posters / *Affiches des étudiants diplômés*

Monogastric

Ruminant

List of Advertisers / *Liste d'annonceurs*

2022 ANIMAL NUTRITION CONFERENCE OF CANADA

ANCC 2022 ANIMAL NUTRITION CONFERENCE OF CANADA

MAY 10 TO MAY 12, 2022 – DELTA SASKATOON DOWNTOWN, SASKATOON, SK

Feed production and formulation technologies: considerations for a sustainable industry

ANCC INFORMATION DESK HOURS

May 10: 2:00 pm to 6:30 pm • May 11: 7:00 am to 5:30 pm • May 12: 7:00 am to 2:15 pm

Conference registration packages will be available at the ANCC information desk.

TUESDAY, MAY 10

Student Networking Event - 3:30 to 4:30 pm

This is an opportunity for students to learn about careers in the feed industry and network with potential employers.

Welcome Cocktail - 4:30 to 6:30 pm

Join us for an evening of networking and animated discussions with industry colleagues. Registration packages will also be available for collection.

WEDNESDAY, MAY 11

PRE-CONFERENCE SYMPOSIUM
EMERGING TRENDS IN ANIMAL NUTRITION
Sponsored by: ADM

ADM Unlocking Nature. Enriching Life.

7:15	Hot breakfast	
8:05	Introduction	
8:10	Influence of antibiotic alternatives on immunity and microflora in support of poultry gut health	**Dr. Morgan Farnell** Texas A&M University
8:50	Intestinal permeability, immune response and inflammation in dairy – nutritional modulations	**Dr. Lance Baumgard** Iowa State University
9:30	Viral viability and infectivity in contaminated feed and opportunities for feed additives to mitigate the challenge	**Dr. Scott Dee** Pipestone Veterinary Services
10:10	Health break	
10:40	Moderating the effect of coccidia and necrotic enteritis challenge using non-pharmaceutical means	**Dr. Milan Hruby** and **Dr. Mohamad Mortada*** ADM Animal Nutrition
11:20	Methane, cows, and climate change: California dairy's path to climate neutrality	**Dr. Frank Mitloehner** University of California Davis
12:00	Lunch	

Opening Plenary

1:05	Opening remarks from ANAC	
1:10	Organizing committee welcome	
1:15	**JM Bell Memorial Lecture**: Prospects for climate neutral beef and dairy production in Canada	**Dr. Karen Beauchemin** Agriculture and Agri-Food Canada
2:15	Sustainable livestock feeding and management: what changes are needed?	**Dr. Donald Broom** University of Cambridge
3:00	Health break	
3:30	On the development of online tools to support the production of sustainable animal feeds	**Dr. Dominique Bureau** University of Guelph
4:15	Feed processing for a profitable and sustainable feed business	**Dr. Menno Thomas** Zetadec - Wageningen University & Research
5:00	**ANAC Scholarship Lecture**: The influence of transition diet energy and protein content on colostrum and early lactation milk composition and bioactive compound concentrations in Holstein dairy cattle	**Amanda Fischer-Tlustos** Graduate Student, University of Guelph
5:15 - 7:15	**ANCC 2022 Reception:** Enjoy an evening of food, drinks and networking. Visit the Industry Partner Showcase and Graduate Student Poster Exhibition.	

Animal Nutrition Association of Canada ANAC Association de nutrition animale du Canada

** Speaker will present virtually.*

Please note at this time three speakers (one pre-conference and two main conference) will be presenting virtually while all others will be at the conference in person.

THURSDAY, MAY 12

SIMULTANEOUS SESSIONS

Please note that the monogastric and ruminant session will be held simultaneously.

Monogastric Session

7:15	Hot breakfast	
8:10	Opening remarks by session chair	
8:15	Intersection of the nutritive value and bioactive potential of two emerging classes of novel feed ingredients: insect and algae products	**Dr. Stephanie Collins** Dalhousie University
9:00	Viral transmission through feed	**Dr. Chad Paulk** Kansas State University
9:45	Health break	
10:15	Appreciating the dynamics of pellet quality improvements to nutrient segregation in poultry houses	**Dr. John Boney** Pennsylvania State University
11:00	The potential for feed processing to reduce ergot toxicity	**Dr. Denise Beaulieu** University of Saskatchewan
11:45	Growth performance, digesta pH and organ weight of weaned pigs fed barley grain differing in fermentable starch and fibre	**Joaquin Sanchez Zannatta** Graduate Student, University of Alberta
12:00	Lunch	

Ruminant Session

7:15	Hot breakfast	
8:10	Opening remarks by session chair	
8:15	Feed formulation for sustainable agriculture	**Dr. Greg Thoma*** University of Arkansas
9:00	Nutritional strategies to mitigate enteric methane emissions from dairy cows: state of knowledge and new perspectives	**Dr. Chaouki Benchaar*** Agriculture and Agri-Food Canada
9:45	Health break	
10:15	Understanding feed efficiency in the feedlot	**Dr. Katie Wood** University of Guelph
11:00	Precision mineral nutrition for dairy cows	**Dr. William Weiss** Ohio State University
11:45	High mycotoxin levels in wheat grain and their effects on beef cattle ruminal fermentation, performance, and carcass traits	**Renée Bierworth** Graduate Student, University of Saskatchewan
12:00	Lunch	

Closing Plenary

1:05	Remarks by session chair	
1:15	Round Table Discussion with Dr. Karen Beauchemin, Dr. Denise Beaulieu, Dr. Donald Broom, Dr. Frank Mitloehner and Dr. Katie Wood	Moderated by: **Dr. Rex Newkirk** University of Saskatchewan
2:00	Feed formulation in the future: quantum change or incremental steps forward	**Dr. John Patience** Iowa State University
2:45	Closing remarks	

MISSED A SESSION?

Recordings of all presentations will be available for on-demand viewing by both virtual and in-person participants.
A password to access the virtual conference platform will be provided by email before the start of the conference.

** Speaker will present virtually.*
Please note at this time three speakers (one pre-conference and two main conference) will be presenting virtually while all others will be at the conference in person.

Technologies de production et de formulation des aliments du bétail : considérations pour une industrie durable

HEURES D'OUVERTURE DU BUREAU D'INFORMATION DU CNAC
10 mai : 14h00 à 18h30 • 11 mai : 7h00 à 17h30 • 12 mai : 7h00 à 14h15
Les trousses d'inscription à la conférence seront disponibles au bureau d'information du CNAC.

LE MARDI 10 MAI

Événement de réseautage pour les étudiants - 15H30 À 16H30
Il s'agit d'une occasion pour les étudiants de se renseigner sur les carrières dans l'industrie de l'alimentation animale et de réseauter avec des employeurs potentiels.

Cocktail de bienvenue - 16H30 À 18H30
Joignez-vous à nous pour une soirée de réseautage et discussions animées avec des collègues de l'industrie.

LE MERCREDI 11 MAI

SYMPOSIUM PRÉ-COLLOQUE
NOUVELLES TENDANCES EN MATIÈRE DE NUTRITION ANIMALE
Commandité par : ADM

7h15	Petit-déjeuner chaud	
8h05	Introduction	
8h10	Influence des solutions de remplacement aux antibiotiques sur l'immunité et la microflore pour favoriser la santé intestinale chez les volailles	**Dr Morgan Farnell** Texas A&M University
8h50	Perméabilité intestinale, réaction immunitaire et inflammation chez les vaches laitières – modulations nutritionnelles	**Dr Lance Baumgard** Iowa State University
9h30	Viabilité et infectiosité virales dans les aliments du bétail contaminés et possibilités d'atténuer le problème grâce aux additifs alimentaires pour le bétail	**Dr Scott Dee** Pipestone Veterinary Services
10h10	Pause-santé	
10h40	Modérer l'effet d'une exposition aux coccidies et à l'entérite nécrotique par l'emploi de solutions non pharmaceutiques	**Dr Milan Hruby** et **Dr Mohamad Mortada*** ADM Animal Nutrition
11h20	Méthane, vaches et changement climatique : La filière laitière californienne sur la voie de la neutralité climatique	**Dr Frank Mitloehner** University of California Davis
12h00	Dîner	

Plénière d'ouverture

13h05	Propos d'ouverture de l'ANAC	
13h10	Bienvenue du comité organisateur	
13h15	**Conférence JM Bell en nutrition animale :** Perspectives d'une production bovine et laitière carboneutre au Canada	**Dre Karen Beauchemin** Agriculture et Agroalimentaire Canada
14h15	Alimentation et gestion durables du bétail : quels sont les changements nécessaires?	**Dr Donald Broom** University of Cambridge
15h00	Pause-santé	
15h30	La question du développement d'outils en ligne pour soutenir la production des aliments durables pour le bétail	**Dr Dominique Bureau** University of Guelph
16h15	La transformation des aliments du bétail pour une entreprise rentable et durable	**Dr Menno Thomas** Zetadec - Wageningen University & Research
17h00	**Conférence de la bourse d'études de l'ANAC** : Influence de la teneur en énergie et en protéines de la ration de transition sur la composition et les concentrations en composés bioactifs du colostrum et du lait de début de lactation chez les vaches laitières Holstein	**Amanda Fischer-Tlustos** Étudiant de cycle supérieur, University of Guelph
17h15-19h15	**Réception du CNAC 2022 :** Venez profiter d'une soirée où gastronomie et réseautage seront à l'honneur. Visitez le salon des partenaires de l'industrie et l'exposition d'affiches pour étudiants diplômés.	

** Ce conférencier fera une présentation virtuelle.*
Veuillez noter qu'à l'heure actuelle, seulement trois conférenciers feront des présentations virtuelles (soit un de la pré-conférence et deux de la conférence principale). Tous les autres conférenciers seront en présentiel lors de la conférence. Toutes les sessions seront présentées en anglais.

DU 10 AU 12 MAI 2022 – DELTA SASKATOON DOWNTOWN, SASKATOON (SK)

LE JEUDI 12 MAI

SÉANCES SIMULTANÉES

Veuillez noter que les séances sur les monogastriques et les ruminants seront offertes simultanément.

Séance sur les monogastriques

7h15	Petit-déjeuner chaud	
8h10	Propos d'ouverture	
8h15	Intersection de la valeur nutritive et du potentiel bioactif de deux classes émergentes d'ingrédients alimentaires nouveaux : les produits à base d'insectes et les produits à base d'algues	**Dre. Stephanie Collins** Dalhousie University
9h00	Transmission virale par l'intermédiaire des aliments	**Dr Chad Paulk** Kansas State University
9h45	Pause-santé	
10h15	Appréciation de la dynamique de l'amélioration de la qualité des granulés dans un contexte de ségrégation des ingrédients dans les poulaillers	**Dr John Boney** Pennsylvania State University
11h00	Possibilité de réduire la toxicité de l'ergot par le traitement des aliments	**Dre Denise Beaulieu** University of Saskatchewan
11h45	Performances de croissance, pH du digesta et poids des organes chez des porcs sevrés nourris avec des grains d'orge de teneurs différentes en amidon et en fibres fermentescibles	**Joaquin Sanchez Zannatta** Étudiant de cycle supérieur, University of Alberta
12h00	Dîner	

Séance sur les ruminants

7h15	Petit-déjeuner chaud	
8h10	Propos d'ouverture	
8h15	Formulation des aliments du bétail pour une agriculture durable	**Dr Greg Thoma*** University of Arkansas
9h00	Stratégies nutritionnelles visant à atténuer les émissions de méthane d'origine entérique chez les vaches laitières : état des connaissances et nouvelles perspectives	**Dr Chaouki Benchaar*** Agriculture et Agroalimentaire Canada
9h45	Pause-santé	
10h15	Comprendre l'efficacité alimentaire au parc d'engraissement	**Dre Katie Wood** University of Guelph
11h00	Nutrition minérale de précision pour bovins laitiers	**Dr William Weiss** Ohio State University
11h45	Teneurs élevées en mycotoxines dans le grain de blé et effets sur la fermentation ruminale, les performances et les caractéristiques de carcasse chez les bovins de boucherie	**Renée Bierworth** Étudiante de cycle supérieur, University of Saskatchewan
12h00	Dîner	

Plénière de clôture

13h05	Annonce des gagnants du concours d'affiches pour étudiants	
13h15	Table ronde avec Dre Karen Beauchemin, Dre Denise Beaulieu, Dr Donald Broom, Dr Frank Mitloehner et Dre Katie Wood	Animée par : **Dr Rex Newkirk** University of Saskatchewan
14h00	La formulation des aliments du futur : changement radical ou avancées progressives	**Dr John Patience** Iowa State University
14h45	Propos de clôture	

VOUS AVEZ MANQUÉ UNE SESSION?

Les enregistrements de toutes les présentations seront disponibles pour un visionnement sur demande par les participants virtuels et en personne. Un mot de passe pour accéder à la plateforme de conférence virtuelle sera fourni par courriel avant le début de la conférence.

** Ce conférencier fera une présentation virtuelle.*

Veuillez noter qu'à l'heure actuelle, seulement trois conférenciers feront des présentations virtuelles (soit un de la pré-conférence et deux de la conférence principale). Tous les autres conférenciers seront en présentiel lors de la conférence. Toutes les sessions seront présentées en anglais.

Pre-conference Symposium
Symposium pré-colloque

ANCC
CNAC

Influence of Antibiotic Alternatives on Immunity and Microflora in Support of Poultry Gut Health

Influence des solutions de remplacement aux antibiotiques sur l'immunité et la microflore pour favoriser la santé intestinale chez les volailles

Morgan Farnell
Department of Poultry Science, Texas A&M University, College Station, TX, USA
morgan.farnell@ag.tamu.edu

Pre-conference sponsored by / Pré-colloque commandité par :

ADM®

Abstract

Poultry are raised intensively to reduce production costs, while creating a wholesome and inexpensive source of animal protein for the consumer. The poultry breeder has primarily focused on advancing genetics and nutrition to achieve these phenomenal performance and economic goals. The sub-therapeutic administration of antibiotics was used to maintain avian gut health and thereby improved performance. However, antibiotic usage in food production animals has been significantly curtailed due to consumer demand and government regulations. Alternatives such as vaccines, probiotics, botanicals, nutraceuticals, and immune modulators are employed to fill this niche. These products may also modify immune function and the microbiome, directly or indirectly affecting animal welfare, sustainability, gut health and food safety. Technology has recently advanced giving researchers tools to better elucidate the mechanism of action of these products. The presentation will give attendees a brief overview of the major components of the avian immune system. We will then discuss the relationship between genetics, nutrition, immunity, and microflora; how they can impact performance and how these systems may be manipulated using antibiotic alternatives.

Résumé

Les volailles sont élevées de manière intensive afin de réduire les coûts de production et de créer une source saine et peu coûteuse de protéines animales pour le consommateur. L'industrie s'est principalement concentrée sur les progrès de la génétique et de la nutrition pour atteindre ces objectifs colossaux de performance et d'économie. L'administration d'antibiotiques à doses sousthérapeutiques a été utilisée pour maintenir la santé intestinale des oiseaux et ainsi améliorer leurs performances. Cependant, l'utilisation d'antibiotiques chez les animaux destinés à la production alimentaire a été considérablement réduite à la suite des pressions exercées par les consommateurs et les gouvernements. D'autres solutions, telles que les vaccins, les probiotiques, les produits botaniques, les nutraceutiques et les modulateurs immunitaires, occupent maintenant ce créneau. Ces produits peuvent également modifier la fonction immunitaire et le microbiome, intervenant ainsi directement ou indirectement sur le bien-être des animaux, la durabilité de la production, la santé intestinale et la salubrité des aliments. Récemment, la technologie a progressé, donnant aux chercheurs des outils pour mieux élucider le mécanisme d'action de ces produits. Cet exposé donnera aux participants un bref aperçu des principaux composants du système immunitaire aviaire. Nous discuterons ensuite de la relation entre la génétique, la nutrition, l'immunité et la microflore, de l'influence de chacun sur les performances et de la possibilité de manipuler ces systèmes en utilisant des solutions de remplacement aux antibiotiques.

Introduction

The poultry industry has experienced unprecedented growth in the past 70 years due to improved genetics, nutrition, housing, water quality and veterinary care driven by post-WWII consumer demands (Havenstein, et al., 2003). Modern-day poultry efficiently convert feed to meat and eggs, but genetic selection for improved performance may have inadvertently selected for a negative correlative response to other traits such as immune function (Cheema et al., 2003). The reader may consider what occurs when they are ill to understand the phenomenon. When an individual is sick, they do not eat, or drink and activity is reduced. Through a process called cachexia, the body uses amino acids and energy from muscles and fats, which is incompatible with growth and reproduction. Therefore, it is reasonable why selecting for a feed efficient bird may not always result in improved immune function.

Poultry are exposed to numerous bacteria, fungi, viruses, protozoa, and parasites throughout their production cycle (Kogut et al. 2020). These microbes can replicate and mutate very quickly, presenting a challenge for the immune system. They have evolved over millions of years and have developed many ways to evade detection and destruction. Further reading on pathogen evasion mechanisms can be reviewed in Finlay and McFadden's manuscript (2006). While many would consider the integument to have the greatest exposure to microbes, human mucosal surfaces are estimated to have a total surface area of approximately four hundred square meters and harbor over 13 trillion bacteria in the hindgut (Murphy et al., 2022). Mucosal surfaces have the greatest exposure to the outside world; therefore, most of the immune system can be found nearby in mucosal or gut associated lymphoid tissues (MALT/GALT). The purpose of this proceedings is to give a brief immunology primer and then discuss the proposed mechanisms of activity of antibiotic alternatives and their potential effects on the microbiota and gut health. Please be reminded that the simple answer is not always the most correct answer, but we will cover the topic at 30,000 feet to give the reader a basic understanding.

Innate Immunity

For ease of understanding, the immune system can be divided into an innate and adaptive immune system. Innate immunity begins with physical barriers to the outside world such as skin, mucus, stomach acid, digestive enzymes, nutrient binding (ovotransferrin), tears (lysozyme), changes in pH, oxygen tension (foregut versus hindgut), competitive exclusion by microflora, antimicrobial peptides (e.g., defensins produced by Paneth cells) and temperature differences. The innate immune system is characterized as being fast acting, non-specific and "what you are born with." It is found in more primitive organisms such as fruit flies and was marginalized until the recent discovery of pattern recognition receptors and the understanding that the innate leads the acquired immune system (Janeway and Medzhitov, 2002). Cells can communicate by producing a multitude of soluble factors, called cytokines, or membrane bound receptors (Kaiser et al., 2005). A subgroup of these cytokines are called chemokines which allow for movement of immune cells to tissues that are damaged or infected. An example of this, would be a gut epithelial cell infected with a bacterium causing it to secrete CXCL-8 (formerly interleukin-8 or IL-8) to promote chemotaxis of immune cells to the site of infection. Microbial associated molecular peptides (MAMPs) are cues immune cells may use to identify a potential pathogen via germ line encoded pattern recognition receptor (PRRs). These MAMPs are typically microbial components that cannot be changed, such as peptidoglycan which gives structure to a bacterial cell membrane. These MAMPs (formerly pathogen associated molecular patterns, PAMPs) allow for a phagocyte to discover, consume, and kill a microbe. In addition to MAMPs, danger associated molecular patterns (DAMPs) are needed for an effective immune response to be mounted. These DAMPs are necessary redundancies. Most microbes, that an individual may encounter, are not harmful but they still exhibit these MAMPs.

Heterophils, macrophages and dendritic cells are white blood cells that eat (or phagocytize) and kill pathogens. Heterophils, the avian equivalent to a mammalian neutrophil, respond to an infection very quickly and kill microbes through oxidative burst, nitric oxide, degranulation and netosis. These phagocytes cause inflammation by releasing cytokines and other immunogenic factors in addition to what is released when the pathogen is killed, such as nucleic acids. Macrophages and dendritic cells arrive shortly after the heterophils. They can kill microbes, like the heterophil, but they also have the advantage of serving as an antigen presenting cell. What this means, is that they can present a small piece or epitope of the pathogen via MHC II (major histocompatibility complex) to other immune cells to coordinate or direct the adaptive immune response. Dendritic cells can cover a large surface area and interact with more cells due to its dendrites, which are analogous to roots of a tree. The natural killer cell is comes from a lymphoid progenitor cell that kills abnormal, or virus infected cells through programmed cell death. These cells are identified by unique PRRs and the absence of MHC I, which is present on all normal nucleated cells. MHC I present endogenous antigens such as a virus fragments as well as self-antigens. This is useful, as a phagocytic cell would not be able to determine that a cell was infected using its PRRs.

Mast cells, basophils and eosinophils are granulocytes that can be activated by the antibody isotype IgE, PRRs and complement. Allergic individuals are called atopic and may have up to 10 times the normal amount as the rest of the population. While severe allergies are harmful, these atopic individuals may have increased resistance to parasitic infections. Mast cells are the primary immune cell that mediates inflammation, releasing factors such as histamine, leukotrienes, and arachidonic acids. Inflammation is characterized by redness, swelling, heat and pain. The process prepares a tissue for battle by allowing for the migration of fluids, nutrients, and white blood cells to an injured or infected area. Eosinophils and basophils primarily focus on multicellular parasites, such as helminths (reviewed by Stone et al., 2010).

Complement may be described as innate humoral immunity. It's a complex system of 35 proteins that may be triggered by MAMPs, or antibody bound to its respective antigen. The ultimate effect is to produce holes in the offending organism, via the membrane attack complex (MAC), allowing for leakage of it internal contents which also will be immunogenic. Complement opsonins may also act similarly to an antibody by binding to a pathogen and serving as a ligand for the complement receptor on phagocytes.

Adaptive Immunity

The adaptive immune system can take longer to respond to infection but is extremely specific and diverse. Estimates vary from 10^{11} to 10^{18} unique epitopes that can be identified by the adaptive immune system, which is an incredible number! While the innate system is always ready to respond quickly, the adaptive immune system can take weeks to months to mount an effective immune response as it goes through a complicated and stringent negative selection process where 95-98% of the lymphocytes are destroyed. It also has the added advantage of creating memory cells, which allows for vaccination to occur once a viable solution has been found.

The cells of the adaptive immune system are lymphocytes called B and T cells. Glick and colleagues discovered B cells by serendipity when they discovered that bursectomized chickens were not able to produce antibodies (1956). It was later discovered that the bursa of Fabricius allowed for a subset of lymphocytes, which produce antibodies, to develop. Its equivalent has not been found in mammals, where B cell maturation occurs in the bone marrow. Antibodies are extremely specific to what they bind to. They may be found on mucosal surfaces, in circulation, in tissues and bound to cells receptors. In a best-case scenario, they may neutralize a pathogen on a mucosal surface allowing it to simply float away in the lumen with no added response required. They may also help cells of the innate immune system to become much more specific by a process called antibody dependent cellular cytotoxicity, in effect serving as a specialized receptor for an innate immune cell.

T cells were later discovered and found to mature in the thymus (Cooper, 2015). These cells primarily produce cytokines to direct, improve or reduce the immune response. The major subtypes are T helper 1, T helper 2, T helper 17, T regulatory and cytotoxic T cells (Th1, Th2, Th17, Treg, and Tc). T helper 1 lymphocytes facilitate cell mediated immunity. These cells produce cytokines that help phagocytic cells to better respond to an infection. T helper 2 cells drive the humoral immune response helping B cells to produce additional and over time, better antibodies. T helper 17 cells produce a cytokine called IL-17 which improves neutrophil/heterophil function. These three cell types would be categorized as mediating a type I, II, or III immune response, respectively. T regulatory cells produce IL-10 and TGF-beta to return cell mediated immunity to baseline conditions prior to an infection. Interestingly, Th1 and Th2 cells produce factors that inhibit the other, so they may act in a regulatory fashion as well. Cytotoxic T cells are like natural killer cells functionally, but they are much more specific searching for a distinct epitope instead of a MAMP or absence of MHC I expression. These cells are particularly important as viruses can be difficult to identify when hiding inside of the host's cell. Cytotoxic T cells make use of MHC I which presents endogenous antigens from within a cell. Some of these may be self-antigen, but it may also present material related to production of a virus. Fortunately, all lymphocytes have been intensively selected to only respond to pathogens and not self-antigen. These cells are especially important when cells become malignant. It is thought that incidence of cancer increases as an individual ages to due to decreased immune surveillance.

Lymphocyte Recirculation/Antigen Trapping

Blood vessels can be considered leaky pipes that transport nutrients to various tissues. When the pressure increases, as seen with hypertension or ascites syndrome more leakage will occur, and tissues swell. Fortunately, these tissues are drained by an efficient lymphatic system that returns the extracellular fluid to the heart via the thoracic duct. There is some similarity between the lymphatic system and a water shed where smaller streams combine forming large rivers. Multiple afferent lymphatic vessels are consolidated at lymph nodes which will have only one efferent vessel. This system provides a very efficient means for white blood cells to travel and to communicate with each other. Interestingly, while the chicken has an effective lymphatic system, it does not have defined lymph nodes. It is thought that other lymphatic tissues found in the Harderian gland, Meckel's diverticulum, Peyer's patches, cecal tonsils, and bursa of Fabricius make up for the lack of avian lymph nodes (Kim and Lillehoj, 2019). Lymphocytes can make copies of themselves within these specialized environments leading to lymph node enlargement that may be observed during an infection. Lymphocyte recirculation is especially important as there may be only one B cell that has to meet only one T cell that are both specific for a single extremely specific epitope to be effective. Once this match has been made, these cells will make many copies of themselves to expand the immune response. They will refine the receptors and antibodies that are produced to mount an improved immune response over time. Further, some of these cells will become memory cells, which may last a lifetime, and are ready to respond quickly upon a secondary infection. Macrophages and dendritic cells also use this recirculation phenomenon, presenting new epitopes to these lymphocytes. Dendritic cells may migrate from the lamina propria of the gut or the epidermis of the skin (Langerhans's cells) to a lymph node to share this information with lymphocytes.

Antibiotic Alternatives

Antibiotics were used in animal production for many years. It is thought that these products fed at a sub-therapeutic level enhanced growth production by reducing microbial competition for nutrients. They also prevent infections, such as caused by *Clostridium perfringens* in necrotic enteritis, which would counter nutrient absorption by the gut (reviewed in Bedford, 2000). Consumer demand and government restrictions have significantly curtailed the use of antibiotic usage in farm animals. It has been contended that these impediments to veterinary care may prevent sick animals from being treated, thereby causing a reduction in animal welfare, and increasing production costs (Singer et al., 2019). Antibiotic alternatives are being looked at closely to fill this void and improve animal health.

Vaccination may be one of the most successful alternatives to antibiotics, but it is not always a perfect solution. Despite significant growth rates and size, broiler chickens are still juveniles. They are harvested at 4-9 weeks of age, while puberty occurs much later at 18-22 weeks. This shortened timeline may not allow for adaptive immunity to develop before the bird is exposed to a pathogen. Maternal antibodies, found in the yolk sac and in circulation, may interfere with vaccination at the hatchery, which is the most convenient time to vaccinate. Microbes have many ways to evade the immune system, changing what it "looks" like making a vaccine ineffective. Poultry appear to be challenged with a constant onslaught of viruses. By the time that the etiology has been decided, the virus has run its course and a new one arises. This obviously makes vaccination difficult, but several of these viruses are also immunosuppressive, which could have a negative effect for vaccinating against another pathogen. Added booster vaccinations are usually needed to achieve maximum protection against a pathogen, but this adds to the cost, labor and handling stress to the animal. Predicting what pathogens will affect a flock can be exceedingly difficult, but a significant amount of time is also needed for the bird's immune system to be properly prepared to defend itself. Vaccinating during an infection won't be helpful and could exacerbate the problem. There are many types of vaccines available. They may be classified as living or not living. Both vaccine types will have the necessary PAMPs to be recognized by PRRs but sometimes that isn't enough. Living vaccines can produce or express DAMPs which are a necessary secondary signal to prevent the immune system from "misfiring". Unfortunately, low dose or weakened live vaccines can also cause the disease in rare cases. Non-living vaccines typically need something else to be noticed by the immune system. A material known as an adjuvant (Latin adjuvere - to help) is used in combination with the vaccine to help the immune system to recognize these PAMPs. These adjuvants cause inflammation and may cause localized tissue damage in extreme circumstances, which could be a negative if it affects muscles consumed as meat.

While a modern hatchery is an environmentally controlled and clean environment, it isolates the neonate from the mother hen thereby interfering with establishment of a healthy microbiota via coprophagy (Nurmi and Rantala, 1973). Prebiotics and probiotics are a means to improve the gut microflora of an individual. Prebiotics are indigestible nutrients to the host but can be used by and favor the growth of certain beneficial bacteria. Probiotics are beneficial bacteria that may be administered via the feed or water to colonize the gut with beneficial bacteria. Both solutions work by a process known as competitive exclusion. Microbes compete for attachment space and nutrients. They produce substances, such as antimicrobial bacteriocins, to kill and out compete other microbes. A major issue with antibiotic usage is that these beneficial bacteria may be indiscriminately killed allowing the gut to be reseeded with bad microbes causing diarrhea and poor nutrient absorption.

Yeast fermentation products were initially produced as a by-product of alcohol production. These products can be fractioned into yeast cell wall fragments, mono-oligosaccharide (MOS), and beta-glucans (reviewed by Sanchez et al., 2021). Yeast beta-glucans have been shown to be immune-stimulatory in chickens (Lowry et al., 2005). Mono-oligosaccharides bind pathogenic bacteria in the gut, thereby neutralizing the pathogen preventing attachment to the epithelium (Peisker, 2017).

Fungi have been shown to be immunogenic and recognized by numerous PRRs as reviewed by Patin and colleagues (2019), but they may also supply nutrients for beneficial bacteria and the host.

Phytogenic feed additives are derived from plants that may have beneficial effects. Essential oils, tannins, herbs and spices are examples of these products. They are reported to have immune-modulatory and antimicrobial properties. El-Hack and colleagues recently published an extensive review concerning this category of antibiotic alternatives (2022)

Bacteriophages are viruses that infect a specific host bacterium. They have shown to be effective when fed to live birds or applied to carcasses but must also be paired with the correct microbe which can be problematic. There are estimated to be more than 10 times as many bacteriophages than bacteria (reviewed by Zbikowska et al., 2020). Therefore, it's possible that a bacteriophage may exist for every or at least many pathogens.

Meat birds typically undergo 8-10 hours of feed withdrawal prior to processing to reduce contamination by ingesta and fecal matter. Unfortunately, the birds are still hungry and begin to eat litter and darkling beetles which may carry foodborne pathogens. They will also consume more water to reach satiety. Products have been administered via the drinker system such as pre/probiotics and acids to counteract this phenomenon. Byrd and colleagues evaluated the administration of acetic, formic and lactic acids in drinking water and found the most improvement with lactic acid reducing Campylobacter and Salmonella contamination in broilers. Pineda and colleagues had some success reducing Salmonella Heidelberg biofilm and cecal colonization using sodium bisulfate in water (2021). While acidification of the foregut is possible, it is quickly countered by the dilution by ingesta and the bicarbonate buffering system. Foodborne bacteria can survive a wide range of pH, which makes them difficult to kill without injuring the host or significantly reducing water consumption.

The importance of nutrition should not be discounted when discussing immunity. Energy and amino acids are needed for an effective immune response, including clonal expansion, production of antibodies, complement and cytokines. While least cost feed formulation allows for meeting all the bird's nutrient needs at the lowest cost, additional factors may need to be considered concerning immunity, microbiota and overall gut health. Cod liver oil may be one of the earliest examples of using a nutraceutical to treat tuberculosis in the 1800's (Grad, 2004). Cod liver oil is rich in omega-3 fatty acids and vitamins A and D which are both touted to affect immunity. Calder reviews the importance of zinc, copper, iron and vitamins A, B6, B9, B12, C, D and E to immune function in his recent paper (2021). While malnutrition will have negative health effects, fasting or reduced consumption of calories can be beneficial. In fact, overeating can cause chronic inflammation which can lead to other maladies such as cancer and heart disease (Collins and Belkaid, 2022).

Enzymes can make nutrients more available to the host and potentially the microflora within the host. Fernandez and colleagues reported that the addition of xylanase to a wheat-based broiler diet reduced mucin viscosity and colonization by *Campylobacter jejuni*. Campylobacter use mucin as a nutrient source and the xylanase increased production of mucin by goblet cells within the gut reducing viscosity and creating a less hospitable environment for the pathogen (2000).

Conclusion

It is amazing to think that trillions of microorganisms are living on and within our bodies trying to eke out an existence. They have had millions of years to adapt and improve through random mutations and shared genes. Most microbes do no harm and are in fact beneficial for our development and nutrient absorption. These organisms can even influence behavior as seen with infections such as rabies and *Toxoplasma gondii*. The immune system is a formidable weapon that keeps these microflorae in check and even encourages growth of the advantageous ones. Production agriculture will continue to use and improve antibiotic alternatives to modulate and improve immune function thereby improving animal welfare, gut health and nutrient absorption, but also decreasing waste which is beneficial for the environment.

References

Bedford, M. 2000. Removal of antibiotic growth promoters from poultry diets: implications and strategies to minimize subsequent problems. World's Poultry Science Journal. 56:347-365.

Byrd, J.A., B.M. Hargis, D.J. Caldwell, R.H. Bailey, K.L. Herron, J.L. McReynolds, R.L. Brewer, R.C. Anderson, K.M. Bischoff, T.R. Callaway, and L.F. Kubena. 2001. Effect of lactic acid administration in the drinking water during preslaughter feed withdrawal on *Salmonella* and *Campylobacter* contamination of broilers. Poultry Science. 80:278-283.

Calder, P.C. 2021. Nutrition and immunity: lessons for COVID-19. European Journal of Clinical Nutrition. 75:1309-1318.

Cheema, M.A., M.A. Qureshi, and G.B. Havenstein. 2003. A comparison of the immune response of a 2001 commercial broiler with a 1957 random bred broiler strain when fed representative 1957 and 2001 broiler diets. Poultry Science. 82:1519-1529.

Collins, N. and Y. Belkaid. 2022. Control of immunity via nutritional interventions. Immunity. 55:210-223.

Cooper, M. 2015. The early history of B cells. Nature Reviews: Immunology. 15:191-197.

El-Hack, M.E.A., M.T. El-Saadony, H.M. Salem, A.M. El-Tahan, M.M. Soliman, G.B.A. Youssef, A.E. Taha, S.M. Soliman, A.E. Ahmed, A.F. El-kott, K.M. Al Syaad, and A.A. Swelum. 2022. Alternatives to antibiotics for organic poultry production: types, modes of action and impacts on bird's health and production. 101:101696.

Fernandez, F., R. Sharma, M. Hinton and M.R. Bedford. 2000. Diet influences the colonization of Campylobacter jejuni and distribution of mucin carbohydrates in the chick intestinal tract. 57:1793-1801.

Finlay, B.B. and G. McFadden. 2006. Anti-Immunology: evasion of the host immune system by bacterial and viral pathogens. Cell. 124:767-782.

Glick, B., T. Chang, R. Jaap, and R. George. 1956. The bursa of Fabricius and antibody production. Poultry Science. 35:224-225.

Grad, R. Cod and the consumptive: a brief history of cod-liver oil in the treatment of pulmonary tuberculosis. Pharmacy in History. 46:106-120.

Havenstein, G.B., P.R. Ferket, and M.A. Qureshi. 2003. Growth, livability, and feed conversion of 1957 versus 2001 broilers when fed representative 1957 and 2001 broiler diets. Poultry Science. 82:1500-1508.

Janeway, C.A. and R. Medzhitov. 2002. Innate immune recognition. Annual Reviews in Immunology. 20:197-216.

Kaiser, P., T.Y. Poh, L. Rothwell, S. Avery, S. Balu, U.S. Pathania, S. Hughes, M. Goodchild, S. Morrell, M. Watson, N. Bumstead., J. Kaufman, and J.R. Young. 2005. A genomic analysis of chicken cytokines and chemokines. Journal of Interferon and Cytokine Research. 24:467-484.

Kim, W.H., and H.S. Lillehoj. 2019. Immunity, immunomodulation, and antibiotic alternatives to maximize the genetic potential of poultry for growth and disease. 250:41-50.

Kogut, M.H., A. Lee and E. Santin. 2020. Microbiome and pathogen interaction with the immune system. 99:1906-1913.

Lamont, S.J. Poultry immunogenetics: which way do we go? 1994. Poultry Science. 73:1044-1048.

Lowry, V.K., M.B. Farnell, P.J. Ferro, C.L. Swaggerty, A. Bahl, and M.H. Kogut. Purified B-glucan as an abiotic feed additive upregulates the innate immune response in immature chickens against *Salmonella enterica* serovar *Enteritidis*. 2005. International Journal of Food Microbiology. 98:309-318.

Murphy, K., C. Weaver, and L. Berg. 2022. Janeway's Immunobiology, 10th edition. W.W. Norton and Company.

Nurmi. E., and M. Rantala. 1973. New aspects of *Salmonella* infection in broiler production. Nature. 241:210-211.

Patin, E.C., A. Thompson, and S. Orr. 2019. Pattern recognition receptors in fungal immunity. Seminars in Cell and Developmental Biology. 89:24-33.

Peisker, M., E. Stensrud, J. Apajalahti and M. Sifri. 2017. Morphological characterization of Pichia guilliermondii and Saccharomyces cerevisiae yeast and their effects on adherence of intestinal pathogens on piglet and chicken epithelium in-vitro. Journal of Animal Research and Nutrition. 2(1):9.

Pineda, M.R., J.A. Byrd, K.J. Genovese, Y.Z. Farnell, D. Zhao, X. Wang, A.C. Milby, and M.B. Farnell. 2021. Evaluation of sodium bisulfate on reducing *Salmonella* Heidelberg biofilm and colonization in broiler crops and ceca. Microorganisms 9:2047.

Sanchez, N.C.B., P.R. Broadway, and J.A. Carroll. 2021. Influence of yeast products modulating metabolism and immunity in cattle and swine. Animals. 11:371.

Singer, R.S., L. J. Porter, D. U. Thomson, M. Gage, A. Beaudoin and J.K. Wishnie. 2019. Raising animals without antibiotics: U.S. producer and veterinarian experiences and opinions. Frontiers in Veterinary Science. 6:452.

Stone, K.D., C. Prussin, and D.D. Metcalfe. 2010. IgE, mast cells, basophils, and eosinophils. Journal of Allergy and Clinical Immunology. 125:S73-S80.

Swaggerty, C.L., C. Bortoluzzi, A. Lee, C. Eyng, G. Dal Pont and M.H. Kogut. 2021. "Potential replacements for antibiotic growth promoters in poultry: interactions at the gut level and their impact on host immunity", pages 145-159 in Recent advances in animal nutrition and metabolism.

Zbikowska, K., M. Michalczuk, and B. Dolka. 2020. The use of bacteriophages in the poultry industry. Animals. 10:872.

Intestinal Permeability, Immune Response and Inflammation in Dairy – Nutritional Modulations

Perméabilité intestinale, réponse immunitaire et inflammation chez les animaux laitiers – modulations nutritionnelles

J. Opgenorth, M.A. Abeyta, B.M. Goetz, S. Rodrigez-Jimenez, E.J. Mayorga, and L.H. Baumgard
Department of Animal Science, Iowa State University, Ames, IA, USA
baumgard@iastate.edu

Pre-conference sponsored by / Pré-colloque commandité par :

ADM®

Abstract

Suboptimal milk yield limits the dairy industry's productive competitiveness, marginalizes efforts to reduce inputs into food production, and increases animal agriculture's carbon footprint. There are a variety of circumstances in a cow's life which result in hindered productivity and these include: heat stress, ketosis, rumen and hindgut acidosis, feed restriction, and psychological stress associated with normal animal practices (i.e., pen changes, weaning, shipping). Although these insults are seemingly very different, they have a common metabolic and inflammatory footprint. It appears that almost all "stressors" negatively affect the gastrointestinal epithelial barrier. The compromised barrier dysfunction allows luminal foreign molecules to infiltrate into local, portal and systemic circulation. The invading antigens stimulate the immune system and inflammation is the response. Once activated, the immune system primarily utilizes glucose for energy and an intensely immune response can utilize > 1 kg of glucose per day. The immune system has a larger priority than growth, milk synthesis and reproduction and the entire body initiates metabolic adjustments to ensure that leukocytes receive the necessary nutrients to resolve the infection. These adaptations come at the cost of productive purposes. Nutritionists and producers can develop strategies to ameliorate the negative consequences of leaky gut.

Résumé

Un rendement laitier sous-optimal limite la compétitivité de l'industrie laitière, marginalise les efforts accomplis pour réduire les intrants nécessaires à la production d'aliments et accroît l'empreinte carbone des activités de production animale. Diverses circonstances dans la vie d'une vache peuvent nuire à sa productivité, notamment : le stress thermique, l'acétonémie, l'acidose du rumen et de l'intestin postérieur, la restriction alimentaire et le stress psychologique associé aux pratiques d'élevage normales (changements d'enclos, sevrage, transport). Bien que ces agressions soient apparemment très différentes, elles ont une signature métabolique et inflammatoire commune. Il semble que presque tous les « facteurs de stress » ont un effet négatif sur la barrière épithéliale gastro-intestinale. La barrière ainsi compromise permet à des molécules luminales étrangères de s'infiltrer dans la circulation locale, porte et générale. Les antigènes envahissants stimulent le système immunitaire et la réponse inflammatoire. Une fois activé, le système immunitaire utilise principalement le glucose comme source d'énergie, et une réponse immunitaire intense peut consommer plus de 1 kg de glucose par jour. Le système immunitaire revendique la priorité sur la croissance, la synthèse du lait et la reproduction, et l'organisme en entier apporte les réglages métaboliques voulus pour que les leucocytes reçoivent les nutriments nécessaires pour juguler l'infection. Ces adaptations se font au détriment des objectifs de production. Les nutritionnistes des animaux et les producteurs laitiers peuvent élaborer des stratégies permettant d'atténuer les effets négatifs associés à la perméabilité de l'intestin.

Introduction

Suboptimal milk yield limits the U.S. dairy industry's productive competitiveness, marginalizes efforts to reduce inputs into food production, and increases animal agriculture's carbon footprint. There are a variety of circumstances in a cow's life which result in hindered productivity including heat stress, rumen and hindgut acidosis, feed restriction, and psychological stress associated with normal animal practices (i.e., pen changes, weaning, shipping). Although these insults have different origins, a commonality among them is increased production of inflammatory biomarkers and markedly altered nutrient partitioning. We and others have generated convincing data strongly implicating intestinally derived lipopolysaccharide (LPS) as sometimes being the culprit in these situations.

Heat Stress

During heat stress (HS), blood flow is diverted from the viscera to the periphery to dissipate heat, and this leads to intestinal hypoxia (Hall et al., 1999). Enterocytes are particularly sensitive to hypoxia and nutrient restriction (Rollwagen et al., 2006), resulting in ATP depletion and increased oxidative and nitrosative stress (Hall et al., 2001). This contributes to tight junction dysfunction and gross morphological changes that ultimately reduce intestinal barrier function (Lambert et al., 2002; Pearce et al., 2013), resulting in increased passage of luminal content into portal and systemic blood (Hall et al., 2001; Pearce et al., 2013). Endotoxin, otherwise referred to as LPS, is a glycolipid embedded in the outer membrane of Gram-negative bacteria, which is abundant and prolific in luminal content, and is a well-characterized potent immune stimulator in multiple species (Berczi et al., 1966; Giri et al., 1990; Tough et al., 1997). Immune system activation occurs when LPS binding protein (LBP) initially binds LPS and together with CD14 and TLR4 delivers LPS for removal and detoxification, thus LBP is frequently used as a biomarker for LPS infiltration (Ceciliani et al., 2012). For a detailed description of how livestock and other species detoxify LPS see our recent review (Mani et al., 2012). Endotoxin infiltration into the bloodstream during HS, which was first observed by Graber et al. (1971), is common among heat stroke patients (Leon,

2007) and is thought to play a central role in heat stroke pathophysiology as survival increases when intestinal bacterial load is reduced or when plasma LPS is neutralized (Bynum et al., 1979; Gathiram et al., 1987). It is remarkable how animals suffering from heat stroke or severe endotoxemia share many physiological and metabolic similarities to HS, such as an increase in circulating insulin (Lim et al., 2007). Intramammary LPS infusion increased (~2 fold) circulating insulin in lactating cows (Waldron et al., 2006). In addition, we intravenously infused LPS into growing calves and pigs and demonstrated >10 fold increase in circulating insulin (Rhoads et al., 2009; Kvidera et al., 2016, 2017c). Interestingly, increased insulin occurs prior to increased inflammation and the temporal pattern agrees with our previous *in vivo* data and a recent *in vitro* report (Bhat et al., 2014) suggesting LPS stimulates insulin secretion, either directly or via GLP-1 (Kahles et al., 2014). The possibility that LPS increases insulin secretion likely explains the hyperinsulinemia we have repeatedly reported in a variety of HS agriculture models (Baumgard and Rhoads, 2013). Again, the increase in insulin during both HS and immunoactivation is energetically difficult to explain as feed intake is severely depressed in both experiments.

Ketosis and the Transition Period

Recently, the concept that LPS impacts normal nutrient partitioning and potentially contributes to metabolic maladaptation to lactation has started to receive attention. Although LPS itself has not been the primary causative focus, general inflammation has been the topic of investigations. Increased inflammatory markers following parturition have been reported in cows (Ametaj et al., 2005; Bionaz et al., 2007; Bertoni et al., 2008; Humblet et al., 2006; Mullins et al., 2012). Presumably, the inflammatory state following calving disrupts normal nutrient partitioning and is detrimental to productivity (Bertoni et al., 2008), and this assumption was recently reinforced when TNFα infusion decreased productivity (albeit without overt changes in metabolism; Yuan et al., 2013; Martel et al., 2014). Additionally, in late-lactation cows, injecting TNFα increased (>100%) liver TAG content without a change in circulating NEFA (Bradford et al., 2009). Our recent data demonstrates increased inflammatory markers in cows diagnosed with ketosis only and no other health disorders (i.e. the inflammation was not apparently due to mastitis or metritis). In comparison with healthy controls, ketotic cows had increased circulating LPS prior to calving and post-partum acute phase proteins such as LBP, serum amyloid A, and haptoglobin were also increased (Abuajamieh et al., 2016). However, even seemingly healthy cows experience some degree of inflammation postpartum (Humblet et al., 2006). The magnitude and persistency of the inflammatory response seems to be predictive of transition cow performance (Bertoni et al., 2008; Bradford et al., 2015; Trevisi and Minuti, 2018). Endotoxin can originate from a variety of locations, and obvious sources in transitioning dairy cows include the uterus (metritis) and mammary gland (mastitis) (Mani et al., 2012). Additionally, we believe intestinal hyperpermeability may also be responsible for periparturient inflammation in dairy cows as many of the characteristic responses (rumen acidosis, decreased feed intake, and psychological stress) occurring during this time can compromise gut barrier function.

As aforementioned, mild inflammation is observed even in cows which seemingly complete the transition period successfully, suggesting that some level of inflammation plays an important role in cow health. In fact, previous reports have demonstrated that blocking endogenous inflammation (via administration of non-steroidal anti-inflammatory drugs [NSAID]) can increase the incidence of negative health outcomes (i.e., fever, stillbirth, retained placenta, metritis) and reduce productivity (Schwartz et al., 2009; Newby et al., 2013, 2017). Beneficial effects of NSAIDs have been observed on production performance (Carpenter et al., 2016a), but inconsistencies exist (Priest et al., 2013; Meier et al., 2014) including how NSAIDs seemingly work better in specific parities (Farney et a., 2013) and interfere with fiber digestion (Carpenter et al., 2016b) and compromise feed intake (Carpenter et al., 2017). Although NSAIDs may be an effective prophylactic strategy during the periparturient period, further research is necessary to determine the timing of administration and type and dose of NSAID that is most effective at improving health. Alternatively, administrating a chemokine (anti or even pro-inflammatory) may hold promise in improving transition cow performance.

Rumen and Hindgut Acidosis

A transitioning dairy cow undergoes a dietary shift from a high forage to a high concentrate ration post-calving. This has the potential to induce rumen acidosis (RA) as increases in fermentable carbohydrates and DMI stimulate the buildup of short chain fatty acids and lactic acid (Nocek, 1997; Enemark, 2008). Rumen acidosis has direct and ancillary consequences accompanied by various production issues (decreased DMI, reduced milk yield, milk fat depression) and health challenges such as laminitis, liver abscesses, and potentially death (Nocek, 1997; Kleen, 2003). The mechanisms linking RA and the development of health disorders are not entirely clear, however, recent literature has indicated that inflammation associated with epithelial damage and consequential LPS translocation are at least partially responsible for production losses associated with RA (Gozho, et al., 2005; Khafipour, et al., 2009). Although many hypothesize LPS translocation occurs at the rumen epithelium directly (Guo et al., 2017; Minuti et al., 2014), others point towards LPS translocation in the hindgut to be a potential source of peripheral inflammation (Li et al., 2012). Interestingly, when RA was induced using either alfalfa pellets or high-grain diets, increased peripheral inflammation was only observed in the high-grain group, irrespective of rumen acidotic conditions being similar between the two treatments (Khafipour et al., 2009a,b). It was hypothesized that the grain supplemented group likely had increased starch flow to the hindgut, and therefore, increased fermentation that could potentially lead to hindgut acidosis and LPS translocation across the large intestine. However, we were unable to recreate production losses and systemic inflammation when we abomasally infused 500 g/d of resistant starch (Piantoni et al., 2018) or even 4 kg/d of purified corn starch (Abeyta et al., 2019). Both of our aforementioned experiments were accompanied with marked reductions in fecal pH so it is unlikely that large intestinal acidosis per se is the specific reason for systemic inflammation described in the previous reports (Li et al., 2012, Khafipour et al., 2009a,b). Regardless, we recently reported that cows with the largest decrease in fecal pH post-calving consumed less feed, produced less milk, had a larger acute phase protein response and had increased NEFA and BHB compared to cows that had a mild decrease in fecal pH following parturition (Rodriguez-Jimenez et al., 2019). Clearly, our current understanding of how hind-gut acidosis impacts the immune system and ultimately periparturient productivity is woeful.

Feed Restriction and Psychological Stress

Stress associated with feed restriction along with several other regular production practices (e.g., heat stress, weaning, transportation, overcrowding, restraint, social isolation/mixing) is frequently encountered in animal agriculture (Chen et al., 2015) and is associated with gastrointestinal hyperpermeability. In fact, we have repeatedly reported reduced intestinal barrier integrity in pigs pair-fed to their HS counterparts (Pearce et al., 2013; Sanz-Fernandez et al., 2014). Furthermore, we recently demonstrated shortened ileum villous height and crypt depth (Kvidera et al., 2017d) as well as increased appearance of the intestinal permeability marker Cr-EDTA (Horst and Baumgard, unpublished), indicating reduced intestinal health in cows fed 40% of ad libitum intake. Recent literature indicates that the corticotropin releasing factor (CRF) system may be the mechanism involved in stress-induced leaky gut (Wallon et al., 2008; Vanuytsel et al., 2014). The CRF and other members of the CRF signaling family including urocortin (1, 2, and 3) and their G-protein couple receptors CRF1 and CRF2, have been identified as the main mediators of the stress-induced intestinal changes including inflammation, altered intestinal motility and permeability, as well as shifts in ion, water, and mucus secretion and absorption (as reviewed by Rodiño-Janeiro et al., 2015). These alterations appear to be regulated in large part by intestinal mast cells (Santos et al., 2000). Mast cells are important mediators of both innate and adaptive immunity and express receptors for the neuropeptides CRF1 and CRF2, which may in part explain the association between emotional stress and intestinal dysfunction (Smith et al., 2010; Ayyadurai et al., 2017). Furthermore, mast cells synthesize a variety of pro-inflammatory mediators (i.e., IFN-γ and TNF-α) that are released upon activation, mainly via degranulation (de Punder and Pruimboom, 2015). Excessive mast cell degranulation plays an important role in the pathogenesis of different intestinal inflammatory disorders (Santos et al., 2000; Smith et al., 2010). A better understanding of the role psychosocial stress plays on the initiation of different intestinal disorders in livestock is of obvious interest for multiple animal agriculture systems.

Metabolism of Inflammation

LPS-induced inflammation has an energetic cost which redirects nutrients away from anabolic processes that support milk and muscle synthesis (see review by Johnson 1997, 1998) and thus compromises productivity. Upon activation, most immune cells become obligate glucose utilizers via a metabolic shift from oxidative phosphorylation to aerobic glycolysis (not anaerobic glycolysis typically learned about in biochemistry classes), a process known as the Warburg effect.

This metabolic shift allows for rapid ATP production and synthesis of important intermediates which support proliferation and production of reactive oxygen species (Calder et al., 2007; Palsson-McDermott and O'Neill, 2013). In an effort to facilitate glucose uptake, immune cells become more insulin sensitive and increase expression of GLUT3 and GLUT4 transporters (Maratou et al., 2007; O'Boyle et al., 2012), whereas peripheral tissues become insulin resistant (Poggi et al., 2007; Liang et al., 2013). Furthermore, metabolic adjustments including hyperglycemia or hypoglycemia (depending upon the stage and severity of infection), increased circulating insulin and glucagon, skeletal muscle catabolism and subsequent nitrogen loss (Figure 1; Wannemacher et al., 1980), and hypertriglyceridemia (Filkins, 1978; Wannemacher et al., 1980; Lanza-Jacoby et al., 1998; McGuinness, 2005) occur. Interestingly, despite hypertriglyceridemia, circulating BHB often decreases following LPS administration (Waldron et al., 2003a,b; Graugnard et al., 2013; Kvidera et al., 2017a). The mechanism of LPS-induced decreases in BHB has not been fully elucidated, but may be explained by increased ketone oxidation by peripheral tissues (Zarrin et al., 2014). Collectively, these metabolic alterations are presumably employed to ensure adequate glucose delivery to activated leukocytes.

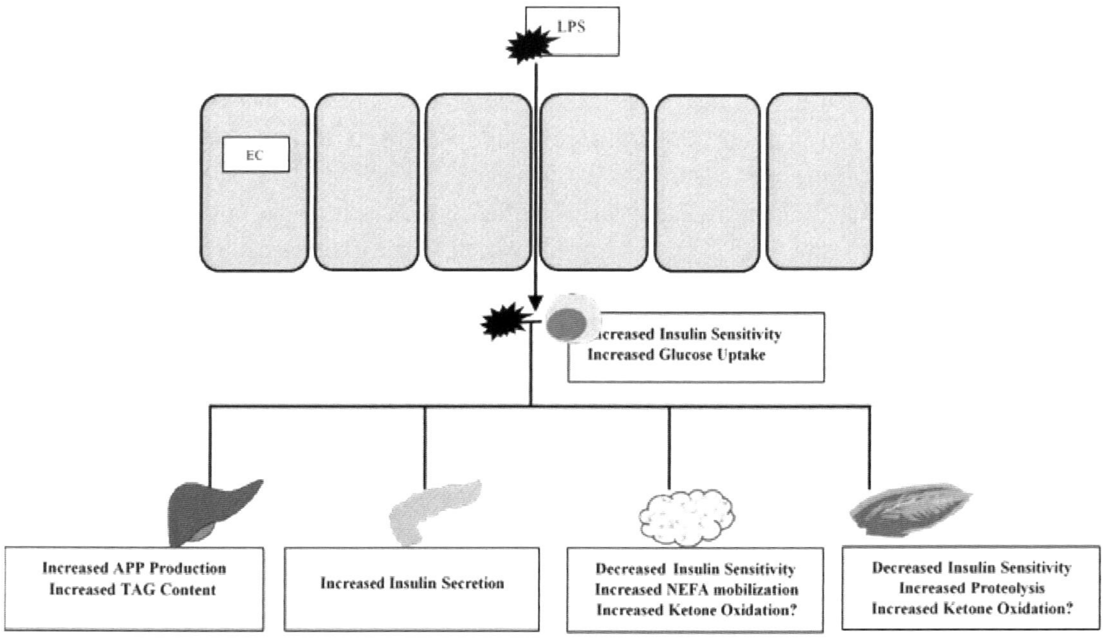

Figure 1. LPS induced alterations in peripheral metabolism.

Energetic Cost of Immune Activation

The energetic costs of immunoactivation are substantial, but the ubiquitous nature of the immune system makes quantifying the energetic demand difficult. Our group recently employed a series of LPS-euglycemic clamps to quantify the energetic cost of an activated immune system. Using this model, we estimated approximately 1 kg of glucose is used by an intensely activated immune system during a 12 hour period in lactating dairy cows. Interestingly, on a metabolic body weight basis the amount of glucose utilized by LPS-activated immune system in mid- and late-lactation cows, growing steers and growing pigs were 0.64, 1.0, 0.94, 1.0, and 1.1 g glucose/kg $BW^{0.75}$/h, respectively; Kvidera et al., 2016, 2017b,c, Horst et al., 2018, 2019). A limitation to our model is the inability to account for liver's contribution to the circulating glucose pool (i.e., glycogenolysis and gluconeogenesis). However, both glycogenolytic and gluconeogenic rates have been shown to be increased during infection (Spitzer et a., 1985; Waldron et al., 2003b) and Waldron et al. (2006) demonstrated that ~87 g of glucose appeared in circulation from these processes. Furthermore, we have observed both increased circulating glucagon and cortisol (stimulators of hepatic glucose output) following LPS administration (Horst et al., 2019) suggesting we are underestimating the energetic cost of immunoactivation. The reprioritization of glucose trafficking during immunoactivation has particular consequences during lactation as it requires ~72 g of glucose for synthesizing 1 kg milk (Kronfeld, 1982).

Increased immune system glucose utilization occurs simultaneously with infection-induced decreased feed intake: this coupling of enhanced nutrient requirements with hypophagia obviously decrease the amount of nutrients available for the synthesis of valuable products (milk, meat, fetus, wool, etc.). We and others have now demonstrated that HS, rumen acidosis, and psychological stress increase circulating markers of endotoxin and inflammation. We believe that the circulating LPS originates from the intestine (small or large) and initiates an immune response. This activated systemic immune response reprioritizes the hierarchy of glucose utilization and milk synthesis is consequently deemphasized.

Nutritional Mitigation Strategies: The Role of Zinc (Zn) Supplementation

Potential dietary mitigation strategies aimed at improving gut health are currently of great interest, especially considering the numerous stressors (i.e., heat stress, feed restriction, acidosis) that potentially impact intestinal permeability. Zinc is an essential nutrient which is crucial for maintaining epithelial integrity (i.e., mammary, uterine, intestinal) and regulating the renewal of damaged epithelium (Alam et al., 1994). Zinc was first demonstrated to improve intestinal "health" in human leaky gut models (Alam et al., 1994; Rodriguez et al., 1996; Sturniolo et al., 2001), and we extended this to improved metrics of intestinal permeability in a variety of farm animal stress models including heat stress (Sanz-Fernandez et al., 2014; Pearce et al., 2015; Opgenorth et al., 2021) and feed restriction (Horst and Baumgard, unpublished using Zn hydroxychloride). Additionally, we observed altered febrile, cytokine, and acute phase protein responses during heat stress (Sanz-Fernandez et al., 2014; Pearce et al., 2015; Mayorga et al., 2018; Opgenorth et al., 2021) and in response to LPS administration (Horst et al., 2019) with dietary Zn supplementation. Presumably the aforementioned changes in inflammatory variables are indicative of a blunted immune response (because of improved intestinal barrier function). Therefore, Zn as a dietary supplement appears to be a promising avenue to improve gut health and to ameliorate alimentary canal associated inflammation.

Conclusion

There are various situations in an animal's life that hinder production performance (i.e., heat stress, feed restriction, rumen acidosis, etc.) and we suggest, based upon the literature and on our supporting evidence, that LPS (of intestinal origin) may be the common culprit in these circumstances. Immune activation in response to LPS markedly alters nutrient partitioning as a means of fueling the immune response. More research is still needed to understand the mechanisms and consequences of intestinal permeability and associated inflammation in order to provide foundational information for developing strategies aimed at maintaining productivity.

*Parts of this manuscript were first published in the proceedings of the 2016, 2017 and 2018 Southwest Nutrition Conference in Tempe AZ and 2019 Cornell Nutrition Conference

References

Abeyta, M. A., E. A. Horst, S. J. Rodriguez-Jimenez, E. J. Mayorga, B. M. Goetz, M. Al-Qasi, P. Piantoni, G. F. Schroeder, H. A. Ramirez-Ramirez, and L. H. Baumgard. 2019. Effects of hindgut acidosis in dairy cows acclimated to a low-starch diet. J. Dairy Sci. (Suppl. 1): 476.

Abuajamieh, M., S. K. Kvidera, M. V. Sanz Fernandez, A. Nayeri, N. C. Upah, E. A. Nolan, S. M. Lei, J. M. DeFrain, H. B. Green, K. M. Schoenberg, W. E. Trout, and L. H. Baumgard. 2016. Inflammatory biomarkers are associated with ketosis in periparturient Holstein cows. Res. Vet. Sci. 109:81-85.

Alam, A. N., S. A. Sarker, and M. A. Wahed. 1994. Enteric protein loss and intestinal permeability changes in children during acute shigellosis and after recovery: effect of zinc supplementation. Gut 35**:**1707–1711.

Ametaj, B. N., B. J. Bradford, G. Bobe, R. A. Nafikov, Y. Lu, J. W. Young, and D. C. Beitz. 2005. Strong relationships between mediators of the acute phase response and fatty liver in dairy cows. Can. J. Anim. Sci. 85:165–175.

Ayyadurai, S., A. J. Gibson, S. D'Costa, E. L. Overman, L. J. Sommerville, A. C. Poopal, E. Mackey, Y. Li, and A. J. Moeser. 2017. Corticotropin-releasing factor receptor subtype 1 is a critical modulator of mast cell degradation and stress-induced pathophysiology. J. Leukoc. Biol. 102:1299-1312.

Baumgard, L. H. and R. P. Rhoads. 2013. Effects of heat stress on postabsorptive metabolism and energetics. Annu. Rev. Anim. Biosci. 1:311–337.

Berczi, I., L. Bertok, and T. Bereznai. 1966. Comparative studies on the toxicity of Escherichia coli lipopolysaccharide endotoxin in various animal species. Can. J. of Microbiol. 12:1070-1071.

Bertoni, G., E. Trevisi, X. Han, and M. Bionaz. 2008. Effects of inflammatory conditions on liver activity in puerperium period and consequences for performance in dairy cows. J. Dairy Sci. 91:3300–3310.

Bhat, U. G., V. Ilievski, T. G. Unterman, and K. Watanabe. 2014. *Porphyromonas gingivalis* lipopolysaccharide (LPS) upregulates insulin secretion from pancreatic beta cells line MIN6. J. Periodontol. 85:1629–1636.

Bionaz, M., E. Trevisi, L. Calamari, F. Librandi, A. Ferrari, and G. Bertoni. 2007. Plasma paraoxonase, health, inflammatory conditions, and liver function in transition dairy cows. J. Dairy Sci. 90:1740-1750.

Bradford, B. J., L. K. Mamedova, J. E. Minton, J. S. Drouillard, and B. J. Johnson. 2009. Daily injection of tumor necrosis factor-α increases hepatic triglycerides and alters transcript abundance of metabolic genes in lactating dairy cattle. J. Nutr. 139:1451–1456.

Bradford, B.J., Yuan, K., Farney, J.K., Mamedova, L.K., Carpenter, A.J., 2015. Invited review: Inflammation during the transition to lactation: New adventures with an old flame. J. Dairy Sci. 98, 6631-6650.

Bynum, G., J. Brown, D. Dubose, M. Marsili, I. Leav, T. G. Pistole, M. Hamlet, M. LeMaire, and B. Caleb. 1979. Increased survival in experimental dog heatstroke after reduction of gut flora. Aviat. Space Environ. Med. 50:816-819.

Calder, P. C., G. Dimitriadis, and P. Newsholme. 2007. Glucose metabolism in lymphoid and inflammatory cells and tissues. Curr. Opin. Clin. Nutr. Metab. Care. 10:531-540.

Carpenter, A. J., C. M. Ylioja, C. F. Vargas, L. K. Mamedova, L. G. Mendonça, J. F. Coetzee, L. C. Hollis, R. Gehring, and B. J. Bradford. 2016. Hot topic: Early postpartum treatment of commercial dairy cows with nonsteroidal anti-inflammatory drugs increases whole-lactation milk yield. J. Dairy Sci. 99:672-679.

Carpenter, A.J., C.M. Ylioja, and B.J. Bradford. 2016. Early postpartum administration of sodium salicylate to multiparous dairy cattle is associated with alterations in feeding behavior up to 120 d in milk. J. Anim Sci. 94. Suppl. 5. 531 (abst).

Carpenter, A.J., C.F.V. Rodriguez, J.A.B. Jantz, and B.J. Bradford. 2017. Sodium salicylate negatively affects rumen fermentation in vitro and in situ. J. Dairy Sci. 100:1935-1939.

Ceciliani, F., J.J. Ceron, P.D. Eckersall, and H. Sauerwein. 2012. Acute phase proteins in ruminants. J. Proteomics. 75:4207-4231.

Chen, Y., R. Arsenault, S. Napper, and P. Griebel. 2015. Models and methods to investigate acute stress responses in cattle. Animals (Basel). 5:1268-1295.

de Punder, K., and L. Pruimboom. 2015. Stress induces endotoxemia and low-grade inflammation by increasing barrier permeability. Front. Immunol. 6:223.

Enemark, J. M. D. 2008. The monitoring, prevention and treatment of sub-acute ruminal acidosis (SARA): a review. Vet. J. 176:32-43.

Farney, J. K., L. K. Mamedova, J. F. Coetzee, J. E. Minton, L. C. Hollis, and B. J. Bradford. 2013. Sodium salicylate treatment in early lactation increases whole-lactation milk and milk fat yield in mature dairy cows. J. Dairy Sci. 96:7709-7718.

Filkins, J. P. 1978. Phases of glucose dyshomeostasis in endotoxicosis. Circ. Shock 5:347-355.

Gathiram, P., M. T. Wells, J. G. Brock-Utne, and S. L. Gaffin. 1987. Antilipopolysaccharide improves survival in primates subjected to heat stroke. Circ. Shock 23:157-164.

Giri, S. N., P. Emau, J. S. Cullor, G. H. Stabenfeldt, M. L. Bruss, R. H. Bondurant, and B. I. Osburn. 1990. Effects of endotoxin infusion on circulating levels of eicosanoids, progesterone, cortisol, glucose and lactic acid, and abortion in pregnant cows. Vet. Microbiol. 21:211-231.

Gozho, G. N., J. C. Plaizier, D. O. Krause, A. D. Kennedy, and K. M. Wittenberg. 2005. Subacute Ruminal Acidosis Induces Ruminal Lipopolysaccharide Endotoxin Release and Triggers an Inflammatory Response. J. Dairy Sci. 88:1399-1403

Graber, C. D., R. B. Reinhold, J.G. Breman, R. A. Harley, and G. R. Hennigar. 1971. Fatal heat stroke. Circulating endotoxin and gram-negative sepsis as complications. JAMA. 216:1195-1196.

Graugnard, D. E., K. M. Moyes, E. Trevisi, M. J. Khan, D. Keisler, J. K. Drackley, G. Bertoni, and J. J. Loor. 2013. Liver lipid content and inflammometabolic indices in peripartal dairy cows are altered in response to prepartal energy intake and postpartal intramammary inflammatory challenge. J. Dairy Sci. 96:918-935.

Guo, J., G. Chang, K. Zhang, L. Xu, D. Jin, M. S. Bilal, and X. Shen. 2017. Rumen-derived lipopolysaccharide provoked inflammatory injury in the liver of dairy cows fed a high-concentrate diet. Oncotarget. 8(29):46769-46780.

Hall, D. M., K. R. Baumgardner, T. D. Oberley, and C. V. Gisolfi. 1999. Splanchnic tissues undergo hypoxic stress during whole body hyperthermia. Am. J. Physiol. 276:G1195-G1203.

Hall, D. M., G. R. Buettner, L. W. Oberley, L. Xu, R. D. Matthes, and C. V. Gisolfi. 2001. Mechanisms of circulatory and intestinal barrier dysfunction during whole body hyperthermia. Am. J. Physiol. Heart Circ. Physiol. 280:H509– H521.

Horst, E. A., S. K. Kvidera, E. J. Mayorga, C. S. Shouse, M. Al-Qaisi, M. J. Dickson, J. Ydstie, H. A. Ramirez Ramirez, A. F. Keating, D. J. Dickson, K. E. Griswold, and L. H. Baumgard. 2018. Effect of chromium on bioenergetics and leukocyte dynamics following immunoactivation in lactating Holstein cows. J. Dairy Sci. 101:5515-5530.

Horst, E. A., E. J. Mayorga, S. L. Portner, M. Al-Qaisi, C. S. McCarthy, M. A. Abeyta, B. M. Goetz, H. A. Ramirez-Ramirez, D. H. Kleinschmit, and L. H. Baumgard. 2019. Effects of dietary zinc on energetic requirements of an activated immune system following lipopolysaccharide challenge in lactating cows. J. Dairy Sci. 102:11681-11700.

Humblet, M. F., H. Guyot, B. Boudry, F. Mbayahi, C. Hanzen, F. Rollin, and J. M. Godeau. 2006. Relationship between haptoglobin, serum amyloid A, and clinical status in a survey of dairy herds during a 6-month period. Vet. Clin. Pathol. 35:188–193.

Johnson, R. W. 1997. Inhibition of growth by pro-inflammatory cytokines: an integrated view. J Anim. Sci. 75: 1244-1255.

Johnson, R. W. 1998. Immune and endocrine regulation of food intake in sick animals. Dome. Animal Endo. 15: 309-319.

Kahles, F., C. Meyer, J. Möllmann, S. Diebold, H.M. Findeisen, C. Lebherz, C. Trautwein, A. Koch, F. Tacke, N. Marx, and M. Lehrke. 2014. GLP-1 Secretion Is Increased by Inflammatory Stimuli in an IL-6–Dependent Manner, Leading to Hyperinsulinemia and Blood Glucose Lowering. Diabetes. 63:3221-3229.

Khafipour, E., D.O. Krause, and J.C. Plaizier. 2009a. A grain-based subacute ruminal acidosis challenge causes translocation of lipopolysaccharide and triggers inflammation. J. Dairy Sci. 92:1060-1070.

Khafipour, E., D. O. Krause, and J. C. Plaizier. 2009b. Alfalfa pellet-induced subacute ruminal acidosis in dairy cows increases bacterial endotoxin in the rumen without causing inflammation. J. Dairy Sci. 92:1712-1724.

Kleen, J. L., G. A. Hooijer, J. Rehage, and J. P. T. M. Noordhuizen. 2003. Subacute ruminal acidosis (SARA): a review. J. Vet. Med. 50:406-414.

Kronfeld, D. S. 1982. Major metabolic determinants of milk volume, mammary efficiency, and spontaneous ketosis in dairy cows. J. Dairy Sci. 65:2204-2212.

Kvidera, S. K., E. A. Horst, M. Abuajamieh, E. J. Mayorga, M. V. Sanz-Fernandez, and L. H. Baumgard. 2016. Technical note: A procedure to estimate glucose requirements of an activated immune system in steers. J. Anim. Sci. 94:4591-4599.

Kvidera, S. K., M. J. Dickson, M. Abuajamieh, D. B. Snider, M. V. Sanz-Fernandez, J. S. Johnson, A. F. Keating, P. J. Gorden, H. B. Green, K. M. Schoenberg, and L.H. Baumgard. 2017a. Intentionally induced intestinal barrier dysfunction causes inflammation, affects metabolism, and reduces productivity in lactating Holstein cows. J. Dairy Sci. 100:4113-4127.

Kvidera, S. K., E. A. Horst, M. Abuajamieh, E. J. Mayorga, M. V. Sanz-Fernandez, and L. H. Baumgard. 2017b. Glucose requirements of an activated immune system in lactating Holstein cows. J. Dairy Sci. 100:2360-2374.

Kvidera, S. K., E. A. Horst, E. J. Mayorga, M. V. Sanz-Fernandez, M. Abuajamieh, and L. H. Baumgard. 2017c. Estimating glucose requirements of an activated immune system in growing pigs. J. Anim. Sci. 95:5020-5029.

Kvidera, S. K., E. A. Horst, M. V. Sanz-Fernandez, M. Abuajamieh, S. Ganesan, P. J. Gorden, H. B. Green, K. M. Schoenberg, W. E. Trout, A. F. Keating, and L. H. Baumgard. 2017d. Characterizing effects of feed restriction and glucagon-like peptide 2 administration on biomarkers of inflammation and intestinal morphology. J. Dairy Sci. 100:9402-9417.

Lambert, G. P., C. V. Gisolfi, D. J. Berg, P. L. Moseley, L. W. Oberley, and K. C. Kregel. 2002. Selected contribution: Hyperthermia-induced intestinal permeability and the role of oxidative and nitrosative stress. J. Appl. Physiol. 92:1750-1761.

Lanza-Jacoby, S., H. Phetteplace, N. Sedkova, and G. Knee. 1998. Sequential alterations in tissue lipoprotein lipase, triglyceride secretion rates, and serum tumor necrosis factor alpha during Escherichia coli bacteremic sepsis in relation to the development of hypertriglyceridemia. Shock 9:46-51.

Leon, L. R. 2007. Heat stroke and cytokines. Prog. Brain Res. 162:481-524.

Li, S., E. Khafipour, D. O. Krause, A. Kroeker, J. C. Rodriguez-Lecompte, G. N. Gozho, and J. C. Plaizier. 2012. Effects of subacute ruminal acidosis challenges on fermentation and endotoxins in the rumen and hindgut of dairy cows. J. Dairy Sci. 95:294-303.

Liang, H., S. E. Hussey, A. Sanchez-Avila, P. Tantiwong, and N. Musi. 2013. Effect of lipopolysaccharide on inflammation and insulin action in human muscle. PLoS One 8:e63983.

Lim, C. L., G. Wilson, L. Brown, J. S. Coombes, and L. T. Mackinnon. 2007. Pre-existing inflammatory state compromises heat tolerance in rats exposed to heat stress. Am. J. Physiol. Regul. Integr. Comp. Physiol. 292:R186-194.

Mani, V., T. E. Weber, L. H. Baumgard and N. K. Gabler. 2012. Growth and development symposium: endotoxin, inflammation, and intestinal function in livestock. J. Anim. Sci. 90:1452-1465.

Maratou, E., G. Dimitriadis, A. Kollias, E. Boutati, V. Lambadiari, P. Mitrou, and S. A. Raptis. 2007. Glucose transporter expression on the plasma membrane of resting and activated white blood cells. Eur. J. Clin. Invest. 37:282-290.

Mayorga, E. J., S. K. Kvidera, E. A. Horst, M. A. Al-Qaisi, M. J. Dickson, J. T. Seibert, S. Lei, Z. J. Rambo, M. E. Wilson, and L. H. Baumgard. 2018. Effects of zinc amino acid complex on biomarkers of gut integrity and metabolism during and following heat stress or feed restriction in pigs. J. Anim. Sci. 96:4173-4185.

Martel, C. A., L. K. Mamedova, J. E. Minton, M. L. Jones, J. A. Carroll, and B. J. Bradford. 2014. Continuous low-dose infusion of tumor necrosis factor alpha in adipose tissue elevates adipose tissue interleukin 10 abundance and fails to alter metabolism in lactating dairy cows. J. Dairy Sci. 97:4897-4906.

McGuinness, O. P. 2005. Defective glucose homeostasis during infection. Annu. Rev. Nutr. 25:9-35.

Meier, S., N. V. Priest, C. R. Burke, J. K. Kay, S. McDougall, M. D. Mitchell, C. G. Walker, A. Heiser, J. J. Loor, and J. R. Roche. 2014. Treatment with a nonsteroidal aniinflammatory drug after calving did not improve milk production, health, or reproduction parameters in pasture-grazed dairy cows. J. Dairy Sci. 97:2932-2943.

Minuti, A., S. Ahmed, E. Trevisi, F. Piccioli-Cappelli, G. Bertoni, N. Jahan, and P. Bani. 2014. Experimental acute rumen acidosis in sheep: Consequences on clinical, rumen, and gastrointestinal permeability conditions and blood chemistry. J. Anim. Sci. 92:3966-3977.

Mullins, C. R., L. K. Mamedova, M. J. Brouk, C. E. Moore, H. B. Green, K. L. Perfield, J. F. Smith, J. P. Harner, and B. J. Bradford. 2012. Effects of monensin on metabolic parameters, feeding behavior, and productivity of transition dairy cows. J. Dairy Sci. 95:1323–1336.

Newby, N.C., D.L. Pearl, S.J. Leblanc, K.E. Leslie, M.A. von Keyserlingk, and T.F. Duffield. 2013. Effects of meloxicam on milk production, behavior, and feed intake in dairy cows following assisted calving. J. Dairy Sci. 96:3682-3688.

Newby, N.C., K. E. Leslie, H. D. Putnam Dingwell, D. F. Kelton, D. M. Weary, L. Neuder, S. T. Millman, and T. F. Duffield. 2017. The effects of periparturient administration of flunixin meglumine on the health and production of dairy cattle. J. Dairy Sci. 100:582-587.

Nocek, J. E. 1997. Bovine acidosis: Implications on laminitis. J. Dairy Sci. 80:1005-1028.

O'Boyle, N. J., G. A. Contreras, S. A. Mattmiller, and L. M. Sordillo. 2012. Changes in glucose transporter expression in monocytes of periparturient dairy cows. J. Dairy Sci. 95:5709-5719.

Opgenorth, J., M. Abuajamieh, E. A. Horst, S. K. Kvidera, J. S. Johnson, E. J. Mayorga, M. V. Sanz-Fernandez, M. A. Al-Qaisi, J. M. DeFrain, D. H. Kleinschmit, P. J. Gorden, and L. H. Baumgard. 2021. The effects of zinc amino acid complex on biomarkers of gut integrity, inflammation, and metabolism in heat-stressed ruminants. J. Dairy Sci. 104:2410-2421.

Palsson-McDermott, E. M. and L. A. O'Neill. 2013. The Warburg effect then and now: from cancer to inflammatory diseases. Bioessays 35:965-973.

Pearce, S. C., V. Mani, T. E. Weber, R. P. Rhoads, J. F. Patience, L. H. Baumgard, and N. K. Gabler. 2013. Heat stress and reduced plane of nutrition decreases intestinal integrity and function in pigs. J. Anim. Sci. 91:5183-5193.

Pearce S. C., M. V. Sanz-Fernandez, J. Torrison, M. E. Wilson, L. H. Baumgard, and N. K. Gabler. 2015. Dietary zinc attenuates heat-stress induced changes in intestinal integrity and metabolism. J. Anim. Sci. 93:4702-4713.

Piantoni, P., M.A. Abeyta, G.F. Schroeder, H.A. Ramirez-Ramirez, H.A. Tucker and L.H. Baumgard. 2018. Induction of leaky gut through feed restriction or abomasal infusion of resistant starch in healthy post-peak lactating cows. J. Dairy Sci. (Suppl. 2): 228.

Poggi, M., D. Bastelica, P. Gual, M. A. Iglesias, T. Gremeaux, C. Knauf, F. Peiretti, M. Verdier, I. Juhan-Vague, J. F. Tanti, R. Burcelin, and M. C. Alessi. 2007. C3H/HeJ mice carrying a toll-like receptor 4 mutation are protected against the development of insulin resistance in white adipose tissue in response to a high-fat diet. Diabetologia 50:1267-1276.

Priest, N. V., S. McDougall, C. R. Burke, J. R. Roche, M. Mitchell, K. L. McLeod, S. L. Greenwood, and S. Meier. 2013. The responsiveness of subclinical endometritis to a nonsteroidal anti-inflammatory drug in pasture-grazed dairy cows. J. Dairy Sci. 96:4323-4332.

Rhoads, M. L., R. P. Rhoads, M. J. VanBaale, R. J. Collier, S. R. Sanders, W. J. Weber, B. A. Crooker, and L. H. Baumgard. 2009. Effects of heat stress and plane of nutrition on lactating Holstein cows: I. Production, metabolism, and aspects of circulating somatotropin. J. Dairy Sci. 92:1986-1997.

Rodiño-Janeiro, B. K., C. Alonso-Cotoner, M. Pigrau, B. Lobo, M. Vicario, and J. Santos. 2015. Role of corticotropin-releasing factor in gastrointestinal permeability. J. Neurogastroenterol. Motil. 21:33-50.

Rodriguez-Jimenez, S., C. S. McCarthy, E. A. Horst, E. J. Mayorga, M. Al-Qaisi, M. A. Abeyta, B. M. Goetz, H. A. Ramirez-Ramirez, and L. H. Baumgard. 2019. Relationships between fecal pH and milk production, metabolism, and acute phase protein response in periparturient dairy cows. J. Dairy Sci. 102 (Suppl. 1): 402.

Rodriguez, P., N. Darmon, P. Chappuis, C. Candalh, M. A. Blaton, C. Bouchaud, and M. Heyman. 1996. Intestinal paracellular permeability during malnutrition in guinea pigs: effect of high dietary zinc. Gut 39:416–422.

Rollwagen, F. M., S. Madhavan, A. Singh, Y. Y. Li, K. Wolcott, and R. Maheshwari. 2006. IL-6 protects enterocytes from hypoxia-induced apoptosis by induction of bcl-2 mRNA and reduction of fas mRNA. Biochem. Biophys. Res. Commun. 347:1094-1098.

Santos, J., M. Benjamin, P. C. Yang, T. Prior, and M. H. Perdue. 2000. Chronic stress impairs rat growth and jejunal eputhelial barrier function: role of mast cells. Am. J. Physiol. Gastrointest. Liver Physiol. 278:G847-G854.

Sanz-Fernandez, M.V., S. C. Pearce, N. K. Gabler, J. F. Patience, M. E. Wilson, M. T. Socha, J. L. Torrison, R. P. Rhoads, and L. H. Baumgard. 2014. Effects of supplemental zinc amino acid complex on gut integrity in heat-stressed growing pigs. Animal. 8:43-50.

Schwartz, G., K. L. Hill, M. J. VanBaale, and L. H. Baumgard. 2009. Effects of flunixin meglumine on pyrexia and bioenergetics variables in postparturient dairy cows. J. Dairy Sci. 92:1963-1970.

Smith, F., J. E. Clark, B. L. Overman, C. C. Tozel, J. H. Huang, J. E. F. Rivier, A. T. Blikslager, and A. J. Moeser. 2010. Early weaning stress impairs development of mucosal barrier function in the porcine intestine. Am. J. Physiol. Gastrointest. Liver Physiol. 298: G352-G363.

Spitzer, J. A., K. M. Nelson, and R. E. Fish. 1985. Time course of changes in gluconeogenesis from various precursors in chronically endotoxemic rats. Metabolism 34:842-849.

Sturniolo, G. C., V. Di Leo, A. Ferronato, A. D'Odorico, and R. D'Incà. 2001. Zinc supplementation tightens "leaky gut" in crohn's disease. Inflamm. Bowel Dis. 7:94–98.

Tough, D. F., S. Sun, and J. Sprent. 1997. T cell stimulation in vivo by lipopolysaccharide (LPS). J. Exp. Med. 185:2089-2094.

Trevisi, E., and A. Minuti. 2018. Assessment of the innate immune response in the periparturient cow. Res. Vet. Sci. 116:47-54.

Vanuytsel, T., S. van Wanrooy, H. Vanheel, C. Vanormelingen, S. Verschueren, E. Houben, S. Salim Rasoel, J. Tóth, L. Holvoet, R. Farré, L. Van Oudenhove, G. Boeckxstaens, K. Verbeke, and J. Tack. 2014. Psychological stress and corticotropin-releasing hormone increase intestinal permeability in humans by a mast cell-dependent mechanism. Gut 63:1293-1299.

Waldron, M. R., B. J. Nonnecke, T. Nishida, R. L. Horst, and T. R. Overton. 2003a. Effect of lipopolysaccharide infusion on serum macromineral and vitamin D concentrations in dairy cows. J. Dairy Sci. 86:3440-3446.

Waldron, M. R., T. Nishida, B. J. Nonnecke, and T. R. Overton. 2003b. Effect of lipopolysaccharide on indices of peripheral and hepatic metabolism in lactating cows. J. Dairy Sci. 86:3447-3459.

Waldron, M. R., A. E. Kulick, A. W. Bell, and T. R. Overton. 2006. Acute experimental mastitis is not causal toward the development of energy-related metabolic disorders in early postpartum dairy cows. J. Dairy Sci. 89:596-610.

Wallon, C., P. C. Yang, A. V. Keita, A. C. Ericson, D. M. McKay, P. M. Sherman, M. H. Perdue, and J. D. Söderholm. 2008. Corticotropin-releasing hormone (CRH) regulates macromolecular permeability via mast cells in normal human colonic biopsies in vitro. Gut 57:50-58.

Wannemacher, R. W., F. A. Beall, P. G. Canonico, R. E. Dinterman, C. L. Hadick, and H. A. Neufeld. 1980. Glucose and alanine metabolism during bacterial infections in rats and rhesus monkeys. Metabolism 29:201-212.

Yuan, K., J. K. Farney, L. K. Mamedova, L. M. Sordillo, and B. J. Bradford. 2013. TNFa Altered Inflammatory Responses, Impaired Health and Productivity, but Did Not Affect Glucose or Lipid Metabolism in Early-Lactation Dairy Cows. PloS One. e80316.

Zarrin, M., O. Wellnitz, H. A. van Dorland, J. J. Gross, and R. M. Bruckmaier. 2014. Hyperketonemia during lipopolysaccharide-induced mastitis affects systemic and local intramammary metabolism in dairy cows. J. Dairy Sci. 97:3531-3541.

Viral Viability and Infectivity in Contaminated Feed and Opportunities for Feed Additives to Mitigate the Challenge

Viabilité et infectiosité virales dans les aliments du bétail contaminés et possibilités d'atténuer le problème grâce aux additifs alimentaires pour le bétail

Scott Dee
Pipestone Veterinary Services, Pipestone, MN, USA
Scott.Dee@pipestone.com

Pre-conference sponsored by / Pré-colloque commandité par :

ADM®

Abstract

The concept of animal feed as a previously overlooked risk factor came to light following the introduction of PEDV into U.S. swine herds in May 2013. Feed ingredients and prepared diets were not considered as potential vehicles for pathogen transport and transmission prior to that time, and no standard biosecurity practices were in place even though swine facilities frequently received new products and supplies on a daily or weekly basis.

Résumé

L'idée selon laquelle l'alimentation animale est un facteur de risque ayant été négligé est apparue à la suite de l'introduction du VDEP dans les troupeaux de porcs américains en mai 2013. Avant cette date, les ingrédients alimentaires et les rations préparées n'étaient pas considérés comme des véhicules potentiels de transport et de transmission d'agents pathogènes, et aucune mesure standard de biosécurité n'était en place, même si les installations porcines recevaient souvent de nouveaux produits et fournitures sur une base quotidienne ou hebdomadaire.

Introduction

The link between diet and disease transmission has raised concerns that U.S. herds could become infected with foreign pathogens through contaminated feed and feed ingredients originating from countries with endemic disease and lax sanitation and quality assurance procedures. Experimental data has already demonstrated that some feed ingredients, particularly soy-based products, can support the viability of at least three significant viral pathogens of swine (i.e., Classical Swine Fever or CSF, ASF, and PRV) (Dee and others, 2018, Niederwerder and others, 2019). ASF survival has been successfully confirmed in a total of 9 distinct feed ingredients, including three soy-based products, choline chloride, three pet diets, pork sausage casings, and complete feed exported from China to the U.S. (Dee et al., 2018). Two key variables that influence the viability of viruses in feed include the specific virus's phenotypic properties and the feed matrix that is produced by specific feed ingredients and additives. Some matrices are more conducive to survival and can support several different viruses at the same time.

New Knowledge

This discovery had added significance in 2018, when ASF was first isolated in Chinese swine herds. As the US imports soy-based products from several ASFV-positive countries, it becomes important to quantity the amount and type of these products that enter the US and to identify at what seaports does this importation occur. Specifically, the type and quantity of soy-based feed ingredients and specific ports of entry (POE) was evaluated during 2018 and 2019 using data from the International Trade Commission Harmonized Tariff Schedule website. In 2018, 104,707 metric tons (MT) of soy-based ingredients (soybeans, soybean meal, soy oil cake and soy oil) were imported to the US with 52.6 % (55,101 MT) originating from China and 42.9% (44,775 MT) originating from the Ukraine. In 2019, 73,331 MT entered the US with 54.7% (40,143 MT) originating from the Ukraine and 8.4% (6182 MT) coming from China. Regarding POEs, approximately 81% of soy-based imports from China entered the US at San Francisco/Oakland, CA (60.36%) and Seattle, WA (20.54%), while 89.4-100% entered from the Ukraine at New Orleans, LA and Charlotte, NC. The approach allows for the identification and quantification of potential channels of foreign disease entry to the US and allows for the focusing of mitigation efforts and resources at high-risk points (Patterson et al, 2020).

Industry Actions

Contaminated feed and feed ingredients are now widely recognized as likely vehicles for the transport and transmission of viral pathogens, highlighting the need for improved biosecurity policies and procedures for imported products intended for use in animal diets. In Canada, restrictions are placed on feed ingredients imported from countries known to be positive for ASF virus that are known to enhance the survival of pathogens, predominantly soy-based products. The Canadian Food Inspection Agency has also established secondary control zones around all national seaports where these high-risk ingredients are received. After arrival, products must be stored under controlled environmental conditions for a specified interval to all adequate time for viral decay prior to distribution to milling facilities. Similar programs have been implemented in Australia and the European Food Safety Authority recently released their scientific opinion that the risk of ASFV movement between countries via feed is "low but cannot be ignored". In their evaluation, the risk of feed was ranked as a higher risk than transport vehicles returning from ASFV-positive countries as well as blood products and bedding.

The situation is complex in terms of challenges, but also presents unique opportunities to advance agricultural biosecurity. Major challenges reflect the vast scope and magnitude of the problem. Even though the risk of pathogen entry through contaminated feed ingredients may be low, the associated consequences associated with a lapse would have extremely negative effects on animal agriculture and commerce. The growing complexity of international trade and expanding movement of feed-related products across multiple borders present constantly changing circumstances that are increasingly difficult to manage and track. However, new, and emerging technologies are developing at an equally rapid pace and can often be leveraged to further enhance the management of these new risk factors at the transboundary and domestic levels. The use of some chemical additives has been shown to have a negative effect on the survival of viral pathogens in products intended for use as animal feed ingredients. A comparison of 15 different additives, which included organic acids, fatty acid blends, formaldehyde-based products, and essential oils, was recently conducted, and the resulting data showed that pigs fed diets with an additive had significantly better health and performance following challenges with PEDV, PRRS, and Seneca virus A as compared to cohorts consuming non-mitigated feeds. (Dee, 2020) Similar results were observed in challenges with ASF (Niederwerder, 2020), suggesting the use of a validated additive may be efficacious in reducing risks associated with viral-contaminated feed ingredients.

Expanding knowledge on the half-lives of viruses found in essential animal feed ingredients has led to science-based protocols in the U.S., which allow these materials to be safely introduced from high-risk countries. This approach is referred to as "Responsible Imports" and relies on a comprehensive risk assessment process that considers: 1) the absolute necessity of importing the material(s); 2) the availability of alternative ingredients that can be obtained through other sources (i.e., countries free of foreign diseases of concern); 3) prevalence of specific virus(es) regarded as credible threats; 4) access to reliable data that describes the half-lives of these agents in designated ingredients and their substrates; 5) projected transport times of feed substrates, from source to end destination; 6) mitigation methods and strategies that can be implemented to reduce viral load during transit; and 7) optimal storage temperatures and times that will eliminate residual virus from various ingredients after receipt and prior to use. The relatively new concept of "feed quarantine" is rapidly evolving, as production companies design storage facilities to safely accommodate incoming products and facilitate secure, long-term trade with a wider range of international partners. (Patterson et al., 2019)

In contrast to the rapid industry-based response to the PEDV epidemic, the risk of feed was immediately downplayed by US governmental agencies, resulting in criticism by the Government Accountability Office (GAO). In their report, the GAO found fault with USDA's lack of response to the outbreaks in 2013 and 2014, and for insufficient actions taken afterward to prevent future outbreaks. In addition, it stated that the government is not fully prepared to track and respond to emerging diseases that could harm animal health, decrease availability of food, and increase food prices. Unfortunately, despite the growing body of scientific evidence in support of the risk as it pertains to other pathogens, the risk of feed ingredients continues to be ignored, thereby continuing to expose US agriculture to the risk of the introduction of foreign animal diseases via contaminated feed ingredients.

Moving Forward

Currently, all parties continue to work together to reach consensus. Research on feed risk mitigation is ongoing and results are being communicated between all parties. Industry stakeholder groups and federal agencies continue to interact via a national task force, focusing on the risk of foreign animal disease entry through contaminated feed. In the end, it is hoped that these efforts will stimulate communication and collaboration between the feed and livestock industries, and governmental agencies, furthering the emerging concept of "global feed biosecurity".

References

Dee, S.A., Bauermann, F.V., Niederwerder, M.C., Singrey, A., Clement, T., de Lima, M, Long, C., Patterson, G., Sheahan, M.A., Stoian, A.M.M., Petrovan, V., Jones, C.K., De Jong. J., Ji, J., Spronk, G.D., Minion, L., Christopher-Hennings, J., Zimmerman, J.J., Rowland, R.R.R., Nelson, E., Sundberg, P. & Diel DG. (2018). Survival of viral pathogens in animal feed ingredients under transboundary shipping models. *PLoS ONE.* 13(3):e0194509. https://doi.org/10.1371/journal.pone.0194509.

Dee S, Niederwerder MC, Edler Roy, Hanson D, Singrey A, Cochrane R, Spronk G, Nelson E. An evaluation of additives for mitigating the risk of virus-contaminated feed using an ice block challenge model (2020). *Transbound Emerg Dis* doi:10.1111/tbed.13749.

Niederwerder MC, Dee S, Stoian AMM, Constance LA, Olcha M, Petrovan V, Patterson G, Cino G, Rowland RRR. (2020). Mitigating the risk of African swine fever virus in feed with antiviral chemical additives. *Transbound Emerg Dis*. doi:10.1111/tbed.13699.

Patterson, G, Niederwerder, M.C. & Dee, S.A. (2019). Risks to animal health associated with imported feed ingredients. Journal of the American Veterinary Medical Association, 1;254(7):790-791. https://doi.org/10.2460/javma.254.7.790.

Patterson G, Niederwerder, MC, Spronk, G and Dee, SA. (2020). Quantification of soy-based feed ingredient entry from ASFV-positive countries to the United States by ocean freight shipping and associated seaports. *Transboundary and Emerging Diseases* doi:10.111/tbed.13881.

ADM
Unlocking Nature. Enriching Life.
Building World-Class Nutrition With You
As a global leader in animal nutrition, ADM partners with you to provide high-quality, sustainable nutrition that meets all of your needs. With our experienced species specialists, a continuously growing product portfolio and access to global learnings, we work with you to advance nutritional performance and develop solutions to drive success today and tomorrow.
Your Edge. Our Expertise.
Lethbridge, AB 403-320-2659
Saint-Hyacinthe, PQ 450-773-7111
Woodstock, ON 519-539-2091

Moderating the Effect of Coccidia and Necrotic Enteritis Challenge Using Non-Pharmaceutical Means

Modérer l'effet d'une exposition aux coccidies et à l'entérite nécrotique par l'emploi de solutions non pharmaceutiques

Mohamad Mortada and Milan Hruby
Archer Daniels Midland Animal Nutrition, Decatur, IL, USA
Milan.Hruby@adm.com

Abstract

One of the challenges facing the global poultry industry is the reduction or complete ban of antibiotic growth promotors (AGPs) and other pharmaceuticals in feed. This situation contributes to increased incidents in enteric diseases, mainly coccidiosis and necrotic enteritis, resulting in detrimental effects to bird's heath and growth performance. Coccidiosis is a disease caused by a protozoan parasite of the genus *Eimeria*. *Eimeria* species are transmitted via the fecal-oral route and enter the intestinal tract, replicate and cause damage to the epithelial layers. Coccidiosis increases the susceptibility of birds to other diseases, mainly affecting the intestinal epithelial barrier, and is mostly associated with necrotic enteritis. This syndrome involves toxicosis caused by *Clostridium perfringens* toxins. Necrotic enteritis is estimated to cause $6 billion per year in losses mostly due to its subclinical form. Enteric diseases generally cause inflammation (clinical or sub-clinical), reduced appetite, decrease nutrient absorption due to "leaky gut" and dysbiosis in the gut microbiota. Feed additives have been employed as AGP alternatives that decrease the host susceptibility to diseases by modulating the innate immunity via exploiting natural mechanisms in the host. Birds rely heavily on their innate immunity, the first line of non-specific and rapid defense against pathogens. Recent studies show that feed additives influence the microbiome composition and modulate gut health. This review will discuss the use of feed additives, such as yeast-based prebiotics, probiotics, phytogenic compounds, organic acids and enzymes, for alleviating coccidiosis and necrotic enteritis.

Key words: Coccidiosis, necrotic enteritis, feed additives, immunity, microbiome

Résumé

L'un des défis auxquels est confrontée l'industrie avicole mondiale est la réduction ou l'interdiction totale de l'utilisation d'antibiotiques promoteurs de croissance (APC) et d'autres produits pharmaceutiques dans les aliments des oiseaux. Cette situation contribue à augmenter les cas de maladies entériques, principalement de coccidiose et d'entérite nécrosante, ce qui a des effets néfastes sur la santé et les performances de croissance des oiseaux. La coccidiose est une maladie causée par un parasite protozoaire du genre Eimeria. Les différentes espèces d'Eimeria sont transmises par la voie fécale-orale et pénètrent dans le tube digestif, se multiplient et causent des dommages aux couches épithéliales. La coccidiose augmente la sensibilité des oiseaux à d'autres maladies, agissant principalement sur la barrière épithéliale intestinale, et est surtout associée à l'entérite nécrosante, un syndrome qui implique une toxicose causée par les toxines de Clostridium perfringens. On estime que l'entérite nécrosante, et principalement sa forme subclinique, entraîne des pertes de 6 milliards $ par année. En général, les maladies entériques provoquent de l'inflammation (clinique ou subclinique), une baisse de l'appétit, une diminution de l'absorption des nutriments (hyperperméabilité intestinale) et d'une dysbiose. Des additifs alimentaires ont été utilisés comme solutions de remplacement aux APC pour diminuer la sensibilité de l'hôte aux maladies en modulant l'immunité innée par l'exploitation de mécanismes naturels chez l'hôte. Les oiseaux dépendent fortement de leur immunité innée, laquelle constitue la première ligne de défense non spécifique et rapide contre les agents pathogènes. Des études récentes montrent que les additifs alimentaires influencent la composition du microbiome et modulent la santé intestinale. Cet exposé portera sur l'utilisation d'additifs alimentaires, tels que les prébiotiques à base de levure, les probiotiques, les composés phytogènes, les acides organiques et les enzymes, pour atténuer la coccidiose et l'entérite nécrosante, et s'attardera principalement aux interactions entre la nutrition, la réponse immunitaire et le microbiome.

Components of Gut Health

The intestine is a complex organ that comprises regions with distinct structural and physiological functions specialized in digestion and nutrient absorption. Simultaneously, the gut represents the primary contact site with foreign antigens and pathogens that can enter, reside and disseminate to the internal organs. For this reason, the gut harbors the majority of immune cells, referred to as gut-associated lymphoid tissues, when compared to other tissues. Oral tolerance to feed particles and commensal bacteria and efficient recognition and elimination of pathogens are two gut-specific functions that affect immunity (Smith et al., 2022). Recently, gut health has been gaining great attention for its role in the general health of animals. By definition, gut health is the number of physiological, microbiological, physical, and immunological functions working together to maintain intestinal homeostasis (Kogut, 2019). This paragraph will briefly discuss gut health moving from the intestinal lumen that contains commensal bacteria, mucosal layer, and various antimicrobial peptides to the epithelial monolayer and finally discusses the gut-associated lymphoid organs in the lamina propria.

Commensal bacteria

Commensal bacteria play a major role in gut health by competing with pathogens for attachment sites and nutrients. One of the clear roles of commensal bacteria has been demonstrated by Nurmi and Rantala (1973) that showed that gut microbiota from healthy adult chickens could protect newly born chicks against *Salmonella* invasion. Commensal bacteria produce different antimicrobial peptides, known as bacteriocins, that have a wide range of antimicrobial activity against various pathogens (Vieco-Saiz et al., 2019). In addition, some commensal bacteria ferment indigestible complex carbohydrates and produce short-chain fatty acids SCFA, primarily acetate, propionate, and butyrate. Short-chain fatty acids have a bactericidal activity mainly against gram-negative bacteria (Ricke, 2003). Butyrate is known to be a source of energy for enterocytes and to modulate gut health in poultry (Bedford and Gong, 2018).

Gut lumen

The first physical barrier against pathogen invasion is mucus, a glycoprotein produced by goblet cells. Mucus traps pathogens that can be expelled from the intestine by the luminal flow. Intestinal IgA is the main antibody class in mucosal secretions and plays a major role in neutralizing gut pathogens and controlling commensal bacteria (Broom, 2018). IgA is in the form of dimer in mucosal secretions and forms a complex with the secretory component from the surface of the epithelial cells that protect the antibody from endogenous proteases (Rautenschlein, 2019). Another major component of the intestinal lumen is host defense peptides (HDP) which are antimicrobial peptides produced by the host and are known to have a wide range of antimicrobial activities against different bacteria, viruses, protozoa, and fungi. Defensins and cathelicidins are the most characterized HDP in chickens (Cuperus et al., 2013).

Epithelial layer

Another physical barrier is the epithelial monolayer after the mucosal lining (lumen). Epithelial cells are held together by complex protein structures known as tight junctions. Villus structures protrude the gut, followed by indentations known as crypts, increasing the gut's surface area. Intestinal epithelial cells play a major role in digestion and their function as a physical barrier. In addition, intestinal epithelial cells have a significant role in innate immunity by expressing pathogen recognition receptors (PRR). PRR monitor and respond to pathogens by recognizing invariant molecular motifs of pathogens known as Pathogen-associated molecular patterns (PAMPs).

Gut-associated Lymphoid Tissue (GALT)

The gut-associated lymphoid tissue is comprised of lymphoid cells distributed in the lamina propria. Chickens lack lymph nodes and rely heavily on lymphoid aggregates within the lamina propria, including Meckel's diverticulum, payer's patches, and cecal tonsils. The GALT is a major immune tissue that harbors most immune cells when compared to other immune tissues (Smith et al., 2022). The epithelial monolayer consists of microfold cells (M cells) that are specialized epithelial cells that sample and deliver contents from the gut lumen to antigen-presenting cells like dendritic cells and macrophages. Antigen-presenting cells delivered antigens to lymphoid follicles rich in B and T lymphocytes. The chicken hindgut is known to harbor more organized lymphoid tissue than the foregut (Smith et al., 2022).

Enteric infections in poultry: The case of coccidiosis and necrotic enteritis

The gut is in constant contact with pathogens, and the mucosal layer and the epithelial layer are the first physical barriers to entry. Gut pathogens can be localized, colonize and persist in the intestinal lumen and cause little damage, such as *Campylobacter*. Another type of pathogens is intracellular pathogens that reside in the intracellular epithelial layer and can cause acute damage, such as *Eimeria*. Other pathogens utilize the intestine to disseminate into internal tissues, and therefore localized immune response is critical for controlling the systemic infection. Innate immunity is the first line of defense activated by pathogen recognition receptors (PRR) such as TLR that are expressed in epithelial cells. These receptors respond to pathogens by recognizing invariant molecular motifs of pathogens known as Pathogen-Associated molecular patterns (PAMPs). In addition, the GALT, such as payer's patches, is lined with M cells that constantly sample the lumen to detect pathogens. Once the innate immune system initiates pathogen recognition, heterophils infiltrate to kill the pathogen or act as antigen-presenting cells (APC) that help recruit other immune cells. NK cells, macrophages, and TCR $\gamma\delta$+ T cells play a major role in limiting infection during the early stages (Smith et al., 2022). After the breakdown and presentation of the pathogen on MHC of APC, the adaptive immune cells play a major role in the resolution of the infection. Adaptive immunity is mainly divided into humoral immune response mediated by B cells that later differentiate into plasma cells and cell-mediated immunity that is mediated by CD4+ helper T cells that help B cells T cells and macrophages, and CD8+ cytotoxic T cells (Rautenschlein, 2019). Immune response against infectious agents does not always lead to the clearance of the pathogen and can often be damaging for the host's tissues.

Coccidiosis

Coccidiosis is a disease caused by a protozoan parasite of the genus *Eimeria*. *Eimeria* species are transmitted via the fecal-oral route and are intracellular parasites that enter the intestinal tract, replicate and cause damages to the epithelial layers (McDougald and Fitz-Coy, 2008). This disruption to the gut barrier, commonly called "leaky gut," results in detrimental effects that include reduction in feed intake, digestibility and nutrient absorption, blood loss, and dehydration. Coccidiosis can be mild or severe depending on the number of ingested oocysts. In poultry, nine *Eimeria* species are known to infect chickens and cause distinct lesions that affect different parts of the GIT (Long, 1985; Pellerdy, 1974). Gross lesion scores of different sections of the GIT are recorded to determine the severity of infection. The most commonly used lesion scoring system uses a graded scale from 0-4, where 0 is for the absence of lesions and 4 is the maximum lesion (Johnson and Reid, 1970). The eimerian life cycle involves oral ingestion of sporulated oocyst (usually sporulate in wet poultry litter), invasion of enterocytes (zoite stage), development of asexual (schizonts) and sexual (gametocytes) reproduction phases, fertilization, and release of oocysts in feces (Smith et al., 2022). Therefore, efficient immune response to *Eimeria* has to be an intracellular response, rapid, and dependent on the primary immune response (Smith et al., 2022). The immune response to *Eimeria* is species-specific and, in some cases, strain-specific such as *E. maxima*. *Eimeria maxima* are the most immunogenic *Eimeria* species. Studies have shown that IFN- -γ has anticoccidial properties and is critical in controlling *Eimeria* infection and cycling. Rothwell et al. (2004) showed that chicken lines more susceptible to *E. maxima* infection have a higher expression of IL-10 mRNA both pre and post-infection in the spleen and higher expression of IL-10 mRNA in the intestine post-infection when compared to resistant lines. Infection with *Eimeria* induces heterophil infiltration, activation of innate NK cells, and antigen-specific B and T cells (Shirley et al., 2005). However, the immune response of chickens to *Eimeria* depends on the age, host genetics, and species of *Eimeria* (Smith et al., 2002). Historically, coccidiosis has been controlled by using different chemotherapeutics and live vaccines. Even though live and attenuated vaccines have been used to decrease the disease severity and oocysts shedding, live vaccines can induce a light infection and decrease body weight and feed conversion. Coccidiosis vaccines rely mainly on inducing immunological memory to protect against secondary infections. A study by Rose and Hesketh (1979) found that B cells play a limited role in the primary infection and re-challenge infection in bersectomised chickens. The authors concluded that CD4+ (helper T cells) but bot CD8+ were essential in resisting *E. maxima* primary infection. Depleting chicken CD4+ helper T cells using anti-CD4 demonstrated that CD4+ cells play a role in *E. tenella* primary infection but not *E. acervulina* (Trout and Lillehoj,1996). *Eimeria* infection, mainly *E. maxima* and *E. acervulina*, has been shown to increase the susceptibility of birds to Necrotic Enteritis that involves the Gram-positive anaerobic bacterium *Clostridium perfringens* (Immerseel et al., 2004).

Necrotic enteritis

Necrotic enteritis is associated with the Gram-positive anaerobic bacterium *Clostridium perfringens*. The disease is mainly caused by *C. perfringens* toxins that induce intestinal necrosis. *Clostridium perfringens* is a ubiquitous and commensal bacterium that is commonly found in the chicken gut. NE is primarily diagnosed by the presence of macroscopic lesions with the presence of Gram-positive *Clostridium perfringens*. Necrosis is not limited to intestinal villi but can also reach internal tissues like the liver, heart, kidney, and bursa. Necrotic enteritis usually causes a sudden spike in mortality. In addition, NE has subclinical forms that cause reduced growth and efficiency, leading to significant economic losses to the poultry industry. *Clostridium perfringens* strains are classified into seven types A – G based on the toxins produced. Poultry pathogenic *C. perfringens* strains belong to types A, C, and G. The pathogenesis on NE is not completely understood. Researchers focused initially on the CPA toxin, and studies later showed that CPA-deleted mutants could still be infectious *in vivo*. The focus shifted to the NetB toxin, and studies showed that NetB positive strains could reproduce NE lesions but not NetB-negative strains. Recently studies showed that NetB is associated with virulent CP strains; however, it is not the only virulent factor since some NetB negative strains could still cause NE lesions experimentally (Rood et al., 2016). Historically, NE was controlled by controlling the growth of *Clostridium perfringens* by the use of antibiotics, namely bacitracin (narrow-spectrum) and virginiamycin (broad-spectrum) activity against Gram-positive bacteria (LaVorgna et al., 2013). In addition, NE was controlled by controlling the exposure to coccidiosis by anticoccidial drugs such as ionophores and chemicals (Opengart and Songer, 2008). Most NE outbreaks occur at around three weeks or later and could be associated with maternal anti-CPA antibodies that usually wane at three weeks of age (Heier et al., 2001). Numerous attempts have been made to develop NE vaccines utilizing different *C. perfringens* toxins and delivery systems (Opengart and Songer, 2008). The immune response of chickens to necrotic enteritis is mainly linked to the NE challenge model utilized in the studies. Table 1 summarizes the various NE challenge models. Necrotic enteritis models rely on either creating an ecological niche or providing nutrients, mainly proteins and amino acids, for the proliferation of *C. perfringens* (Antonissen et al., 2016). Necrotic enteritis challenge models induce bacterial dysbiosis in the gut that is observed by a reduced abundance of immune-modulating segmented filamentous bacteria and lactic acid and butyrate-producing bacteria (Antonissen et al., 2016).

Table 1. Published necrotic enteritis challenge models.

NE model	Reference
High NSP in diet	Annett et al., 2002
Fish meal	Drew et al., 2004
Mycotoxins (deoxynivalenol and fumonisins)	Antonissen et al., 2015; Antonissen et al., 2014
Coccidiosis (*Eimeria maxima* and *Eimeria acervulina*	Immerseel et al., 2004
Dietary calcium source and particle size	Zanu et al., 2020
Meat and bone meal	Zanu et al., 2020

Non-Antibiotic Growth Promoters for alleviating coccidiosis and necrotic enteritis

The poultry industry has been under immense pressure to reduce the reliance on antibiotics in feeds. Non-antibiotic growth promotors (non-AGPs), such as feed additives, have been tested for their use as on-farm interventions that can alleviate the detrimental effects of coccidiosis and necrotic enteritis. Research feed additives against coccidiosis and necrotic enteritis include probiotics, prebiotics, phytogenic compounds, enzymes, organic acids, and organic trace minerals. Table 2 summarizes the hypothesized mechanism of action for every feed additive intervention and brief literature results. It is worth noting that feed additive combinations also provide an advantage for providing synergistic biological activity. However, combinations should also be economical (below $2/short ton). Product combinations can also be beneficial for decreasing pathogen resistance by avoiding the repetitive use of a single product. One clear example of the benefit of feed additive combinations could be the bioshuttle coccidiosis control programs in no-antibiotics-ever (NAE) production.

Table 2. Non-pharmaceutical interventions for alleviating necrotic enteritis and coccidiosis in poultry.

Intervention	Description	Hypothesized Mechanisms	Challenge	Brief results	References
Probiotics	Spore-formers (*Bacillus*); Lactic acid-producing bacteria; yeast	Competitive exclusion, immune modulation, microbiome, SCFA, bacteriocins	Coccidiosis	Probiotic mixture improved growth performance. *Bacillus subtilis* increased Villi height and *Lactobacillus reuteri* increase in beneficial commensal bacteria.	Giannenas et al., 2012
			Necrotic enteritis	*Bacillus subtilis* improved intestinal health and reduced *C. perfringens.*	Jayaraman et al., 2013
Prebiotics	Yeast-derived, Plant-derived	Microbiome; immune modulation (PAMPs)	Coccidiosis	*Pichia guilliermondii* in feed improved production parameters, reduced oocyst shedding, increased IL-1 and decreased IL-10 gene expression in the ceca, reduced *E. coli*, and *Salmonella* in ceca.	Shanmugasundaram et aal., 2013 a,b
			Necrotic enteritis	*Saccharomyces cerevisiae* prevented the NE-induced performance decline.	M'Sadeq et al., 2015
Phytogenic compounds	Plant extracts	Antimicrobial, antioxidant, immune modulation, metabolism, microbiome	Coccidiosis	Carvacrol, cinnamaldehyde, and capsicum oleoresin improved coccidiosis resistance through immune modulation and reduced oocyst shedding.	Lillehoj et al., 2011
			Necrotic enteritis	Carvacrol, cinnamaldehyde, and capsicum oleoresin improved BWG and reduced lesion scores in NE-challenged birds.	Lee at al., 2013
Enzymes	Phytase	Reduce dietary Ca, increase nutrient digestibility	Necrotic enteritis	High phytase reduced NE-induced drop in performance by increasing nutrient digestibility.	Zanu et al., 2020
	Xylanase	Prebiotic effect, increase nutrient digestibility	Coccidiosis	Xylanase improved growth performance, nutrient digestibility, ileal NSP concentration, and SCFA.	Craig et al., 2020
	Protease	Microbiome	Commercial conditions	Protease improved livability in turkeys and enhanced gut health by reducing avian pathogenic *E. coli.*	Kim et al., 2021
	Protease + Xylanase	Microbiome, SCFA, Immune modulation	Coccidiosis	Exogenous enzymes alleviated *Eimeria*-induced immune gene expression and unfavorable cecal fermentation pattern.	Lin and Olukosi, 2021
Organic acids	SCFA and MCFA	Antimicrobial, SCFA, microbiome, epithelial cell proliferation (butyrate)	Necrotic enteritis	Butyrate and lauric acid blend improved gut barrier function, gut microbiota, and SCFA production.	Kumar et al., 2022
Organic trace minerals	Zinc, Selenium	Gut health, antioxidant, immune modulation	Necrotic enteritis	Organic Zn reduced intestinal permeability and alleviated enteric inflammation induced by NE.	Bortoluzzi et al., 2019

Necrotic enteritis and coccidiosis interventions beyond feed additives

Poultry producers today rank restrictions on antibiotics as a key factor impacting the cost of production (Watt Poultry Survey, 2021). A widespread reduction of antibiotic usage increased broiler health and, consequently, also production challenges not seen before implementing these changes. It might be natural to consider only nutrition to help with these issues since antibiotics and ionophore coccidiostats were mostly added via feed, which means that the replacement strategies should also be feed-based. However, the issue is larger than just nutrition. Experience has shown us that nutritional interventions alone cannot fully compensate for performance losses associated with approaches like NEA (no antibiotics ever). Management, in particular, is an important factor since much needs to be done to control coccidiosis and necrotic enteritis. Issues such as bird density (Tsiouris et al.,2015a), the occurrence of wet litter (Hermans and Morgan, 2007), the use of coccidia vaccine (Williams, 2002), and temperature stress (Tsiouris et al., 2015b) can all impact NE. Limiting exposure to infectious agents through biosecurity might be a tool to reduce the incidence of NE and improve gut health (Shawkat et al., 2015). Among the general nutrition topics impacting NE and its severity are mycotoxin contaminations (Antonissen et al., 2014) but also other water and feed quality parameters such as the condition of oils and fats, presence of antinutrients in soybean meal, and retrograded starch and reduction of protein solubility in corn (Oviedo-Rondon, 2019). Grain source (Branton et al., 1987), type and level of protein (Drew et al., 2004), and mineral levels (Paiva et al., 2013) can also impact NE.

In conclusion, necrotic enteritis and coccidiosis are the top two enteric challenges facing as poultry production moves towards no-antibiotics ever production systems (NAE). Research methods are constantly advancing, especially in the field of omics that provide a valuable tool for the comprehensive understanding of challenges in the biological system (Karahalil, 2016). Therefore, research aimed at controlling coccidiosis and necrotic enteritis and understanding the functionality of omics data will elucidate the complex host-microbiome-pathogen interactions. This effort would require cross-functional teamwork between poultry producers, veterinarians, nutritionists, microbiologists, immunologists, and bioinformaticians.

References

Annett, C. B., Viste, J. R., Chirino-Trejo, M., Classen, H. L., Middleton, D. M., & Simko, E. (2002). Necrotic enteritis: effect of barley, wheat and corn diets on proliferation of *Clostridium perfringens* type A. *Avian Pathology*, *31*(6), 598-601.

Antonissen G, Van Immerseel F, Pasmans F, Ducatelle R, Haesebrouck F, et al. (2014) The Mycotoxin Deoxynivalenol Predisposes for the Development of *Clostridium perfringens*-Induced Necrotic Enteritis in Broiler Chickens. PLOS ONE 9(9).

Antonissen, G., Croubels, S., Pasmans, F., Ducatelle, R., Eeckhaut, V., Devreese, M., ... & Van Immerseel, F. (2015). Fumonisins affect the intestinal microbial homeostasis in broiler chickens, predisposing to necrotic enteritis. *Veterinary Research*, *46*(1), 1-11.

Antonissen, G., Eeckhaut, V., Van Driessche, K., Onrust, L., Haesebrouck, F., Ducatelle, R., ... & Van Immerseel, F. (2016). Microbial shifts associated with necrotic enteritis. *Avian Pathology*, *45*(3), 308-312.

Antonissen, G., Van Immerseel, F., Pasmans, F., Ducatelle, R., Haesebrouck, F., Timbermont, L., ... & Croubels, S. (2014). The mycotoxin deoxynivalenol predisposes for the development of *Clostridium perfringens*-induced necrotic enteritis in broiler chickens. *PLoS One*, *9*(9), e108775.

Bedford, A., & Gong, J. (2018). Implications of butyrate and its derivatives for gut health and animal production. *Animal Nutrition*, *4*(2), 151-159.

Bortoluzzi, C., Lumpkins, B., Mathis, G. F., França, M., King, W. D., Graugnard, D. E., ... & Applegate, T. J. (2019). Zinc source modulates intestinal inflammation and intestinal integrity of broiler chickens challenged with coccidia and *Clostridium perfringens*. *Poultry science*, *98*(5), 2211-2219.

Branton, S. L., Reece, F. N., & Hagler Jr, W. M. (1987). Influence of a wheat diet on mortality of broiler chickens associated with necrotic enteritis. *Poultry Science*, *66*(8), 1326-1330.

Broom, L. J. (2018). Gut barrier function: effects of (antibiotic) growth promoters on key barrier components and associations with growth performance. *Poultry science*, *97*(5), 1572-1578.

Craig, A. D., Khattak, F., Hastie, P., Bedford, M. R., & Olukosi, O. A. (2020). The similarity of the effect of carbohydrase or prebiotic supplementation in broilers aged 21 days, fed mixed cereal diets and challenged with coccidiosis infection. Plos one, 15(2), e0229281.

Cuperus, T., Coorens, M., van Dijk, A., & Haagsman, H. P. (2013). Avian host defense peptides. *Developmental & Comparative Immunology*, *41*(3), 352-369.

Drew, M. D., Syed, N. A., Goldade, B. G., Laarveld, B., & Van Kessel, A. G. (2004). Effects of dietary protein source and level on intestinal populations of *Clostridium perfringens* in broiler chickens. *Poultry science*, *83*(3), 414-420.

Drew, M. D., Syed, N. A., Goldade, B. G., Laarveld, B., & Van Kessel, A. G. (2004). Effects of dietary protein source and level on intestinal populations of *Clostridium perfringens* in broiler chickens. *Poultry science*, *83*(3), 414-420.

Giannenas, I., Papadopoulos, E., Tsalie, E., Triantafillou, E. L., Henikl, S., Teichmann, K., & Tontis, D. (2012). Assessment of dietary supplementation with probiotics on performance, intestinal morphology and microflora of chickens infected with *Eimeria tenella*. *Veterinary parasitology*, *188*(1-2), 31-40.

Heier, B. T., Lovland, A., Soleim, K. B., Kaldhusal, M., & Jarp, J. (2001). A field study of naturally occurring specific antibodies against *Clostridium perfringens* alpha toxin in Norwegian broiler flocks. *Avian Diseases*, 724-732.

Immerseel, F. V., Buck, J. D., Pasmans, F., Huyghebaert, G., Haesebrouck, F., & Ducatelle, R. (2004). *Clostridium perfringens* in poultry: an emerging threat for animal and public health. *Avian pathology*, *33*(6), 537-549.

Jayaraman, S., Thangavel, G., Kurian, H., Mani, R., Mukkalil, R., & Chirakkal, H. (2013). *Bacillus subtilis* PB6 improves intestinal health of broiler chickens challenged with *Clostridium perfringens*-induced necrotic enteritis. *Poultry science*, *92*(2), 370-374.

Johnson, J., & Reid, W. M. (1970). Anticoccidial drugs: lesion scoring techniques in battery and floor-pen experiments with chickens. *Experimental parasitology*, *28*(1), 30-36.

Karahalil, B. (2016). Overview of systems biology and omics technologies. Current medicinal chemistry, 23(37), 4221-4230.

Keyburn, A. L., Sheedy, S. A., Ford, M. E., Williamson, M. M., Awad, M. M., Rood, J. I., & Moore, R. J. (2006). Alpha-toxin of *Clostridium perfringens* is not an essential virulence factor in necrotic enteritis in chickens. *Infection and immunity*, *74*(11), 6496-6500.

Kim, E.J., Mussini, F., Gruber, J., Perry, M., & Remus, JC. (2021). Field evaluation of exogenous protease in commercial turkey diets. Poultry Science Association 2021 annual meeting.

Kogut, M. H. (2019). The effect of microbiome modulation on the intestinal health of poultry. *Animal feed science and technology*, *250*, 32-40.

Kumar, A., Toghyani, M., Kheravii, S. K., Pineda, L., Han, Y., Swick, R. A., & Wu, S. B. (2022). Organic acid blends improve intestinal integrity, modulate short-chain fatty acids profiles and alter microbiota of broilers under necrotic enteritis challenge. *Animal Nutrition*, *8*(1), 82-90.

LaVorgna, M., Schaeffer, J. L., Bade, D., Dickson, J., Cookson, K., & Davis, S. W. (2013). Performance of broilers fed a broader spectrum antibiotic (virginiamycin) or a narrower spectrum antibiotic (bacitracin methylene disalicylate) over 3 consecutive grow-out cycles. *Journal of Applied Poultry Research*, *22*(3), 574-582.

Lee, S. H., Lillehoj, H. S., Jang, S. I., Lillehoj, E. P., Min, W., & Bravo, D. M. (2013). Dietary supplementation of young broiler chickens with Capsicum and turmeric oleoresins increases resistance to necrotic enteritis. *British Journal of Nutrition*, *110*(5), 840-847.

Lillehoj, H. S., Kim, D. K., Bravo, D. M., & Lee, S. H. (2011, December). Effects of dietary plant-derived phytonutrients on the genome-wide profiles and coccidiosis resistance in the broiler chickens. In *BMC proceedings* (Vol. 5, No. 4, pp. 1-8). BioMed Central.

Lin, Y., & Olukosi, O. A. (2021). Exogenous Enzymes Influenced *Eimeria*-Induced Changes in Cecal Fermentation Profile and Gene Expression of Nutrient Transporters in Broiler Chickens. *Animals*, *11*(9), 2698.

Long, P. (1985). The biology of Coccidia. In *The biology of coccidia* (pp. 502-502).
McDougald LR, Fitz-Coy SH. Coccidiosis (Chapter 28-Protozoal Infections). Diseases of Poultry (12th Ed.). Saif YM et al.(ed.). Wiley-Blackwell Publishing, Ames, Iowa. 2008.

M'Sadeq, S. A., Wu, S. B., Choct, M., Forder, R., & Swick, R. A. (2015). Use of yeast cell wall extract as a tool to reduce the impact of necrotic enteritis in broilers. *Poultry science*, *94*(5), 898-905.

Nurmi, E., & Rantala, M. (1973). New aspects of *Salmonella* infection in broiler production. *Nature*, *241*(5386), 210-211.

Opengart, K., & Songer, J. G. (2008). Necrotic enteritis. *Diseases of poultry*, *1*, 972-976.
Oviedo-Rondón, E. O. (2019). Holistic view of intestinal health in poultry. *Animal Feed Science and Technology*, *250*, 1-8.

P. G. Hermans & K. L. Morgan (2007) Prevalence and associated risk factors of necrotic enteritis on broiler farms in the United Kingdom; a cross-sectional survey, *Avian Pathology*, 36:1, 43-51.

Paiva, D. M., Walk, C. L., & McElroy, A. P. (2013). Influence of dietary calcium level, calcium source, and phytase on bird performance and mineral digestibility during a natural necrotic enteritis episode. *Poultry Science*, *92*(12), 3125-3133.

Patterson, J. A., & Burkholder, K. M. (2003). Application of prebiotics and probiotics in poultry production. *Poultry science*, 82(4), 627-631.

Pellerdy, L. (1974). Anseriformes. *Coccidia and: coccidiosis*, 158-170.

R. B. Williams (2002) Anticoccidial vaccines for broiler chickens: Pathways to success, *Avian Pathology*, 31:4, 317-353.

Rautenschlein, S. (2019). The Avian Immune System. *Diseases of Poultry*, 80.

Ricke, S. C. (2003). Perspectives on the use of organic acids and short chain fatty acids as antimicrobials. *Poultry science*, *82*(4), 632-639.

Rood, J. I., Keyburn, A. L., & Moore, R. J. (2016). NetB and necrotic enteritis: the hole movable story. *Avian Pathology*, *45*(3), 295-301.

Rose, M. E., & Hesketh, P. A. T. R. I. C. I. A. (1979). Immunity to coccidiosis: T-lymphocyte- or B-lymphocyte-deficient animals. *Infection and immunity*, *26*(2), 630-637.

Rothwell, L., Young, J. R., Zoorob, R., Whittaker, C. A., Hesketh, P., Archer, A., ... & Kaiser, P. (2004). Cloning and characterization of chicken IL-10 and its role in the immune response to *Eimeria maxima*. *The Journal of Immunology*, *173*(4), 2675-2682.

Shanmugasundaram, R., Sifri, M., & Selvaraj, R. K. (2013). Effect of yeast cell product (CitriStim) supplementation on broiler performance and intestinal immune cell parameters during an experimental coccidial infection. *Poultry Science*, *92*(2), 358-363.

Shanmugasundaram, R., Sifri, M., & Selvaraj, R. K. (2013). Effect of yeast cell product supplementation on broiler cecal microflora species and immune responses during an experimental coccidial infection. *Poultry science*, *92*(5), 1195-1201.

Shawkat A. M'Sadeq, Shubiao Wu, Robert A. Swick, Mingan Choct (2015). Towards the control of necrotic enteritis in broiler chickens with in-feed antibiotics phasing-out worldwide. Animal Nutrition, Volume 1, Issue 1, Pages 1-11, ISSN 2405-6545.

Shirley, M. W., Smith, A. L., & Tomley, F. M. (2005). The biology of avian *Eimeria* with an emphasis on their control by vaccination. *Advances in parasitology*, *60*, 285-330.

Smith, A. L., Hesketh, P., Archer, A., & Shirley, M. W. (2002). Antigenic diversity in *Eimeria maxima* and the influence of host genetics and immunization schedule on cross-protective immunity. *Infection and Immunity*, *70*(5), 2472-2479.

Smith, A. L., Powers, C., & Beal, R. (2022). The avian enteric immune system in health and disease. In *Avian immunology* (pp. 303-326). Academic Press.

Trout, J. M., & Lillehoj, H. S. (1996). T lymphocyte roles during *Eimeria acervulina* and *Eimeria tenella* infections. *Veterinary Immunology and Immunopathology*, *53*(1-2), 163-172.

V. Tsiouris, I. Georgopoulou, C. Batzios, N. Pappaioannou, R. Ducatelle & P. Fortomaris (2015a) High stocking density as a predisposing factor for necrotic enteritis in broiler chicks, Avian Pathology, 44:2, 59-66.

V. Tsiouris, I. Georgopoulou, C. Batzios, N. Pappaioannou, R. Ducatelle & P. Fortomaris (2015b) The effect of cold stress on the pathogenesis of necrotic enteritis in broiler chicks, *Avian Pathology*, 44:6, 430-435.

Vieco-Saiz, N., Belguesmia, Y., Raspoet, R., Auclair, E., Gancel, F., Kempf, I., & Drider, D. (2019). Benefits and inputs from lactic acid bacteria and their bacteriocins as alternatives to antibiotic growth promoters during food-animal production. *Frontiers in microbiology*, *10*, 57.

Zanu, H. K., Kheravii, S. K., Bedford, M. R., & Swick, R. A. (2020). Dietary calcium and meat and bone meal as potential precursors for the onset of necrotic enteritis. *World's Poultry Science Journal*, *76*(4), 743-756.

Zanu, H. K., Kheravii, S. K., Morgan, N. K., Bedford, M. R., & Swick, R. A. (2020). Over-processed meat and bone meal and phytase effects on broilers challenged with subclinical necrotic enteritis: Part 1. Performance, intestinal lesions and pH, bacterial counts and apparent ileal digestibility. *Animal Nutrition*, *6*(3), 313-324.

Methane, Cows, and Climate Change: California Dairy's Path to Climate Neutrality

Méthane, vaches et changement climatique : La filière laitière californienne sur la voie de la neutralité climatique

Frank Mitloehner
CLEAR Center, University of California, Davis, CA, USA
fmmitloehner@ucdavis.edu

Pre-conference sponsored by / Pré-colloque commandité par :

ADM®

Abstract

Climate change is a global issue that requires comprehensive and far-reaching solutions across all economic and demographic jurisdictions. The Paris Climate Agreement, adopted in 2015, sets out a global framework to address harmful climate impacts by limiting additional global warming to well below 2 degrees Celsius (°C) (1.5 °C goal).

The U.S. dairy industry recently announced efforts to address climate change, boldly aiming for carbon neutral or better (net zero climate impact) by 2050 (Innovation Center for U.S. Dairy, 2020). As part of these important efforts, California's dairy farms are leading change and making significant progress in reducing the amount of GHG emissions released into the environment. Producing a glass of milk from a California dairy cow generates 45 percent less GHG emissions today than it did 50 years ago. This finding, recently published in the Journal of Dairy Science, comes from a life-cycle assessment of California dairy farms in 1964 and 2014, conducted by researchers at the University of California, Davis (Naranjo et al., 2020). Significant advancements in farming efficiency, feed crop yields, veterinary care, sustainable feed practices, and animal nutrition have helped reduce the environmental footprint of individual cows. Building on these gains, more can be done to lower the climate footprint of milk production in the coming decade.

Résumé

Le changement climatique est un problème mondial qui nécessite des solutions globales et de grande envergure dans toutes les juridictions économiques et démographiques. L'Accord de Paris sur le climat, adopté en 2015, définit un cadre mondial pour lutter contre les effets néfastes sur le climat en limitant le réchauffement de la planète à bien moins de 2 degrés Celsius (°C) (objectif de 1,5 °C).

L'industrie laitière américaine a récemment annoncé des mesures pour contrer le changement climatique et vise l'objectif ambitieux de la carbo-neutralité ou mieux (impact climatique net nul) d'ici 2050 (Innovation Center for U.S. Dairy, 2020). Ainsi, les exploitations laitières californiennes sont à l'avant-garde du changement et réalisent des progrès considérables dans la réduction de la quantité de GES rejetés dans l'environnement. La production d'un verre de lait d'une vache californienne génère 45 % moins d'émissions de GES aujourd'hui qu'il y a 50 ans. Cette constatation, récemment publiée dans le Journal of Dairy Science, provient d'une évaluation du cycle de vie des exploitations laitières californiennes en 1964 et 2014 réalisée par des chercheurs de l'Université de Californie, Davis (Naranjo et al., 2020). Des avancées significatives en matière d'efficacité agricole, de rendement des cultures fourragères, de soins vétérinaires, de pratiques alimentaires durables et de nutrition animale ont permis de réduire l'empreinte environnementale individuelle des vaches. En s'appuyant sur ces progrès, il est possible de faire davantage pour réduire l'empreinte climatique de la production laitière au cours de la prochaine décennie.

Introduction

Climate change is a global issue that requires comprehensive and far-reaching solutions across all economic and demographic jurisdictions. The Paris Climate Agreement, adopted in 2015, sets out a global framework to address harmful climate impacts by limiting additional global warming to well below 2 degrees Celsius (°C) (1.5 °C goal). The accord recognizes regional differences and the need for specific actions across all jurisdictions, including developed economies providing leadership and assistance to developing nations in their climate mitigation efforts.

California continues to lead the United States and world in implementing measures to achieve emissions reductions of greenhouse gases (GHGs) that advance climate change. Toward this end, California has established ambitious goals for reducing GHG emissions (Senate Bill 32) by 40 percent by 2030 and 80 percent by 2050. Senate Bill 1383 (2016) also established specific goals for reducing short-lived climate pollutants (SLCPs), such as methane, by 40 percent from 2013 levels. Ultimately, California is working toward a goal of "net-zero" carbon emissions by 2045 (Executive Order B-55-18).

The U.S. dairy industry recently announced efforts to address climate change, boldly aiming for carbon neutral or better (net zero climate impact) by 2050 (Innovation Center for U.S. Dairy, 2020). As part of these important efforts, California's dairy farms are leading change and making significant progress in reducing the amount of GHG emissions released into the environment. Producing a glass of milk from a California dairy cow generates 45 percent less GHG emissions today than it did 50 years ago. This finding, recently published in the Journal of Dairy Science, comes from a life-cycle assessment of California dairy farms in 1964 and 2014, conducted by researchers at the University of California, Davis (Naranjo et al., 2020). Significant advancements in farming efficiency, feed crop yields, veterinary care, sustainable feed practices, and animal nutrition have helped reduce the environmental footprint of individual cows. Building on these gains, more can be done to lower the climate footprint of milk production in the coming decade.

California's dairy farmers are working closely with the California Department of Food and Agriculture (CDFA) and the California Air Resources Board (CARB) to further reduce dairy methane emissions. As the efforts continue, it is also important to improve our understanding of how methane and other GHGs contribute to climate impacts, as we seek to limit warming. Leading climate scientists are now recognizing that moderately reducing methane emissions can quickly stabilize the climate pollutant's powerful impact, and further reductions can actually offset the far more damaging impacts of carbon dioxide (CO2), which accumulate in the atmosphere for hundreds of years.

California's Greenhouse Gas Emissions

California, the fifth largest economy in the world, is responsible for about 1 percent of all global GHG emissions. More than 80 percent of California's emissions come from the transportation (41 percent), industrial (23 percent) and electrical (16 percent) sectors. Even though California is the United States' largest agricultural producer—producing fruits, vegetables, nuts, livestock, and other commodities for much of the U.S. and world—the sector's GHG contribution is only 8 percent of the state's total. California's largest-in-the-nation dairy sector accounts for about half of the agricultural share, or 4 percent of the state's total GHG emissions. The U.S. dairy sector accounts for 2 percent of the nation's total GHG emissions.

While CO2 is the primary GHG driving climate warming, methane (CH4), nitrous oxide (N2O), and refrigerants are also important GHGs in California. According to CARB, carbon dioxide accounts for about 83 percent of California's GHG inventory. In comparison, methane accounts for 9 percent, and N2O accounts for about 3 percent. In addition to knowing how much of each gas is being emitted, understanding how each gas causes actual warming is most critical to fully understanding and addressing climate change. Recent work by leading climate scientists at the Oxford Martin School and Environmental Change Institute at Oxford University has shed light on important differences among these GHGs and their impact on climate change (Lynch, 2019).

Methane emissions are generated by a number of processes, both those resulting from human related activity (anthropogenic) and natural (biogenic). Fossil- fuel methane (more commonly known as "natural gas") results from the process of extracting coal or oil, or from leakage during the extraction, storage, or distribution of natural gas for homes and businesses. Fossil methane is largely converted to CO2 when we burn natural gas in our homes, factories, buildings, and other businesses.

Biogenic methane emissions are created by wetlands, rice cultivation, and ruminant livestock, as well as the waste sector, when microbes digest organic matter in our landfills and sewage treatment plants. Animal agriculture activity (all livestock) in California represents the largest source of biogenic methane emissions, accounting for roughly 55 percent of all human-related methane emissions in the state. California is the largest dairy state, producing roughly 18.5 percent of the nation's milk (USDA, 2019). The dairy livestock sector accounts for about 45 percent of all methane emitted in the state (CARB, 2015), primarily from two sources. Roughly half (55 percent) of dairy methane emissions come from manure management (storage, handling, and utilization), and the remaining 45 percent comes from enteric emissions.

In ruminant animals, methane is produced during manure decomposition as well as during enteric fermentation, where microbes decompose and ferment plant materials in the first compartment of their stomach, known as the rumen. This methane is expelled by the animal through belching.

Fossil Methane vs. Biogenic Methane

Fossil methane impacts the climate differently than biogenic methane. Fossil methane, such as natural gas, is carbon that has been locked up in the ground for millions of years and is extracted and combusted in homes and businesses. The burning of fossil methane directly transfers carbon that was stored in the ground (geologic carbon) into the atmosphere as CO2. That carbon continues to accumulate and persist in the environment, contributing to climate change for hundreds of years. Bottom line: Fossil methane increases the total amount of carbon in the atmosphere, which drives warming.

Biogenic methane from cows is part of a natural carbon cycle, where after about 12 years it is removed from the atmosphere. As part of photosynthesis, plants capture CO2 from the atmosphere, absorbing the carbon and releasing oxygen. That carbon is converted into carbohydrates in the plant, which are then consumed by the cows, digested, and released from the cows as methane (CH4). After about 12 years in the atmosphere, that methane is oxidized and converted into CO2. These carbon molecules are the same molecules that were consumed by cows in the form of plants. As part of the biogenic carbon cycle, the carbon originally utilized by the plant is returned to the atmosphere, contributing no net gain of CO2.

Global Warming Potential of California's Primary Greenhouse Gases

Each GHG captures and retains heat at a unique rate, known as its global warming potential or GWP (as shown in Table 1 as GWP 100). For example, CH4 has 28 times the warming potential of CO2 over a 100-year period. Understanding how emissions impact global climate; however, requires consideration of not just the potency, but also how long each type of GHG will last in the atmosphere (atmospheric lifetime).

This is particularly important for methane, as it is a SLCP, with emissions breaking down after about 12 years (Farlie 2019; Lynch, 2019). In contrast, a significant proportion of CO2 emissions are expected to persist in the atmosphere for hundreds of years, or even longer (Farlie, 2019; Lynch, 2019). As a result, the treatment of all GHGs as CO2 equivalent (CO2e) using GWP—and failure to consider the atmospheric removal of SLCPs—misrepresents the impact of methane on future warming (Frame et al., 2018; Cain, 2018). Recognizing this shortcoming, leading climate scientists expanded on GWP and developed GWP* (GWP-Star), which quantifies a GHG's actual warming potential, instead of just its CO2 equivalence, by factoring in how much more or less methane is being emitted from a source over a period of time. GWP* appropriately builds on the conventional GWP approach employed in typical reporting of GHG emissions (Lynch, 2019). GWP* recognizes the rate and degradation of methane emissions, in addition to the total amount of CO2 and other long-lived gases emitted (Lynch, 2019; Cain, 2018; Frame et al., 2018).

Climate Impact Potential/GWP* (GWP-Star)

Recognizing the important differences in how methane and carbon dioxide affect climate change is critical to quantifying their actual climate impacts. GWP* was developed to better and more completely account for the warming impacts of short- and long-lived gases and better link emissions to warming (Cain, 2018). GWP* is still based on GWP, but recognizes how different gases such as methane affect warming (Cain, 2018).

Because CO2 emissions last in the atmosphere for so long, they can continue to impact warming for centuries to come. New emissions are added on top of those that were previously emitted, leading to increases in the total atmospheric stock or concentration of CO2. As a result, when additional CO2 is emitted, additional global warming occurs (Frame et al., 2018).

In contrast, methane emissions degrade in the atmosphere relatively quickly, after about 12 years, and do not act cumulatively over long periods of time. For a constant rate of methane emissions, one molecule in effect replaces a previously emitted molecule that has since broken down. This means that for a steady rate of methane release—as emitted by a constant number of dairy cows, for example—the amount of methane in the atmosphere (concentration) stays at the same level and does not increase. As a result, when a steady amount of methane is emitted for more than 12 years, no additional global warming occurs (Frame et al., 2018).

This improved understanding of how short-lived versus long-lived emissions affect climate differently is critical to addressing further global warming. Limiting climate change requires that we bring emissions of CO2 and other long-lived GHGs down to net-zero (Frame et al., 2018). For methane, however, it is possible to have steady ongoing emissions that do not result in additional warming (Frame et al., 2018).

This does not mean that methane can or should be ignored. Increasing methane emissions would result in significant warming. Because of its short-lived atmospheric lifetime, reducing methane emissions can lead to a drop in atmospheric concentration relatively quickly. So, reducing methane emission rates presents an important mitigation opportunity, which could reverse some of the warming the planet has already experienced (Lynch, 2019). Put simply, a reduction in methane emissions has climate cooling effects (Cain, 2018).

Climate-Neutral Dairy: Achievable in California's Near Future

Understanding how methane impacts global warming is critical to understanding the role of dairy production as a contributor to climate change. California's dairy sector is an excellent case in point. It is no longer growing and expanding production. The number of milk cows raised in the state reached a peak in 2008, around the same time that California passed its first climate policy (2006). Since then, the number of cows has declined by a little more than 7 percent (CDFA, 2017). Total milk production has also decreased in recent years. As a result, the amount of methane in the atmosphere contributed by California milk production is less today than in 2008, as more methane is being removed from the atmosphere each year through its natural breakdown process (biogenic methane cycle) than is created by fewer dairy cows.

California dairy farms are also taking important, voluntary steps to further reduce methane from farms by installing anaerobic digesters designed to capture methane. Other projects, such as compost pack barns and solid separators, are designed to reduce methane production on farms. More than 213 dairy methane reduction projects have been incentivized with state funds to date (CDFA, 2019). These efforts alone are expected to achieve more than 2.2 million additional metric tons of GHG reduction each year, as the projects continue to be implemented (CDFA, 2019). Hundreds of additional dairy methane reduction projects are expected in future years.

As discussed earlier, enteric emissions (belching) from cows account for a significant share (45 percent) of total dairy methane emissions in California. Identifying solutions to reduce these emissions will also be necessary to meet state goals. While research into enteric emission mitigation is being conducted, and some feed additives show promise, commercially proven and cost-effective solutions are not yet available (Webinar on CARB's Analysis of Progress Toward Achieving Methane Emissions Target from Dairy and Livestock Sector, 2020).

Dairy farms also create other GHGs, such as CO2 and nitrous oxide (N2O), from the use of farm equipment for dairy management and the utilization of manure for growing crops. These emissions account for about 20 percent of all GHGs produced by the dairy production sector (Naranjo et al., 2020). Reducing or offsetting these emissions will also be necessary for the state's dairy production sector to achieve climate neutrality, or the point at which operations and resulting emissions are stable and no longer adding to global warming (no net global warming impact). California dairies are also reducing the amount of CO2 they emit into the atmosphere through the adoption of solar energy and electrification of feed mixing and water pumping operations. Fossil fuel use per unit of milk produced has dropped by 58.5 percent from 1964 to 2014 (Naranjo et al., 2020). As dairy methane emissions are reduced further below current levels, then resulting cooling effects can offset some of the remaining CO2 and other gases contributed by dairy production.

Conclusions

A continued focus on methane is necessary, as it is a powerful GHG and an important contributor to climate change. Under all scenarios, methane is significant, second only to carbon dioxide in terms of its overall contribution to global, human-driven climate change (Lynch, 2019). Over the last decade, global methane concentrations have increased (Lynch, 2019). Agriculture, including animal agriculture, is partially responsible for the increase, as dairy and meat production and consumption continue to expand globally, particularly in low- and middle-income countries. That notwithstanding, evidence is growing that shale gas production is a larger source of methane emissions than previously assumed (Howarth, 2019). Like every sector of the global economy, agriculture must do its part if we are to succeed in achieving the overarching goal of limiting global warming. Equally important, California acting alone cannot accomplish significant global dairy methane emission reductions.

Recognizing how methane impacts global climate is also critical to assessing whether the state and world are on track to meet the goals of the Paris Agreement and limit warming to well below 2°C. Comparing GHGs with each other using GWP* preserves the link between emissions and warming or cooling of the atmosphere (Schleussner et al., 2019). It also provides an informative and better suited way to assess the relative merits of different options for reducing GHG emissions, especially in ambitious mitigation scenarios (Cain, 2019). More accurate expression of mitigation efforts in terms of their direct contribution to future warming also better informs burden-sharing and long-term policies and measures in pursuit of ambitious global temperature goals (Allen, 2018; Schleussner et al., 2019).

Reducing methane emissions and achieving climate neutrality is no small undertaking. California is among the most efficient producers of milk and dairy products, and its life-cycle carbon footprint (per gallon of milk produced) is among the lowest of any region in the world. Achieving these or similar levels of production efficiency (more milk with fewer cows) is a critical first step for other dairy regions to begin stabilizing methane emissions and work toward climate neutrality. The impact of such an accomplishment would have profound climate effects. Attaining California's level of production efficiency in all global dairy production regions could reduce total global GHG emissions by as much as 1.73 percent (E. Kebreab, calculations based on Naranjo et al., 2020 and FAO & GDP, 2018).

A full understanding of the potential climate impact of all greenhouse gases is also important in ensuring effective policies are developed to address methane and other flow pollutants in line with their effects. Dairy production primarily produces flow emissions (80 percent is methane) with smaller amounts of stock emissions, such as CO_2 and N_2O (Naranjo et al., 2020). Policy or consumption decisions that trade off and result in greater concentrations of CO_2 and N_2O, while reducing methane, may ultimately leave a warmer planet behind in the long term (Frame et al., 2018).

Adopting sustainable farming practices to vastly improve production efficiency is probably the single-most important step other dairy- producing countries can take to begin to stabilize regional and global methane emissions and begin to achieve climate neutrality. The United Nations Food and Agriculture Organization (FAO) estimates that improved management practices alone could reduce net global methane emissions by 30 percent (FAO, 2019). These efforts will be critical to reduce livestock methane emissions and present important opportunities for reaching global climate mitigation targets. Further reductions in methane emissions will lead to atmospheric concentrations falling relatively quickly, which could reduce some of the warming already experienced (Lynch, 2019).

Case Study: California Dairy Methane Reduction

Fully understanding the climate cooling potential of dairy methane reduction efforts in California is critical for state regulators and policymakers. California is seeking to reduce dairy methane emissions by roughly 7.2 million metric tons (MMT) per year by 2030 (40% reduction). What will this mean for California's overall emissions reduction goal of being "net zero" by 2045?

Achieving the state's goal of reducing dairy methane emissions by 7.2 MMTCO2e annually will provide about 20 MMT of annual reduction (cooling) equivalent each year from 2030 to 2045. These reductions will be critical to mitigate continually accumulating CO_2 emissions from other sectors of the economy, and the achievement of the state's "net zero" long-term goal. In the race to manage global warming, reducing methane can provide fast returns.

This analysis using GWP* shows the true value of the state's dairy methane reduction efforts and programs such as CDFA's Dairy Digester Research and Development Program (DDRDP) and Alternative Manure Management Program (AMMP), which are expected to incentivize more than half of the 7.2 MMT of methane reduction. This analysis also underscores the importance of continuing to fully fund these California Climate Investment Programs at a minimum of $85 million per year. (CARB Preliminary Analysis of Dairy Methane Reduction Progress, May 2020).

References

Allen, M.R., Fuglestvedt, J.S., Shine, K.P., Reisinger, A., Pierrehumbert, R.T., & Forster P.M. (2016). New use of global warming potentials to compare cumulative and short-lived climate pollutants. Nature Climate Change. 6. 773–6. Retrieved from https://www.nature.com/articles/nclimate2998?cacheBust=1508877188307

Allen, M.R., Shine, K.P., Fuglestvedt, J.S., Millar, R.J., Cain, M., Frame, D.J., & Macey, A.H. (2018). A solution to the misrepresentations of CO2-equivalent emissions of short-lived climate pollutants under ambitious mitigation. npj Climate and Atmospheric Science. 1(16). Retrieved from https://www.nature.com/articles/s41612-018-0026-8

California Air Resources Board. (2019, August 12). California 2017 Greenhouse Gas Inventory. Retrieved from https://www.arb.ca.gov/cc/inventory/data/tables/ghg_inventory_bygas.pdf

California Air Resources Board. (2015). California's methane inventory based on the 2015 edition the CARB greenhouse gas inventory. Retrieved from https://www.arb.ca.gov/ghg-slcp-inventory

California Department of Food and Agriculture. (2019, September 18). CDFA Awards Nearly $102 Million for Dairy Methane Reduction Projects [Press release]. Retrieved from https://www.cdfa.ca.gov/egov/ Press_Releases/Press_Release.asp?PRnum=19-085

California Department of Food and Agriculture Dairy Marketing, Milk Pooling, and Milk and Dairy Foods Safety Branches. (2017). [California dairy cows and milk productions]. Unpublished raw historical data.

Cain, M. (2018). Guest post: A new way to assess 'global warming potential' of short-lived pollutants. Carbon Brief. Retrieved from https://www.carbonbrief.org/guest-post-a-new-way-to-assess- globalwarming-potential-of-short-lived-pollutants

Cain, M., Lynch, J., Allen, M.R., Fuglestedt, D.J. & Macey, A.H. (2019). Improved calculation of warming- equivalent emissions for short-lived climate pollutants. npj Climate and Atmospheric Science. 2(29). Retrieved from https://www.nature.com/articles/s41612-019-0086-4

Dairy Cares. (2019, August 28). Cows vs Cars? [Video file]. Retrieved from https://www.youtube.com/ watch?v=RW8BclS27aI&vl=en

Dairy Industries International. (2019). Sustainability project aims for net zero climate impact in US Dairy. Retrieved from https://www.dairyindustries.com/news/32149/sustainability-project-aims-for-net-zero- climate-impact-in-us-dairy/

Fairlie, S. (2019). A Convenient Untruth. Resilience. Retrieved from https://www.resilience.org/stories/2019-05-10/a-convenient-untruth/

FAO. (2019). Five practical actions towards low-carbon livestock. Rome. Retrieved from http://www.fao.org/documents/card/en/c/ca7089en/

FAO and GDP. (2018). Climate change and the global dairy cattle sector – The role of the dairy sector in a low-carbon future. Rome. 36 pp. Licence: CC BY-NC-SA- 3.0 IGO. Retrieved from https://dairysustainabilityframework.org/wp-content/uploads/2019/01/Climate-Change-and-the-Global-Dairy-Cattle-Sector.pdf

Frame, D., Macey, A.H., & Allen, M. (2018). Why methane should be treated differently compared to long-lived greenhouse gases. The Conversation.
Retrieved from https://theconversation.com/why-methane-should-be-treated-differently-compared-to- long-lived-greenhouse-gases-97845

Howarth, R. W. (2019). Ideas and perspectives: is shale gas a major driver of recent increase in global atmospheric methane?. Biogeosciences, 16, 3033–3046. Retrieved from https://doi.org/10.5194/bg-16- 3033-2019

Lynch, J. (2019). Agricultural methane and its role as a greenhouse gas. Food Climate Research Network, University of Oxford. Retrieved from https://foodsource.org.uk/building-blocks/agricultural- methane-and-its-role-greenhouse-gas

Naranjo, A., Johnson, A., Rossow, H., & Kebreab, E. (2020). Greenhouse gas, water, and land footprint per unit of production of the California dairy industry over 50 years. Journal of Dairy Science. 103, 3760-3. Retrieved from https://www.journalofdairyscience.org/article/S0022-0302(20)30074-6/fulltext

Schleussner C., Nauels, A., Schaeffer, M., Hare, W., & Rogelj, J. (2019). Inconsistencies when applying novel metrics for emissions accounting to the Paris agreement. Environmental Research Letters. 14(12). Retrieved from https://iopscience.iop.org/article/10.1088/1748-9326/ab56e7/meta

United States Department of Agriculture, National Agricultural Statistics Service. (2019). Milk Production, Disposition, and Income 2018 Summary. Retrieved from https://usda.library.cornell.edu/ concern/publications/4b29b5974

Conference

Colloque

ANCC
CNAC

Prospects for Climate Neutral Beef and Dairy Production in Canada

Perspectives pour une production bovine et laitière sans impact climatique au Canada

Karen Beauchemin
Agriculture and Agri-Food Canada, Lethbridge Research and Development Centre, Lethbridge, AB
karen.beauchemin@agr.gc.ca

JM Bell Lectureship in Animal Nutrition / Conférence JM Bell en nutrition animale

UNIVERSITY OF SASKATCHEWAN

Abstract

Canada has committed to cut its greenhouse gas (GHG) emissions by 40-45% below 2005 levels by 2030, joining over 120 countries committed to net-zero emissions by 2050. Likewise, numerous agribusinesses have set ambitious voluntary targets to achieve net-zero emissions. Net-zero is defined as a stage when anthropogenic GHG emissions are balanced by their removal. With over 70% of the GHGs from beef and milk occurring prior to the farm gate, livestock producers will need to aggressively adopt practices that decrease GHG emissions to continue to sell their products. The carbon footprints of beef and milk to the farm gate in Canada are both less than 50% of global averages and continue to decrease due to advances in nutrition, genetics, health, manure management, feed production and farm management. However, continued improvements in production efficiency to lower carbon footprints will not achieve the deep reductions in absolute GHGs emissions needed. This paper discusses opportunities for decreasing GHG emissions from Canadian beef and dairy farms, and in particular the innovation required to substantially decrease enteric methane emissions. While net-zero ruminant products may not be achievable in the near future, climate neutral beef and dairy may be a more realistic goal. This goal could be achieved by aggressively adopting best practices to lower on-farm emissions, accounting for soil carbon sequestered in grasslands, and using the new global warming potential star (GWP*) metric. This metric recognizes that methane from ruminants is short-lived in the atmosphere and part of a biological cycle unlike CO_2 from fossil fuels, and therefore does not contribute to temperature warming when animal numbers remain constant as is the case in Canada.

Key words: carbon footprint, greenhouse gases, methane, sustainability

Résumé

Le Canada s'est engagé à réduire ses émissions de gaz à effet de serre (GES) de 40 à 45 % par rapport aux niveaux de 2005 d'ici 2030 et s'est joint à plus de 120 pays qui ont promis d'atteindre la carboneutralité d'ici 2050. Dans ce contexte, de nombreuses entreprises agroalimentaires se sont volontairement fixé des objectifs ambitieux qui vont dans le même sens. La carboneutralité est un stade auquel les émissions anthropiques de GES sont compensées par une élimination équivalente. Étant donné que plus de 70 % des GES associés à la production de bœuf et de lait sont émis à la ferme, les éleveurs devront redoubler d'ardeur pour adopter des pratiques de réduction des émissions de GES afin de continuer à vendre leurs produits. Au Canada, les empreintes carbones du bœuf et du lait à la ferme sont inférieures à 50 % de la moyenne mondiale et continuent de diminuer grâce aux progrès réalisés en matière de nutrition, de génétique, de santé, de gestion du fumier, de production d'aliments et de gestion agricole. Cependant, l'amélioration de l'efficacité de la production pour diminuer les empreintes carbones ne permettront pas d'atteindre les objectifs de réduction des émissions absolues de GES requises. Le présent document traite des possibilités de réduire les émissions de GES des exploitations bovines et laitières canadiennes et, tout particulièrement, des innovations nécessaires pour diminuer considérablement les émissions de méthane entérique. Bien qu'il ne sera pas possible d'obtenir des produits issus de ruminants à zéro émissions nettes dans un avenir proche, il pourrait être plus réaliste de viser la production de bœuf et de produits laitiers sans impact climatique. Cet objectif pourrait être atteint en adoptant de manière rigoureuse les pratiques exemplaires pour réduire les émissions à la ferme, en comptabilisant le carbone du sol séquestré dans les prairies et en appliquant la nouvelle méthode de calcul du potentiel de réchauffement global (PRG*) pour reconnaître que le méthane provenant des ruminants a une courte durée de vie dans l'atmosphère et fait partie d'un cycle biologique, contrairement au CO2 issu des combustibles fossiles, et qu'il ne contribue donc pas au réchauffement des températures lorsque le nombre d'animaux demeure constant, comme c'est le cas au Canada.

Introduction

Governments, the private sector, and consumers are increasingly aware of the potential impacts of elevated greenhouse gases (**GHG**) in the atmosphere and their impacts on climate change. All segments of society are challenged with reducing GHG emissions to limit global warming to 1.5°C above pre-industrial levels by 2050. While no government legislation or carbon tax has been imposed on livestock emissions in Canada, there is increasing pressure on the ruminant industries to lower emissions. The focus is mainly on enteric methane (**CH_4**), which is produced from the fermentation of plant material in the rumen and escapes into the atmosphere through eructation (belching) and respiration.

The concern is that the world's increasing demand for animal sourced protein products will cause enteric CH_4 emissions to increase unless mitigation is adopted. Decreasing enteric CH_4 emissions from ruminants is an extremely challenging goal and limited mitigation options are currently available to producers. The objective of this paper is to provide an overview of the key issues related to GHGs from ruminant production in Canada, discuss some potential approaches for CH_4 mitigation, and provide insight into the potential for carbon neutral beef and dairy production.

How Much do Ruminants Contribute to GHG Emissions?

There is a lot of misinformation in the popular press on the contribution of livestock production and more specifically, ruminants, to GHG emissions. We've all heard comments like: "I am a vegan because I want to do what's right for the environment" and "Cow farts are ruining are planet". The Hollywood movie Cowspiracy claimed that 51% of global GHGs are created by livestock. No wonder consumers are confused.

Let's look at the numbers. The main GHGs from livestock agriculture are: CH_4 from ruminant animals (enteric) and manure, nitrous oxide (**N_2O**) from soils due to the use of organic and inorganic fertilizers for feed production, and carbon dioxide (**CO_2**) from the use of fossil fuels in machinery used on farms. In addition to these emissions, there are also emissions and removals of CO_2 from land use change in the form of carbon loss and storage in soils, respectively (more about that later). The GHGs are usually compared on a CO_2-equivalent (**CO_2e**) basis using their 100-year global warming potentials (**GWP_{100}**) (CH_4, 25 × CO_2; N_2O, 298 × CO_2; ECCC, 2021). For example, a steer producing 100 g/day of enteric CH_4 (100 g/day × 25 GWP_{100}) produces 2,500 g/day of CO_2e.

Using GWP_{100}, the International Panel on Climate Change estimated that globally, enteric CH_4 accounts for 3 to 5% of the total emissions by all sectors (Smith et al., 2014). If GHGs from all livestock production are considered, the contribution is 5.6 to 7.5 Gt CO_2, or 11 to 15% of total GHG emissions. The Food and Agriculture Organization estimated that all livestock production accounts for 14.5% of global emissions (7.1 Gt CO_2); 45% from feed production including land use change (e.g., deforestation in the Amazon), 39% from enteric CH_4, 10% from manure storage and processing, and the remainder attributed to processing and transportation of animal products (Gerber et al., 2013).

For Canada, 2019 National Inventory Report from Environment and Climate Change (ECCC, 2021) gives a breakdown of the nation's total GHG emissions (730 Mt of CO_2e) by sector. Emissions from agriculture (excluding fuel use and land use change) were 59 Mt, or 8.1% of total emissions (Fig. 1). Inclusion of CO_2 from fuel would bring agriculture's contribution to 10%. Emissions from land use change for feed production are minimal in Canada. Compare GHG from agriculture to emissions associated with transportation, power generation, and industrial activity, which combined account for nearly 90% of Canada's GHG emissions.

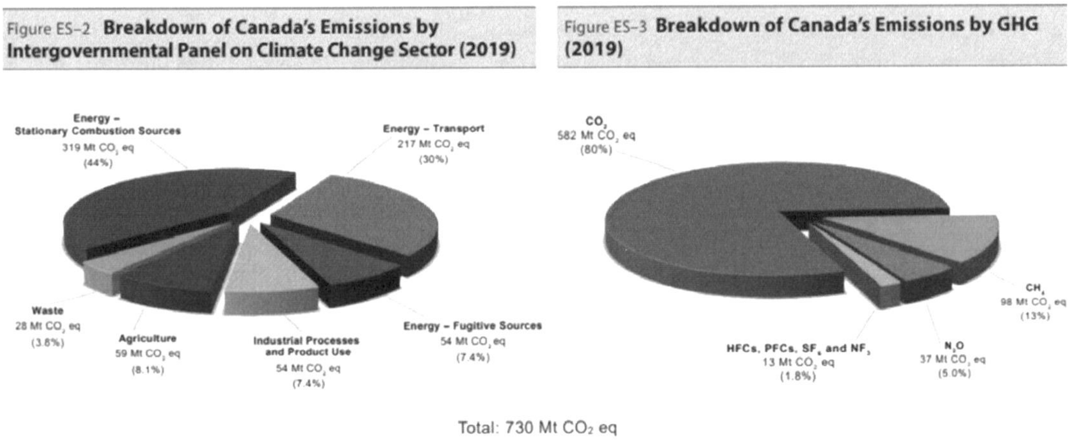

Figure 1. Canada's 2019 emissions by sector (left) and by GHG (right) on a CO_2e basis (ECCC, 2021).

A further breakdown of agriculture by source indicates that enteric CH_4 contributes 41% of agricultural emissions (Fig. 2), or 3.3% of Canada's total GHG emissions. Beef production accounts for 81%, and dairy production 15%, of enteric CH_4 emissions.

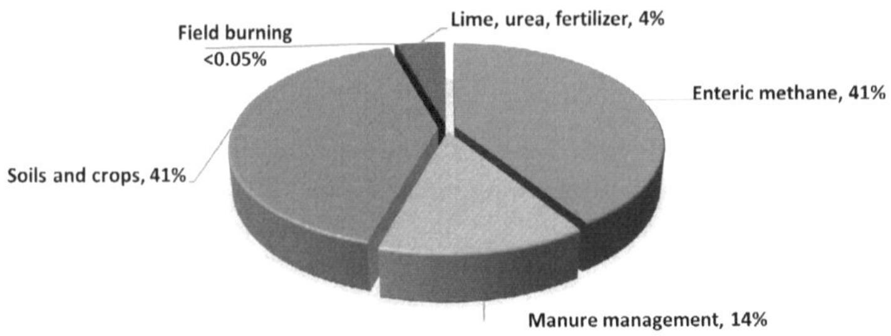

Figure 2. Canada's 2019 agricultural GHG emissions on a CO_2e basis by source (ECCC, 2021).

GHG Reduction Targets

Last year at the COP26 meeting, Canada declared its goal of reducing GHG emissions by 40 to 45% by 2030 compared with 2005 levels, with a focus on reducing emissions from the use of fossil fuels. If this target is applied to agriculture it would require cutting current emissions in half over the next 8 growing seasons. Canada, along with over 120 countries, also signed the Global Methane Pledge to reduce CH_4 by 30% from 2020 levels by 2030 (www.globalmethanepledge.org). The intent is to minimize fugitive emissions from oil and gas production, but given that 28% of total CH_4 emissions is enteric fermentation in Canada, one can foresee increasing pressure on ruminants.

Agribusinesses involved in the meat and dairy supply chains have also set ambitious voluntary GHG reduction targets, many setting goals to achieve net-zero emissions. With over 70% of the GHGs from beef and milk occurring prior to the farm gate, livestock producers will need to aggressively adopt practices that decrease GHG emissions to continue to sell their products. In response to mounting pressure, the Canadian Roundtable for Sustainable Beef (https://crsb.ca/) set a target of reducing GHG per kilogram of beef by 33% by 2030, while Dairy Farmers of Canada has a goal to reach net-zero GHG emissions from farm-level dairy production by the year 2050.

These goals can be confusing. Achieving net-zero means that anthropogenic GHG emissions are balanced by their removal or are offset (e.g., carbon storage in trees and soils, biogas replacing fossil fuel energy, etc.). Government goals are to reduce actual GHG emissions into the atmosphere, while many industries focus on GHG intensity (e.g., carbon footprint) defined as kg of GHG or CH_4 per kg of meat or milk produced. Intensity goals allow the industry to continue to expand, albeit more efficiently, but absolute emissions may not decrease if the increase in animal population offsets the decrease in intensity. Globally, agricultural CH_4 emissions are projected to increase by 30% relative to 2010 by 2050, due to large increases in meat and milk production required to meet human population demands. In Canada, enteric emissions have not increased in the past decade because cattle populations have remained relatively stable, and that trend is likely to continue for some time (Fig. 3).

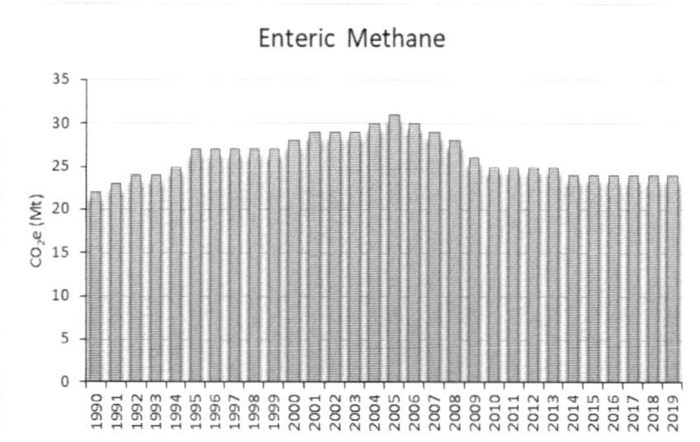

Figure 3. Enteric methane emissions in Canada over time (ECCC, 2021).

Net-zero and Climate Neutral Beef and Dairy

While net-zero ruminant products may not be achievable in the near future, climate neutral beef and dairy may be a more realistic goal. The Paris Agreement states that countries must report their emissions using GWP_{100}, but a number of scientists have made the case that this metric does not reflect the true warming potential of the various gases. GWP_{100} treats all gases as pulse gases, and does not account for the atmospheric removal of CH_4. What does this mean? CO_2 is a pulse gas, because once it is released it lasts in the atmosphere for hundreds and even thousands of years. In contrast, CH_4 is a short-lived flow gas, and only lasts in the atmosphere for about a decade. A flow gas means that as CH_4 is emitted, CH_4 is also being destroyed. Thus, CH_4 produced by ruminants only contributes to increased warming compared with present day if animal numbers increase, which is a trend in developing countries but not in Canada. The converse is also true, if CH_4 decreases, then it contributes to cooling relative to present day.

Net-zero requires that emissions must either be completely eliminated or offset with additional CO_2e removals. However, this is only necessary for CO_2, due to its cumulative impacts over time. It has been argued that for short-lived gases such as CH_4, a climate impact equivalent to 'net-zero CO_2' can be achieved with some ongoing emissions, because CH_4 is relatively quickly eliminated from the atmosphere. This presents an opportunity for ruminants – adopting CH_4 mitigation practices could act as an offset to fuel based emissions and provide additional revenue to producers.

To overcome the limitations of GWP_{100}, the metric of GWP star (**GWP***) has been proposed (Allen et al., 2016). GWP* allows emissions of short-lived and long-lived GHGs to be expressed within a single metric; a change in the emission rate of CH_4 is compared with an equivalent single emissions pulse of CO_2. GWP* indicates how a decrease in CH_4 relative to a baseline (usually the previous 20 years) can lead to cooling. Use of GWP* would credit a cooling effect to ruminant systems that use mitigation strategies to decrease CH_4 production. This could help achieve climate neutral beef and dairy production, which Place and Mitloehner (2021) define as animal production having no net contribution to additional temperature increases.

Furthermore, when calculating GHG emissions from ruminant production, CO_2 removal through carbon storage in soils is typically not considered. Forage based ruminant production can enhance soil carbon reserves, thereby withholding CO_2 from the air. Compared with annual crops, perennial forages return a greater portion of plant biomass to the soil, sustaining greater amounts of soil carbon. The amount of additional carbon that can be stored in forage systems depends on prior management practices. Lands that have been under well-managed grass for decades may already have large stores of carbon that cannot be further increased. However, planting grasses onto lands previously used for annual cropping can significantly increase soil carbon. Similarly, soil carbon can be increased by introducing perennial forages into rotations with annual crops (especially with manure) or by adopting better practices on previously mismanaged grazing lands. The Canadian government is developing an Enhanced Soil Organic Carbon protocol that may allow farmers to participate in GHG offset markets and receive additional revenue by applying good grazing management practices. Enhanced soil carbon strategies would contribute significantly to carbon neutral beef and dairy production.

Sources of GHG from Meat and Milk Production

Life cycle assessment (**LCA**) of milk production in Canada to the farm gate indicates a national average GHG intensity of 1.01 kg CO_2e/kg fat and protein corrected milk (Dairy Farmers of Canada), which is less than half the global average. There is significant variation among farms indicating opportunity for improvement. For example, a survey of 142 dairy farms in Ontario reported a range of 0.44 to 1.73 kg CO_2e/kg fat and protein corrected milk (Jayasundara et al., 2019). In dairying, approximately 40 to 50% of the farm emissions is enteric CH_4, 15 to 25% is from manure management, and 20 to 25% is from feed production (note that this is lower than globally because of the lack of deforestation), and 10% to 15% is from energy use.

The national average GHG intensity of beef production in Canada is 11.4 kg of CO_2e/kg of live weight at the farm gate (https://crsb.ca/). The carbon footprint is dominated by enteric emissions (55 to 70%), followed by manure management (20 to 30%), feed production (15 to 20%), and energy use (< 5%). The GHG intensities do not typically include changes in soil carbon stocks, but had the effects of land use been included, the GHG intensity would be reduced from 11.4 to 10.5 kg CO_2e/kg of live weight. Approximately 70 to 80% of beef GHGs originate from the cow-calf sector, while 20 to 30% is from the feedlot sector, depending upon the calf finishing system.

Intensification of Ruminant Production Lowers GHG Intensity

Research in the area of reducing GHGs from ruminants has grown exponentially in the last two decades (Beauchemin et al., 2020). Intensification of production is regarded as the most effective option for GHG reduction, and thus the feed industry can play an important role in lowering GHGs by helping farms improve productivity and efficiency. Intensification is especially effective in decreasing GHG intensity (GHG/kg product) in low-income countries where carbon footprint of animal products is very high because farms have low productivity. However, the importance of intensification for high income countries should not be overlooked. Between 1991 and 2011, the GHG intensity of milk produced in Canada decreased by 22% (Jayasundara and Wagner-Riddle, 2014), while between 1981 and 2011, the GHG intensity of beef decreased by 14% (Legesse et al., 2016). These decreases in GHG intensity over time were due to improvements in production efficiency – better nutrition, reproduction, genetics, health, crop yields and management. Emission intensity decreases with increased farm productivity because fewer animals are required to produce a given amount of product. However, total emissions per animal may increase due to greater feed intake and manure output, along with greater upstream emissions from feed and farm inputs. Thus, absolute emissions from the farm or sector only decrease if animal numbers decrease, as is true for the Canadian dairy sector. This is not the case globally where livestock populations are increasing to meet the demand of a growing population.

As farms continue to adopt new technologies, further decreases in GHG intensity of meat and milk are expected (approx. -2%/year for dairy and -0.5%/year for beef in Canada). The magnitude of these improvements is far below government and industry GHG reduction targets. Therefore, to make a major impact on reducing GHG from milk and meat production, pronounced decreases in enteric CH_4 production are also needed, particularly from the cow-calf sector.

Are Pronounced Decreases in Enteric Methane Emissions Possible?

The most promising ways to substantially decrease enteric CH_4 are the dietary inclusion of: 1) the chemical inhibitor 3-nitrooxypropanol (**3-NOP**; DSM Nutritional Products) and 2) the bromoform-containing red algae *Asparagopsis* spp. (Ungerfeld et al., 2022). A thorough analysis of CH_4 mitigation strategies is provided by Beauchemin et al. (2020) with a brief summary below.

3-NOP. A comprehensive review of 3-NOP was published by Yu et al. (2021). 3-NOP targets methyl-coenzyme M reductase, the enzyme that catalyzes the last step of methanogenesis in the rumen. Within hours of ingestion, 3-NOP is rapidly hydrolyzed in the rumen to nitrate, nitrite and 1,3-propandiol, a carbon source of low toxicity that is used in gluconeogenesis. At typical inclusion levels in beef and dairy diets, 3-NOP decreased CH_4 production by an average of 30%, although our lab has reported reductions up to 82% in some feedlot finishing studies. The effects of 3-NOP are dose and diet dependent; CH_4 decreases linearly with increasing 3-NOP concentration in the diet and the effectiveness of 3-NOP is inversely proportional to dietary concentration of neutral detergent fiber (**NDF**).

Long term inhibition of enteric CH_4 production by 3-NOP has been reported in numerous studies, although a couple of studies have shown a decline in effectiveness when a low dose of 3-NOP was fed. 3-NOP causes a shift in rumen fermentation from acetate to propionate with no negative effects on digestibility. Most studies indicate no consistent benefits in animal productivity when feeding 3-NOP. However, our group has noted improvements in feed conversion efficiency of up to 5% in several feedlot studies with 3-NOP.

Because dietary concentration of 3-NOP is very low (40 to 200 mg/kg DM), CO_2 emissions during manufacture and transport are also very low. Effects on manure emissions appear to be non-existent or minimal. 3-NOP was recently approved for use in Brazil, Chile, and the E.U., while dossiers have been submitted for approval in Canada and the U.S, where the approval process is lengthy because any claim for CH_4 invokes the drug regulatory process.

In its present form, 3-NOP is limited to confinement systems using formulated diets, as it needs to be fed as part of a ration. However, research is ongoing to develop a slow release form that might offer potential for grazing cattle.

***Asparagopsis* seaweed.** *Asparagopsis* is a red seaweed that grows in tropical waters. *Asparagopsis taxiformis* and *A. armata* accumulate halogenated compounds of which bromoform is the most abundant. The halogenated compounds reduce methanogenesis by blocking methyl transfer. Dose and diet dependent decreases in CH_4 production as high as 98% have been reported for sheep and cattle fed diets supplemented with *Asparagopsis*. Similarly to 3-NOP, *Asparagopsis* appears to be more effective at decreasing CH_4 production with high concentrate diets than with high forage diets. Effects of *Asparagopsis* on animal performance have been investigated in only a few small scale studies. Feed intake was reduced in most studies, leading to increased feed efficiency.

A number of issues will need to be resolved before farmers can use *Asparagopsis* as a feed additive in Canada. Safety will need to be addressed, as the U.S. EPA classifies bromoform as a probable human carcinogen. At the concentrations used, bromoform residues have not been detected in milk or meat, but further study is required. *Asparagopsis* also contains very high levels of iodine, which can accumulate in milk and meat. Wide use of *Asparagopsis* will depend on the ability to sustainably grow it in aquaculture with consistent concentration of the active compounds. The CO_2e emissions of growing, harvesting, processing (drying, extracting), and transporting seaweeds at a large scale will need to be considered. *Asparagopsis* is not listed on the Canadian Feed Inspection Agency's Schedule IV or V. It is possible that it could be approved as a feedstuff given that some seaweeds are already approved, although a CH_4 reduction claim may require a more complex approval process. There is tremendous interest in *Asparagopsis* across many countries because it is viewed as natural and is considered "generally recognized as safe' by some regulatory authorities.

Currently Available Enteric Methane Mitigation Solutions

Although 3-NOP and *Asparagopsis* are not approved in Canada, a range of enteric CH_4 mitigation strategies are currently available (as reviewed by Beauchemin et al., 2020). Although none of these strategies limits enteric CH_4 production to the same extent as 3-NOP and *Asparagopsis,* there is possibility to combine strategies with different modes of action for additive mitigation effects (Table 1). All strategies listed in Table 1 require additional research to establish conditions for deployment on farms, as most have only been evaluated at the research scale. Most strategies require dietary supplementation, and thus are not applicable to grazing cow-calf herds. Prior to recommending any mitigation strategy, a LCA that considers all sources of emissions, including those from feed production, manure, machinery use, and changes in soil carbon need to be considered.

Lipids. The most well-researched mitigation approach is dietary supplementation with non-rumen protected lipids (oils, oilseeds). Lipids elicit a CH_4 mitigating effect through: toxicity of certain fatty acids against methanogens and protozoa, biohydrogenation of unsaturated fatty acids serving as a minor alternative hydrogen sink, shifting rumen fermentation to promote the production of propionate, and lowering of diet fermentability by replacing carbohydrates with lipids and in some cases decreased fiber digestibility. Various meta-analyses indicate a decrease in CH_4 yield (g/kg dry matter intake, **DMI**) between 3.5 to 5% per 10 g/kg DM supplemental fat (Beauchemin et al., 2020). Medium-chain fatty acids such as myristic acid and polyunsaturated fatty acids found in fish oil, sunflower, linseed and canola are the most effective for reducing CH_4 emissions. Most oilseeds need to be processed prior to feeding to ensure availability of the lipids in the rumen, and oils are typically more effective than crushed oilseeds. The inhibitory effect of lipids on CH_4 is greater with concentrate diets compared with forage diets. Care must be used when supplementing diets with fats as high concentrations of dietary fat (> 6% total fat) can have detrimental effects on rumen fermentation, feed digestion and animal performance. The high cost of fat sources limits their scope for adoption.

Concentrates. Increasing the concentrate proportion of the diet decreases fiber intake, increases propionate production, increases rumen outflow rate, and lowers rumen pH – factors that decrease CH_4 production. The efficacy of increasing level of concentrate is variable. Based on an intercontinental database for beef cattle, van Lingen et al. (2019) reported a CH_4 yield of 20.7 g/kg DM (range: 6.29 to 35.1) for high- (≥25%) forage diets compared with 15.2 g/kg DM (7.50 to 30.9) for low- (≤18%) forage diets. The magnitude of CH_4 production from grain sources follows the order: wheat < corn < barley, with the ranking highly dependent on the extent of processing of the grain and the resulting rumen pH. Diets containing steam-flaked corn rather than dry-rolled or ground grains lower CH_4 production by 10 to 20%.

Although higher concentrate diets reduce CH_4 production, there can be increased emissions due to the use of nitrogen (**N**) fertilizers during feed production, and soil carbon is lost during the conversion of pastureland to cropland. Increased feeding of grains should be promoted as a CH_4 mitigation strategy only after assessment using LCA. In Canada, the potential for increasing concentrate use is low because grain use in ruminant diets is already near maximum. Furthermore, grain feeding ignores the importance of ruminants in converting fibrous feeds, unsuitable for human consumption, to high-quality protein sources (i.e., milk and meat).

Forage Digestibility. Increasing forage digestibility usually increases DMI and improves animal performance, which decreases CH_4 yield (g/kg DMI) and intensity (g/kg product). Digestibility of forages conserved as hay or silage can be maximized by harvesting at a vegetative stage, and in pastoral systems digestibility can be enhanced by optimizing grazing management to decrease forage maturity (e.g., adjusting stocking rates, ensuring pre-grazing herbal mass is not excessive). Absolute CH_4 production (g/day) due to increased forage digestibility usually remains constant or increases due to greater DMI and increased fermentation in the rumen.

Perennial Legumes. At the same physiological stage of maturity, legume forages contain less NDF than grasses. Although fiber in legumes is more lignified, the decline in fiber digestibility with advancing maturity is greater for grasses than legumes, especially in warmer locations. In addition, some legumes contain secondary compounds that decrease CH_4 production. Animal performance is often increased with dietary inclusion of legumes, which decreases CH_4 intensity. It is difficult to quantify the mitigation effect of legumes because it depends on the quality of the forages being compared. Perennial legume forages biologically fix N, which reduces the amount of fertilizer used and associated emissions. The high crude protein concentration of legumes can decrease the use of purchased supplements and associated emissions, and fuel-based CO_2 emissions from the use of farm equipment can be less for perennial compared with annual forages.

High Starch Forages. Use of high-starch forages such as corn silage and small-grain cereals can increase starch and decrease NDF concentrations of diets. The resulting rumen fermentation promotes propionate production, which competes with methanogenesis for metabolic hydrogen, and can lower rumen pH, inhibit methanogens, and decrease CH_4 production. With some diets, incorporating high-starch forages increases digestible energy intake of animals and enhances animal performance, thereby decreasing CH_4 intensity. However, Little et al. (2017) showed that although replacing alfalfa silage with corn silage in the diet of lactating dairy cows lowered enteric CH_4 production by 10%, differences in GHG emission intensity between the two forage systems were minimal when change in soil carbon was accounted for. Thus, feeding high-starch forages to reduce enteric CH_4 emissions is not recommended unless substantiated by LCA that includes soil carbon changes. Greatest potential of high-starch forages to reduce GHG emissions is when replacing another annual forage crop.

Pastures and Grazing Management. Grazing systems manage pastures to balance livestock nutritional requirements with herbage availability and quality while promoting rapid pasture re-growth and long-term pasture resilience. Grazing management can enhance herbage quantity and quality leading to increased soil carbon stocks and decreased CH_4 intensity. Grazing management for CH_4 mitigation considers pre-grazing and post-grazing sward height and biomass to maximize herbage nutritional quality as a means of lowering CH_4 intensity, but daily CH_4 production is not expected to change or it may increase if DMI is increased. The extent to which grazing management lowers CH_4 intensity is extremely variable depending upon the production system and local conditions.

Ionophores. In a meta-analysis, Appuhamy et al. (2013) reported average decreases in daily CH_4 production of 3.6% and 10.7% in dairy cows and beef steers, respectively, when administered ionophores. Additionally, monensin increases feed conversion efficiency, which decreases CO_2e emissions from feed production needed to sustain animal production. Ionophores are already used extensively in Canada to improve feed efficiency, thus scope for further use is limited.

Nitrate. Nitrate draws electrons away from methanogenesis by incorporating them into alternative pathways. In the rumen nitrate is reduced to nitrite and then ammonia, which can be incorporated into microbial protein. Nitrite can be absorbed through the rumen wall and react with hemoglobin to form methaemoglobin, which cannot transport oxygen. This condition can be fatal, although it is possible to gradually adapt the rumen to nitrate supplementation. Most studies indicate that nitrate supplementation of dairy and feedlot cattle diets (1.5 to 2% of DM) reduces enteric CH_4 by 15 to 20%. While effective, nitrate should only be recommended in production systems where feed intake is closely managed. Calcium nitrate is often more than twice as expensive compared with urea as a source of non-protein N. No nitrate product is currently approved in Canada.

Tannins. Tannins are polyphenolic plant compounds with affinity to bind to proteins and other compounds, and are classified as either condensed or hydrolysable. Both types have been shown to exert anti-methanogenic effects by directly inhibiting some methanogens and indirectly by decreasing protozoal numbers, which symbiotically host methanogens. Some of the decrease in CH_4 can also be due to a decline in DMI and nutrient digestibility, which can negatively impact animal production. A meta-analysis of *in vivo* studies indicated a linear decrease in CH_4 yield of 3.65% ($r^2 = 0.47$) with each 10 g/kg DM addition of tannin (Jayanegara et al., 2012). However, the decrease in CH_4 yield was accompanied by a decrease in organic matter digestibility of 2.6% per 10 g/kg DM addition. Most tannin-containing legumes grown in Canada (e.g., sainfoin, birdsfoot trefoil, cicer milk vetch) contain relatively low concentrations (<30 g/kg DM) of condensed tannins, thus CH_4 reductions are relatively small. Therefore, tannin extracts from shrubs and trees (e.g., *Acacia mearnsii,* chestnut, quebracho) have been examined as an alternative. Another important environmental benefit of dietary tannins is their ability to improve N utilization in ruminants. Tannins interact with dietary proteins in the gastrointestinal tract, and can improve N utilization, decrease urinary N losses, and decrease ammonia and N_2O emissions from excreta.

Table 1. Summary of existing enteric methane mitigation strategies.

Strategy	Expected CH_4 decrease range		Adoption Potential			Comments
	g/day	g/kg meat or milk	Dairy & Feedlots	Cow-calf grazing	Cow-calf winter feeding	
Increased animal productivity	I	L	Y	Y	Y	Widely applicable, variable potential among farms
Lipids	M	M	Y	Y	Y	Can impact product quality
Concentrates	Nc or I	L	Y	N	N	Limited scope for further increase, LCA needed
Forage digestibility	I	L to M	Y	Y	Y	Widely applicable

Perennial legumes	L	L to M	Y	Y	Y	Widely applicable
High starch forages	L	L	Y	Y	Y	Limited to certain locations, LCA needed
Pasture management	Nc to I	L to M	N	Y	N	Important for cow-calf sector
Ionophores	Nc to L	Nc to L	Y	N	N	Limited ability to expand
Nitrate	M	M	Y	Y	Y	No product approved, high risk
Tannins	L	L	Y	Y	Y	Can alter N utilization

H = high decrease (≥25%); I = increase; L = small decrease (≤15%); M = medium decrease (15-24%); N = no; Nc = no change; Y = yes.

Enteric Methane Mitigation Solutions Under Development

The global commitment to curbing CH_4 emissions is driving significant investment and innovation by the private and public sectors. A number of CH_4 mitigation strategies are under development although the timeframe for market availability is unknown (Table 2). An example is genetic selection programs for low-CH_4 emitting animals. One of the main challenges for selecting animals with low CH_4 production is the difficulty of measuring CH_4 on a large number of animals for weeks at a time. As animal breeding is one of the few anti-methanogenic strategies that can be applied to extensive production systems where animals are not supplemented, there is tremendous interest in pursuing this line of research. Other possibilities are the development of feed additives, such as essential oils, direct fed microbials, and plant extracts (saponins, polyphenols) that alter rumen fermentation in a manner that decreases CH_4 production or that have direct toxic effects on methanogens. Another line of research is development of an anti-methanogenic vaccine that stimulates the immune system of animals to produce antibodies against methanogens. This research was initiated over 2 decades ago; however, progress has been slow and the research has proven to be challenging. Effects *in vivo* on CH_4 production have so far been small indicating lack of broad spectrum effectiveness against the rumen methanogenic community. Another approach is to program the rumen microbiome during early life in manner that decreases CH_4 emissions later in life. Research on early life intervention is at an early stage with a few very promising results when 3-NOP was dosed to newborn calves.

Recent advances in characterizing the rumen microbiome, genome sequencing of rumen methanogens, and an in-depth analysis of the enzymatic pathways involved in methanogenesis are leading to new microbial-based CH_4 mitigation approaches. One avenue being explored is archaeal phages that produce lytic enzymes that breakdown pseudomurein, the principal cell wall component of rumen methanogens. Others are taking an engineering approach by exploring the possibility of developing a face mask for cattle that oxidizes CH_4 as it is released from the animal's nostrils (www.ZELP.co). It is clear that CH_4 mitigation options will continue to expand in future years.

Table 2. Some enteric methane mitigation strategies under development.

Strategy	Expected CH_4 decrease range		Adoption Potential			Comments
	g/day	g/kg meat or milk	Dairy & Feedlots	Cow-calf grazing	Cow-calf winter feeding	
3-NOP	H	H	Y	N	N	Not approved in Canada, highly effective, need slow release version
Asparagopsis	H	H	Y	N	Y	Not approved in Canada, highly effective
Genetic selection	L	L	Y	Y	Y	Long-term mitigation
Essential oils	L	L	Y	N	Y	None with CH_4 claim yet, can shift rumen fermentation
Direct-fed microbials	L	L	Y	N	Y	None with CH_4 claim yet, can shift rumen fermentation
Saponins	L	L	Y	Y	Y	Can alter N utilization
Vaccine	M	M	Y	Y	Y	Still > 5 years away
Early life programming	L	L	Y	Y	Y	Studies using 3-NOP have shown good results

H = high decrease (≥25%); L = small decrease (≤15%); M = medium decrease (15-24%); N = no; Y = yes.

Conclusions

There is increasing pressure on ruminant livestock producers to lower GHG emissions from meat and milk production, both in terms of absolute daily emissions and emissions relative to animal product outputs. The focus is on enteric CH_4 because it is the largest GHG source in ruminant production. A number of enteric CH_4 mitigation solutions are now available, but continued innovation is needed to develop additional technologies that accommodate the large variation in ruminant production systems. Intensification of animal production through improved nutrition, genetics, and management is the most immediate and widely applicable means of decreasing CH_4 emissions relative to animal products, but absolute CH_4 emissions may not always decrease.

While research continues to provide mitigation options to producers, many challenges limit adoption on farms. Although the cow-calf sector is the largest source of enteric CH_4 in Canada, mitigation is particularly challenging for pasture-based systems because it is difficult to provide feed additives and supplements at the required dose. Another major barrier is affordability; farmers need greater information on the cost of mitigation and impacts on animal productivity. Incentives may be needed to encourage adoption because in most cases decreased CH_4 production is not expected to increase animal performance. Regulatory approval requirement for some promising feed additives may slow their adoption. Before adopting an enteric CH_4

mitigation practice, the net impact on total GHG emissions from the entire production system needs to be considered.

Net-zero emissions dairy and beef products may not be achievable in the near future, but climate neutral beef and dairy may be possible by aggressively adopting practices that lower GHG emissions and bolster soil carbon reserves. Adoption of enteric CH_4 mitigation solutions and protocols that enhance soil carbon offer farmers the opportunity to diversify their income by producing branded products and participating in carbon offset markets. As the new global warming potential star (GWP*) metric becomes more universally accepted, CH_4 from ruminants will be recognized as being part of a biological cycle unlike CO_2 from fossil fuels. Ruminants may no longer be seen as the villain, but rather an opportunity to offset GHG emissions, especially in Canada where ruminant numbers are not increasing and deforestation for livestock production does not occur.

References

Allen, M.R., J.S. Fuglestvedt, K.P. Shine, A. Reisinger, R.T. Pierrehumbert and P.M. Forster. 2016. New use of global warming potentials to compare cumulative and short-lived Nature Climate Change 6:773–776. https://doi.org/10.1038/nclimate2998

Appuhamy, J.A.D.R.N., A.B. Strathe, S. Jayasundara, C. Wagner-Riddle, J. Dijkstra, J. France, and E. Kebreab. 2013. Antimethanogenic effects of monensin in dairy and beef cattle: a meta-analysis. J. Dairy Sci. 96:5161–5173. https://doi.org/10.3168/jds.2012-5923

Beauchemin, K.A., E.M. Ungerfeld, R. Eckard and M. Wang. 2020. REVIEW: Fifty years of research on rumen methanogenesis - lessons learned and future challenges for mitigation. Anim. 14:S1, s2–s16. https://doi.org/10.1017/S1751731119003100

ECCC. 2021. https://www.canada.ca/en/environment-climate-change/services/climate-change/greenhouse-gas-emissions/inventory.html

Gerber, P.J., H. Steinfeld, B. Henderson, A. Mottet, C. Opio, J. Dijkman, A. Falcucci and G. Tempio. 2013. Tackling climate change through livestock – A global assessment of emissions and mitigation opportunities. Food and Agriculture Organization of the United Nations (FAO), Rome.

Jayanegara, A., F. Leiber, and M. Kreuzer. 2012. Meta-analysis of the relationship between dietary tannin level and methane formation in ruminants from in vivo and in vitro experiments. J. Anim. Physiol. Anim. Nutr. 96:365–375. https://doi.org/10.1111/j.1439-0396.2011.01172.x

Jayasundara, S. and C. Wagner-Riddle. 2014. Greenhouse gas emissions intensity of Ontario milk production in 2011 compared with 1991. Can. J. Anim. Sci. 94, 155–173, https://doi.org/10.4141/CJAS2013-127.

Jayasundara, S., D. Worden, A. Weersink, T. Wright, A. VanderZaag, R. Gordon and C. Wagner-Riddle. 2019. Improving farm profitability also reduces the carbon footprint of milk production in intensive dairy production systems. J. Cleaner Prod. 229:1018–1028, https://doi.org/10.1016/j.jclepro.2019.04.013.

Legesse, G., K.A. Beauchemin, K.H. Ominski, E.J. McGeough, R. Kroebel, D. MacDonald, S. M. Little and T.A. McAllister. 2016. Greenhouse gas emissions of Canadian beef production in 1981 as compared with 2011. Anim. Prod. Sci. 56:153-168. https://doi.org/10.1071/AN15386

Little S.M., C. Benchaar, H.H. Janzen, R. Kröbel, E.J. McGeough and K.A. Beauchemin. 2017. Demonstrating the effect of forage source on the carbon footprint of a Canadian dairy farm using whole-systems analysis and the Holos model: alfalfa silage vs. corn silage. Climate 5:87. https://doi.org/10.3390/cli5040087.

Place, S.E and F.M. Mitloehner. 2021. Pathway to climate neutrality for U.S. beef and dairy cattle production. https://clear.ucdavis.edu/sites/g/files/dgvnsk7876/files/inline-files/CLEAR%20Center%20Climate%20Neutrality%20White%20Paper_2.pdf

Smith, P., M. Bustamante, H. Ahammad, H. Clark, H. Dong, E. A. Elsiddig, H. Haberl, R. Harper, J.I. House, M. Jafari, O. Masera, C. Mbow, et al. 2014. Agriculture, Forestry and Other Land Use (AFOLU). In O. Edenhofer et al. (Eds.), *Climate Change 2014: Mitigation of Climate Change. Contribution of Working Group III to the Fifth Assessment Report of the Intergovernmental Panel on Climate Change,* Cambridge, University Press. https://doi.org/http://mitigation2014.org/

Ungerfeld, E.M., K.A. Beauchemin and C. Muñoz. 2022. Current perspectives on achieving pronounced enteric methane mitigation from ruminant production. Hypothesis and theory article. Front. Anim. Sci., 03 January 2022. https://doi.org/10.3389/fanim.2021.795200

van Lingen, H. J., M. Niu, E. Kebreab, S. C. Valadares Filho, J. A. Rooke, et al. 2019. Prediction of enteric methane production, yield and intensity of beef cattle using an intercontinental database. Agric. Ecosyst. Environ. 283:106575. https://doi.org/10.1016/j.agee.2019.106575

Yu, G., K.A. Beauchemin and R. Dong. 2021. A review of 3-nitrooxypropanol for enteric methane mitigation from ruminant livestock. Animals 11:3540. https://doi.org/10.3390/ani11123540

Sustainable Livestock Feeding and Management: What Changes are Needed?

Alimentation et gestion durables en production animale : quels changements sont nécessaires?

Donald Broom
Department of Veterinary Medicine, University of Cambridge, United Kingdom
dmb16@cam.ac.uk

Sponsored by / Commandité par :

O&T Farms
Naturally Better

Abstract

World economics has been changing. 'Push production', driven by producers, is being replaced by 'pull production', driven by consumers who demand sustainable systems and products and think that people should be less human-centred. A system or procedure is sustainable if it is acceptable now and if its expected future effects are acceptable, in particular in relation to resource availability, consequences of functioning and morality of action.

There are many components of sustainability. A food production system might be unsustainable because of inefficient usage of world food resources; adverse effects on human welfare, including health; poor welfare of production animals or other animals; harmful environmental effects, such as greenhouse gas production, low biodiversity or insufficient conservation; unacceptable genetic modification; not being "fair trade", in that producers in poor countries are not properly rewarded; or damage to rural communities. A scoring method based on scientific information has been developed and shows that the best beef production systems are much more sustainable than the worst.

In future, consumers choosing efficient usage of world resources and avoiding causing harms will change food production. Likely changes include: increased plant food consumption; avoidance of some plant production methods; increased use of animals consuming leaves; greater use of plants producing high protein leaves; more mixed herb, shrub, and tree systems; less feeding of grain to farmed animals; more use of human food waste to feed farmed animals after treatment to avoid disease; more use of herbivorous fish, insects and mammals; and more cell-cultured meat.

Résumé

L'économie mondiale a changé. La production « à flux poussé », axée sur les producteurs, est remplacée par la production « à flux tiré », axée sur les consommateurs qui exigent des systèmes et des produits durables et qui considèrent que les gens devraient être moins centrés sur l'individu. Un système ou un processus est durable s'il est acceptable aujourd'hui et si ses effets futurs attendus sont acceptables, notamment en ce qui a trait à la disponibilité des ressources, aux conséquences du fonctionnement et la moralité de l'action.

La durabilité est constituée de nombreux éléments. Un système de production alimentaire peut être jugé non durable pour différentes raisons : utilisation inefficace des ressources alimentaires mondiales; effets néfastes sur le bien-être humain, y compris la santé; mauvais traitements infligés aux animaux de production ou à d'autres animaux; effets environnementaux néfastes, tels que production de gaz à effet de serre, diminution de la biodiversité ou gaspillage; modifications génétiques inacceptables; commercialisation non équitable dans la mesure où les producteurs des pays pauvres ne sont pas correctement rémunérés; dommages causés aux communautés rurales. Une méthode de notation basée sur des données scientifiques a été mise au point et elle montre que les meilleurs systèmes de production de viande bovine sont beaucoup plus durables que les pires.

À l'avenir, les consommateurs qui opteront pour une utilisation efficace des ressources mondiales et qui ne cause pas de dommages modifieront la production alimentaire. Parmi les changements probables, citons : consommer plus d'aliments d'origine végétale; éviter certaines méthodes de production végétale; utiliser davantage les animaux qui consomment des feuilles; utiliser davantage de plantes qui produisent des feuilles à haute teneur en protéines; favoriser les systèmes mixtes d'herbes, d'arbustes et d'arbres; donner moins de grains aux animaux d'élevage; donner plus de déchets alimentaires humains aux animaux d'élevage après les avoir traités pour éviter les maladies; utiliser plus de poissons, insectes et mammifères herbivores; utiliser davantage de viande produite à partir de cultures cellulaires.

Introduction

The greater availability of accurate information about products and production systems of various kinds is increasingly changing many aspects of the world economy from a 'push economy' to a 'pull economy'. While producers of food and other goods purchased by consumers formerly determined methods of production, consumers are now exerting more control on production methods (Broom 2014, 2017a). 40-50 years ago, many consumers stopped buying some goods on moral grounds, for example carpets whose production involved child labour were ostracised. The World Trade Organization (WTO) permitted countries to ban imports of such goods on public morality grounds. At this time, and until about 20 years ago, the public wanted food products but farmers determined production methods. Then many retail companies such as supermarkets and fast-food chains were forced by consumers to sell Fair Trade and Welfare-Friendly products (Bennett et al 2002, Broom 2010). The demand pull by consumers has become more detailed as information has become more available (Kim and Lee, 2009, Antonelli and Gehringer, 2015). Retail food companies were also pressurised to publicly state their standards in a number of areas by consumers who said that they would refuse to buy anything from the companies if they continued to sell, for example, pig meat if sows were confined in stalls or tethers, or eggs from hens kept in battery cages. The WTO extended the use of the public morality reason for blocking imports, to include seal skins when the killing method was often not humane, and many governments changed legislation about livestock production (Broom 2016, 2017a).

The concepts of 'one biology', 'one health' and 'one welfare' emphasize the great similarities between humans and other animal species and are changing human attitudes (Tarazona et al 2020). Most people regulate their actions because they consider that each person has obligations to avoid causing harms to humans, other animals and aspects of the world environment. As a consequence, a key question about any system is whether or not it is sustainable (Aland and Madec 2009). Sustainability is now a broader concept than it was when first used (Herrero et al 2010). A system or procedure is sustainable if it is acceptable now and if its expected future effects are acceptable, in particular in relation to resource availability, consequences of functioning and morality of action (Broom 2014, modified after Broom 2001). Sustainability is demanded in relation to: travel, erecting buildings, heating or cooling buildings, producing clothes and all other goods, disposal of waste, etc. Examples given here are mainly from food production.

The quality of food, as assessed by many consumers now, is not just about taste and nutritional value but also includes its impacts on consumer health and various aspects of any perceived negative effects of its production. Moral issues concerning food production emphasise the close links between sustainability and food quality. A palatable, nutritious food item whose production method is unacceptable to the general public for any reason is not regarded as being of good quality and the production system is not sustainable. Some consumers would not purchase certain items whatever the price whilst others might buy them if the price was low so the market size was reduced. The rejection of food items produced in certain ways now varies less from country to country than was formerly the case.

Sustainability components

Sustainability has many components such as: adverse effects on human welfare, including human health, no fair reward for producers in poor countries, and not preserving rural communities; poor welfare of animals used in production; unacceptable genetic modification; harmful environmental effects such as causing climate change or biodiversity reduction; and inefficient use of world resources. In order to evaluate systems of production and products, the source of all raw materials and product modifiers and the fate and impact of all of the product after sale or usage must be taken into account. The life-cycle approach (Day et al 1981) requires evaluation of all environmental changes resulting from the system (Ciambrone 1997). Another terminology used is to say that all externalities of the functioning of the system must be measured, e.g those of motor vehicle use (Delucchi 2000) or of farming systems (Balmford et al 2012, 2018). There have been attempts to express all of the changes in terms of money, or carbon usage, or energy usage but none of these can be used for all sustainability components. The scoring method (Broom 2021a) must be based on scientific information. In order to be able to compare systems, in the example shown here, a score of any negative effects is allocated ranging from 0, for no negative effect, to -5 for the worst effect described in the scientific literature to occur. The scoring could also be positive. Where there is evidence that some consumers completely avoid a product because of the negativity of the particular component, this is indicated by Z.

How do we measure each component of sustainability that affects a system? Some impacts of food products on the human health component are positive. For ingestion of meat, saturated fats may increase the risk of disorders so consumer mortality or morbidity rate can be measured. Public opinion surveys may also provide quantitative information. Fair trade labels require good traceability of products for the main measure of the extent of the effect is what consumers buy. Job satisfaction on farms can be measured by asking people doing the work and considering ease of getting staff. Impacts on rural populations can be measured by human population movement and, to some extent, by use of questionnaires.

The welfare of production animals is an important sustainability component for livestock products. In Brazil, consumers stated that for beef, the welfare of the production animal is the most important sustainability attribute and that traceability is important (Burnier et al 2021). There is now much good quality scientific evidence about the welfare of production animals. Wild animal welfare and mortality rates are also affected by plant production but there is less scientific evidence about this.

Some genetic modification, including gene editing, can negatively change animal functioning in substantial ways. A range of genetic modification effects can be measured, as can public acceptances of GM products.

Environmental effects of livestock production systems can be measured by: greenhouse gas output, biodiversity reduction and pollution of water etc. per unit of product. Inefficient usage of world food resources is likely to become much more important as a factor affecting plant and animal production. One aspect is the amount of food waste produced and what happens to it. Another major factor is the energy loss between plant product and human consumption. Animals that eat leaves and other food that humans cannot eat, such as ruminants and herbivorous fish, will become more important than animals that are carnivores or that eat grain. When wheat, maize, soya etc. are produced, it is much more efficient for humans to eat it rather than to feed it to pigs or poultry. Even worse is to feed it to ruminants as they can eat food that we cannot eat. If dairy cows are fed grain, food resources are wasted. The dairy production has a negative overall energy balance if more than 30% of cow diet is grain. Ruminants, herbivorous fish and insects are expected to be important sources of human food for the future.

When the sustainability component land area per unit of meat is to be measured, the land needed to grow all food given to the animals must be measured, as well as land actually occupied by the animals (Broom 2019). The other ways in which the land could be used is relevant to the efficiency of use of world resources mentioned above. Similarly, the largest water cost is often that used in growing feed. Conserved water is usually a better measure of water cost than total water, including rainfall, while purified water has costs additional to the water costs themselves.

Beef sustainability

What do we know about the sustainability of beef production systems? Recent and current world production of beef has high greenhouse gas production (Steinfeld 2006). Extensive beef uses a lot of land so there is less land available for conservation but more highly productive beef systems have environmental advantages over low production systems Balmford et al 2012, Balmford 2021). There have been improvements in water usage, land usage and greenhouse-gas production as compared with 40 years ago (Capper 2011). However, there are large differences across current beef production systems in the amount of land and water needed per kilo of beef (Broom 2019). Beef and lamb can be produced using plant material that humans cannot eat. It would therefore be better to stop feeding substantial amounts of grain to ruminants (Eisler et al 2014, Wilkinson and Lee 2017). Given this background, an analysis of the following widely-used beef production systems was carried out: extensive pasture degraded or not degraded; fertilised irrigated pasture with and without concentrate feeding; feedlots preceded by fertilised irrigated pasture or by extensive pasture; indoor housing throughout life or preceded by fertilised irrigated pasture or by extensive pasture; and semi-intensive silvopastoral system. The sustainability components for which there were some differences across these systems were: human health; welfare of production animals; efficiency of use of world resources: land usage; efficiency of use of world resources: land area per kg meat (conservation); efficiency of use of world resources: amount of conserved water per kg meat; greenhouse gas production per unit of product; extent of water pollution and nitrogen/phosphorus cycle disruption; biodiversity decline; and reduction in carbon sequestration (Broom 2021a). For example, degraded extensive pasture used most land per kilo of beef, semi-intensive silvopastoral used the least land and feedlot was intermediate. The totals of all sustainability components are shown in Table 1.

Table 1. Totals of sustainability components for each beef production system (modified after Broom 2021a)

Beef production system	
Extensive pasture, degraded	-26,ZZ
Extensive pasture, not degraded	-12
Fertilised irrigated pasture, plus concentrates	-23
Fertilised irrigated pasture, no concentrates	-16
Fertilised irrigated pasture, then feedlot	-25,Z
Extensive pasture, then feedlot	-25,Z
Fertilised irrigated pasture, then indoor housing.	-26,Z
Extensive pasture, then indoor housing	-26,ZZ
Indoor rearing, then indoor housing	-29,Z
Semi-intensive silvopastoral	-5

As a result of this analysis, Conclusions 1-4 at the end of this paper concern beef production systems.

Sustainability comparisons of beef with other meat and with plant production

No comparable scientific analysis of all components of the sustainability of other meat and of plant production has been conducted. I hope that this work will be done. However, the data for lamb from extensively reared sheep would be similar to those from extensively-reared beef cattle. Estimates for pork, chicken and cultured meat are made here. Pork and chicken would score worse than beef for efficiency of use of world resources and significantly worse for production animal welfare but a little better for land area, water use, greenhouse gas and biodiversity decline. The overall sustainability score would be slightly worse (-2 estimated) for pork and chicken than for beef. This estimate contradicts the widely stated view, based largely on greenhouse gas production, that beef is less sustainable than pork and chicken. How will future judgements about sustainability be made? If efficiency of use of world resources or animal welfare is the most important component - beef and lamb are better than pork or chicken. If greenhouse gas, land area used, or biodiversity decline are the most important components - pork and chicken are better than beef or lamb. For each meat, the best systems are much more sustainable than the worst.

I have endeavoured to obtain information from three cultured meat producers that would enable me to do an analysis like that carried out for beef. None of the producers would give me sufficient information to allow the analysis to be carried out. Publications in this area are mainly written by those with a commercial interest in cultured meat production. My estimates for cultured meat suggest that it would be better than beef in land area, water use, biodiversity decline, and production animal welfare and not substantially worse in any components. Hence it could be more sustainable than beef, pork or chicken. However, a reliable replacement for fetal calf serum as an initial nutrient, a usable main nutrient source and safeguards for the spread of human and farm animal disease are needed. Tentative conclusions 5 and 6 follow the estimated comparisons of meat production systems.

How does the sustainability of eating meat compare with that of eating plant-based food? The efficiency of use of world resources is better by a factor of between 3 and 15 if plants are eaten rather than feeding the same plant material to animal species and then eating those animals. Plants can provide all essential human nutrients. A lot of protein in the US diet already comes from plants (8.7% from various breads, 7.2% from pasta, rice, potatoes and other vegetables, 4.5 from breakfast cereals etc. – a total of 32% as compared with 68% from animal sources (Pasiakos et al 2015). Comparable totals for a study in France were 31% of protein from plants, 69% from animals (de Gavelle et al 2017). However, as mentioned above, humans cannot eat most leaves, branches and roots and only some fruits and seeds. Leaves make up most of the plant food that is available to animals. As a consequence, mammals, birds, fish, insects etc. that eat leaves are more important as human food than seed or fruit eaters. World resource use is likely to rapidly become an important issue throughout the world and will have a major impact on livestock feeding. Plant cultivation systems also vary in sustainability. Some involve clearing natural vegetation and almost all involve killing native plants and animals.

What is likely to change in agriculture?

Changing farm animal diet, both that fed to livestock and that foraged by the animals, can reduce greenhouse-gas output and water use. However, the effects are relatively small. Feed additives, such as 3-hydroxy propanol and providing certain forage plants can reduce gut methanogen numbers or activity by 2-12% (White et al 2014, Feng and Kebreab 2020). Changing farm animal diet should be a rapid response in the production industry and there is likely to be greater reduction of output of methane and other greenhouse gases possible but not enough to solve the greenhouse gas problem. In addition, any changes must take account of reduction in anti-microbial resistance and be transparent and traceable (Maia de Souza et al 2017). System change is also needed for a fully sustainable future. A lot of studies focus on greenhouse-gas production (Kamilaris et al 2020) but all components of sustainability should be considered. It is likely that the components now widely considered to be of key importance: greenhouse-gas reduction and avoidance of poor welfare of livestock, will become even more important in the next few years. Using world resources efficiently by minimising waste of food and reducing use of livestock that compete directly with humans for food is very likely to become much more important. This would reverse the recent trend of increase in pig and poultry production. At the same time there would be increase in ruminant production, in production of herbivorous fish, and in production of some insects and other leaf eaters for human consumption. Pigs and poultry might return to their earlier role of consuming human food waste after it has been properly treated to avoid disease spread (zu Ermgassen et al 2016).

Biodiversity loss is an increasingly important factor for many of the general public who are driving these changes. Extensive pasture systems have higher biodiversity than irrigated pasture and most crops such as maize and soya have less biodiversity than pasture. Biodiversity of insects and birds is much higher on silvopastoral systems than on pasture areas in the same locality (Murgueitio et al 2008).

Biodiversity decline on farmed land has been greater in the last 15-20 years than ever before in world history. Much of the decline is a consequence of widespread herbicide use and some of it is a consequence of pesticide use. This decline involves the death of enormous numbers of animals and plants. The general questions of conservation and biodiversity decline have led to discussion about future aims. Should conservation be aimed at keeping islands of natural vegetation in a relatively barren world of agriculture (land-sparing) or should all land be biodiverse (land-sharing)? High productivity per unit of land allows more land to be spared for conservation (Balmford et al 2018, Balmford 2021). My personal view is that there should be some conserved areas managed to have only native communities of plants and animals but that the consequences that we see now of high production agriculture with very low biodiversity are too negative for wild animals and plants and for humans. Reducing biodiversity by causing the death of very many animals and other living organisms is morally wrong. Hence, both land-sparing and land-sharing are needed. Plant production methods need much more rigorous analysis so that unsustainable practices can be avoided. This might well be the greatest change in farming in forthcoming years.

In some areas, the use of semi-intensive silvopastoral systems, together with some entirely natural vegetation areas, is a good solution. Plant production from a mixture of herbs, shrubs and trees is much greater than from a single level pasture system. The use of nitrogen-fixing shrubs such as *Leucaena* as part of the forage for livestock, or as cut fodder, in semi-intensive silvopastoral systems can be economically successful and sustainable. As indicated in the analyses above there is typically: greater production and biodiversity; less pollution run-off because of water-holding properties of soil; less methane production per kg of meat; better carbon sequestration; less disease and better welfare in semi-intensive silvopastoral systems (Murgueitio et al 2008, Broom et al 2013, Broom 2017b).

Should you be vegetarian or vegan?

The answer to the question of whether or not an individual should be vegetarian or vegan depends on the reason for being, or becoming, vegetarian or vegan (Broom 2018). Reasons A, B, and C are briefly considered here.

A. "I cannot eat something that was once a live animal." This is an aesthetic reason and is likely to persist if all food-production systems become sustainable.
B. "Eating animals is poor usage of world resources." This is true for animals that compete with humans for food. Hence it is better to eat more plant food, to stop eating animals

that eat food that humans can eat, or that comes from land where the plants humans eat can be grown, but to eat animals whose food is not edible for humans.

C. "I do not agree with killing animals in order that human food can be produced." Very large numbers of animals are killed during crop production, during harvesting and during storage. For some crops the numbers are higher than for some animal production systems. Hence this is not a logical argument for being vegetarian or vegan.

Food from the sea and other open water

A final point concerns other changes in food production and other human activities, especially in relation to marine and other aquatic habitats. The chemical industry, building industry, clothing production industry, agriculture, human sewage and other human activities all cause pollution of the seas and open water that kills animals. There should be studies of the number of deaths of any animal per unit of pollutant and fines at a deterrent level linked to the number of deaths. A few examples are presented here. There has been a big decline in puffins, guillemots and terns because sand-eels have been fished for use as fertiliser. Albatrosses, dolphins etc. are killed in drift nets and other nets and are called "by-catch". Several fishing methods are non-selective. They catch many animals that are not the target animals. Non-selective catching methods are morally condemned or are not permitted for most forms of vertebrate trapping. One of the worst fishing methods in this respect is trawling. While a few of the non-target animals caught are returned to the sea, very large numbers are injured or killed. Bottom-dwelling animals are particularly vulnerable. Plastic pollution in the sea has small effects on some species but very large effects on others. Legal redress is needed. Unwanted fishing gear also has large effects, killing many marine birds and mammals as well as fish and invertebrates. Penalties for leaving the fishing gear in the sea should be high but the identification of people who dump plastic or leave fishing gear in the sea is often difficult. As with food production on land, world resources are better used if people eat more aquatic plants.

Future livestock food production and feeding; some predictions

As the current trend to seriously consider the sustainability of human products and actions accelerates, what consequences for livestock food and feeding can be expected? Systems of production that cause poor welfare of the production animals, such as high production rates in broiler chickens and high milk yield in dairy cows, will be modified greatly and close confinement in systems that do not provide for the needs of sows, calves, etc., described by (Broom 2021b, 2022) will cease. The feeding of grain, soya, and other foods that humans can eat, to cattle and other ruminants will cease. Ruminant diets that reduce methane production will be refined and become universal. Systems for the management of livestock involving the animals foraging, or being fed cut forage from edible high protein leaves of trees, shrubs and pasture plants will increase. Poultry, pigs and other monogastrics will be fed less grain, soya and other human-edible material. Human food-waste treated to prevent disease spread will be fed to poultry and pigs. The overall effect will be a substantial decline in poultry and pig production. The farming of herbivorous fish will increase even faster then it has in recent years and fish and insect protein will become more important in the human diet. Cell-cultured food will become more widespread if no disease-risk or other sustainability issue prevents this.

Conclusions

1. Beef production systems have a wide sustainability range with the best much better than the worst.

2. The least sustainable beef production systems are extensive grazing that causes land degradation and the use of feedlots or indoor housing with grain feeding.

3. The most sustainable beef production systems are semi-intensive silvopastoral systems. Well-managed pasture-fed beef from land where crop production is uneconomic is also sustainable.

4. For all meat and meat-like products, consumers need reliable sustainability labels taking account of all sustainability components. These would allow them to avoid purchase unless there is a sustainability label and to avoid the least sustainable meat.

5. Consumers should not avoid beef and lamb on grounds of overall sustainability, they should probably prefer beef and lamb to poultry and pig-meat.

6. Cultured meat may be more sustainable than beef, pork and chicken but reliable evidence is lacking.

7. The use of high-protein pasture plants, shrubs and trees should increase.

8. Use semi-intensive silvopastoral systems when possible.

9. Increase efficient use of extensive pasture.

10. Stop feeding maize, wheat, other cereals and soya to livestock, especially cattle.

11. Stop using feedlots and indoor housing of beef cattle.

12. Change dairy production to avoid high-producing cows and closely-confined calves.

13. Reduce pig and poultry production and increase sustainable beef and lamb.

14. Increase use of treated, unwanted human food for pigs, poultry, farmed fish.

15. Minimise ploughing and other soil-damaging activity.

16. Subsidise carbon-sequestration.

17. Switch fish-farming to the use of herbivorous fish: Tilapia, carp etc. Do not farm salmon, trout and other predatory species unless waste food is used.

18. Regulate human exploitation to prevent harms to world marine and other open water habitats. Develop exploitation of aquatic plants for human consumption.

References

Aland, A. and Madec, F. (eds) 2009. *Sustainable Animal Production*. Wageningen Academic Publishers, Wageningen, Netherlands.

Antonelli, C. and Gehringer, A. 2015. The competent demand-pull hypothesis. In: *The Economics of Knowledge, Innovation and Systemic Technology Policy*, ed. F.Crespi & F. Quatraro, 48–69. Routledge, London & New York

Balmford, A. 2021. Concentrating versus spreading our footprint: how to meet humanity's needs at least cost to nature. *J. Zool.* 315, 79-109. doi:10.1111/jzo.12920

Balmford, A., Amano, T., Bartlett, H., Chadwick, D., Collins, A., Edwards, D., Field, R., Garnsworthy, P., Green, R., Smith, P., Waters, H., Broom, D.M., Chará, J., Finch, T., Garnett, E., Gathorne-Hardy, A., Hernandez-Medrano, J, Herrero, M., Hua, F., Latawiec, A., Misselbrook, T., Phalan, B., Simmons, B., Takahashi, T., Vause, J., zu Ermgassen, E. and Eisner, R. 2018. The environmental costs and benefits of high-yield farming. *Nature Sustainability,* 1 477- 485. doi 10.1038/s41893-018-0138-5

Balmford, A., Green, R. and Phalan, B. 2012. What conservationists need to know about farming. *Proc. Roy. Soc. B,* 279 2714-2724.

Bennett, R.M., Anderson, J. and Blaney, R.J.P. 2002. Moral intensity and willingness to pay concerning farm animal welfare issues and the implications for agricultural policy. J. Agric. Environ. Ethics, 15 187–202.

Broom, D.M. 2001. The use of the concept Animal Welfare in European conventions, regulations and directives. *Food Chain 2001*, 148-151, Uppsala: SLU Services.

Broom, D.M. 2010. Animal welfare: an aspect of care, sustainability, and food quality required by the public. J. Vet. Med. Educ., 37 83-88.
doi: 10.3138/jvme.37.1.83

Broom, D.M. 2014. *Sentience and Animal Welfare.* Wallingford, U.K. CABI.

Broom, D.M. 2016. International animal welfare perspectives, including whaling and inhumane seal killing as a public morality issue. In *Animal Law and Welfare – International Perspectives*, 45-61, (eds) D.Cao and S. White. Springer International Publishing, Switzerland. Book DOI 10.1007/978-3-319-26818-7.

Broom, D.M. 2017a. *Animal Welfare in the European Union*. (pp 75). Brussels: European Parliament Policy Department, Citizen's Rights and Constitutional Affairs. ISBN 978-92-846-0543-9 doi: 10-2861/891355.

Broom, D.M. 2017b. Components of sustainable animal production and the use of silvopastoral systems. *Rev. Bras. Zootecn.,* 46 683-688. doi.org/10.1590/S1806-92902017000800009

Broom, D.M. 2018. The scientific basis for action on animal welfare and other aspects of sustainability. In: D'Silva, J. and McKenna, C. (eds) *Farming, Food and Nature: respecting*

animals, people and the environment, 93-100. London and New York: Earthscan, Routledge. ISBN: 978113854141

Broom, D.M. 2019. Land and water usage in beef production systems. *Animals*, 9 286. doi.org/10.3390/ani9060286

Broom, D.M. 2021a. A method for assessing sustainability, with beef production as an example. *Biol. Rev.*, 96 1836-1853. doi.org/10.1111/brv.12726

Broom, D.M. 2021b. Dairy cattle welfare and other aspects of sustainability. In: Endres, M. ed. *Understanding the Behaviour and Improving the Welfare of Dairy Cattle*, 1-13. Cambridge, UK, Burleigh Dodds Science Publishing.

Broom, D.M. 2022. *Broom and Fraser's Domestic Animal Behaviour and Welfare, 6th edn* (pp.545). CABI.

Broom, D.M., Galindo, F.A. and Murgueitio, E. 2013. Sustainable, efficient livestock production with high biodiversity and good welfare for animals. *Proceedings of the Royal Society B, 280 2013-2025.*

Burnier, P.C., Spers, E.E. and de Barcellos, M.D., 2021. Role of sustainability attributes and occasion matters in determining consumers' beef choice. *Food Quality Pref.*, 88 p.104075.

Capper, J.L., 2011. The environmental impact of beef production in the United States: 1977 compared with 2007. *J. Anim. Sci.*, 89, 4249-4261.

Ciambrone, D. F. 2018. *Environmental Life Cycle Analysis*, pp 160. Boca Raton: CRC Press.

Day, G.S. 1981. The product life cycle: analysis and applications issues. *J. Marketing,* 45 60-67.

De Gavelle, E., Huneau, J.F., Bianchi, C.M., Verger, E.O. and Mariotti, F. 2017. Protein adequacy is primarily a matter of protein quantity, not quality: modeling an increase in plant: animal protein ratio in French adults. *Nutrients*, 9 1333.

Delucchi, M.A., 2000. Environmental externalities of motor-vehicle use in the US. *Journal of Transport Economics and Policy*, 34 135-168.

Eisler, M.C., Lee, M.R., Tarlton, J.F., Martin, G.B., Beddington, J., Dungait, J.A., Greathead, H., Liu, J., Mathew, S., Miller, H. and Misselbrook, T. 2014. Agriculture: steps to sustainable livestock. *Nature*, 507 32-34.

Feng, X. and Kebreab, E. 2020. Net reductions in greenhouse gas emissions from feed additive use in California dairy cattle. *Plos one*, 15 p.e0234289.

Herrero, M., Thornton, P.K., Notenbaert, A.M., Wood, S., Msangi, S., Freeman, H.A., Bossio, D., Dixon, J., Peters, M., van de Steeg, J. and Lynam, J. 2010. Smart investments in sustainable food production: revisiting mixed crop-livestock systems. *Science* 327 822–825.

Kamilaris, C., Dewhurst, R.J., Sykes, A.J. and Alexander, P. 2020. Modelling alternative management scenarios of economic and environmental sustainability of beef finishing systems. *Journal of Cleaner Production*, 253 119888.

Kim, W. and Lee, J.D. 2009. Measuring the role of technology-push and demand-pull in the dynamic development of the semiconductor industry: the case of the global DRAM market. *J. Appl. Econ.,* 12 83–108.

Maia de Souza, D., Petre, R., Jackson, F., Hadarits, M., Pogue, S., Carlyle, C.N., Bork, E. and McAllister, T., 2017. A review of sustainability enhancements in the beef value chain: state-of-the-art and recommendations for future improvements. *Animals*, 7 26.

Murgueitio, E., Cuartas, C. A. and Naranjo, J. F. 2008. *Ganadería del Futuro*. Fundación CIPAV, Cali.

Pasiakos, S.M., Agarwal, S., Lieberman, H.R. and Fulgoni, V.L. 2015. Sources and amounts of animal, dairy, and plant protein intake of US adults in 2007–2010. *Nutrients*, *7* 7058-7069.

Tarazona, A.M., Ceballos, M.C. and Broom, D.M. 2020. Human relationships with domestic and other animals: one health, one welfare, one biology. *Animals,* 10 43 (pp. 23) doi:10.3390/ani10010043

White, R. R., Brady, M., Capper, J.L. and Johnson, K.A. 2014. Optimizing diet and pasture management to improve sustainability of US beef production. *Agric. Systems*, 130 1-12.

Wilkinson, J.M. and Lee, M.R.F., 2018. Use of human-edible animal feeds by ruminant livestock. *Animal*, 12 1735-1743.

Zu Ermgassen, E. K., Phalan, B., Green, R. E. and Balmford, A. 2016. Reducing the land use of EU pork production: where there's swill, there's a way. *Food Policy,* 58 35–48.

On the Development of Online Tools to Support the Production of Sustainable Animal Feeds

Sur le développement d'outils en ligne pour supporter la production durable d'aliments pour animaux

Dominique P. Bureau
Department of Animal Biosciences, University of Guelph, Guelph, ON
dbureau@uoguelph.ca

Sponsored by / Commandité par :

LALLEMAND

LALLEMAND ANIMAL NUTRITION

Abstract

Animal production relies on the supply of nutritious feeds supporting high growth, welfare and health of animals and the production of high quality products in an environmentally sustainable and profitable manner.

Feed manufacturers rely least-cost feed formulation programs and an array of chemical analysis technique and equipment to support the production of cost-effective feeds of stable and appropriate quality. However, sustained interactions with industry stakeholders worldwide suggest that many are relying on relatively static, potentially outdated, information on the nutritive value of ingredients, the nutritional requirements of animals and may have limited insights on how to optimize nutritional specifications for specific production conditions. Animal production also relies on a "nutritional supply chain", that includes the production and procurement of feed ingredients, formulation and manufacturing of different feeds and the feeding of these feeds to groups of animals. Tracking this nutritional supply chain is regarded as increasingly important for different reasons. For many medium and small-size feed manufacturers developing or acquiring and implementing ERP systems to track and manage information is complicated and costly.

Several user-friendly, innovative and economical online platforms are being developed and these could help feed manufacturers of any size with:

1) Better prediction of the nutritive value of ingredients

2) Optimization of nutritional specifications on demand

3) Improvement of Quality Assurance and Quality Control (QA&QC) and feed traceability procedures

4) Implementation of effective feedback loop enabling continuous improvement of feed quality

Résumé

La production animale repose sur un approvisionnement en aliments nutritifs favorisant un rythme de croissance élevé, le bien-être et la santé des animaux et la production rentable et durable sur le plan environnemental de produits de bonne qualité.

Les fabricants d'aliments pour animaux s'appuient sur des programmes de formulation d'aliments à moindre coût et sur un ensemble de techniques et d'équipements d'analyse chimique pour soutenir la production lucrative d'aliments de qualité stable et adéquate. Cependant, des échanges réguliers avec des acteurs de l'industrie du monde entier suggèrent que beaucoup d'entre eux s'appuient sur des données relativement statiques, potentiellement dépassées, se rapportant à la valeur nutritive des ingrédients et aux besoins nutritionnels des animaux et que ces intervenants pourraient ne pas disposer de toutes les connaissances requises pour optimiser les spécifications nutritionnelles selon les conditions de production spécifiques. La production animale repose également sur une « chaîne d'approvisionnement nutritionnelle » qui comprend la production et la fourniture d'ingrédients alimentaires, la formulation et la fabrication de différents aliments et l'administration de ces aliments à des groupes d'animaux. Il est considéré comme de plus en plus important d'assurer la traçabilité de cette chaîne d'approvisionnement nutritionnelle pour différentes raisons. Pour de nombreux fabricants d'aliments de petite et moyenne taille, la création ou l'acquisition et la mise en œuvre de systèmes de planification des ressources de l'entreprise permettant de suivre et gérer l'information sont compliquées et coûteuses.

Plusieurs plateformes en ligne conviviales, innovantes et peu coûteuses sont en cours de développement et pourraient aider les fabricants d'aliments pour animaux, quelle que soit leur taille :

1) à mieux prévoir la valeur nutritive des ingrédients;

2) à optimiser les spécifications nutritionnelles à la demande;

3) à améliorer les procédures d'assurance-qualité, de contrôle de la qualité et de traçabilité des aliments pour animaux;

4) à mettre en œuvre d'une boucle de rétroaction efficace permettant une amélioration continue de la qualité des aliments pour animaux.

Introduction

Animal production relies on the supply of nutritious feeds supporting high growth, welfare and health of animals and the production of high-quality products in an environmentally sustainable and profitable manner. With feed cost representing 50 to 80% of the total production cost, feed manufacturers have a very important role in the overall economic viability of animal agriculture enterprises. Animal production involves a complex "nutritional supply chain", that includes the production and procurement of a large variety of feed ingredients, the formulation and manufacturing of different feeds and the feeding of these feeds to different animal populations (Figure 1). Tracking this nutritional supply chain is regarded as increasingly important and feed manufacturers obviously play a vital role in this process.

Variations in feed quality due to variation in the quality of ingredients can have profound impacts on the performance of animals, the quality and safety of the final products, the amount of waste outputs and the associated potential environmental impacts and the profitability of agricultural enterprises. Variations in feed quality may be attributable to variations in the quality of the feed ingredients or due to issues encountered during the manufacturing and handling of the finished feeds. Feed manufacturers need to be able to ensure traceability of their products and be able to trace back problems to their source. The strong and growing emphasis by food distributors and consumers place on product safety and traceability, animal welfare and environmental sustainability makes it important to ensure that feed manufacturers track many different metrics related to the feed ingredients they source, the feeds they produce and the animal populations consuming these feeds. In parallel, the high price of agricultural commodities and the slim profit margins achieved on most animal agriculture enterprises and the important competition that exists between feed manufacturers in many sectors result in a need to keep feed cost competitive. Feeds need to be formulated to nutritional specifications that are not excessive but that result in high performance, health and well-being of the animals and enable producers (clients) to make a profit. This is not an easy feat given the great diversity of conditions and market demands encountered in animal production. Sustained interactions with industry stakeholders worldwide suggest that many are relying on relatively static, potentially outdated, information on the nutritive value of ingredients, the nutritional requirements of animals and may have limited insights on how to optimize nutritional specifications for specific production conditions.

Animal feed manufacturers rely on an array of tools to support the production of cost-effective feeds of stable and appropriate quality and track the nutritional supply chain. These range from standalone least-cost cost feed formulation programs to paper forms to a wide array of spreadsheets and reporting or accounting programs. Years of working with animal feed manufacturers globally indicate that many feed manufacturers are relying on idiosyncratic processes that are tedious to integrate and prone to error. Many of the tools may not adequately address the multitude of needs of the animal feed manufacturers. A growing number of online resources and cloud-based programs have been developed and maintained as part by public or commercial endeavors. Many of these resources are free or highly affordable and could help feed manufacturers of any size address several of their needs, including:

1) Better prediction of the nutritive value of ingredients
2) Estimation of optimal nutritional specifications for different species, genetics, life stages production environment and market demands
3) Dynamic adjustment of feed formulations to changing raw material prices and characteristics
4) Improvement of Quality Assurance and Quality Control (QA&QC) and feed traceability
5) Implementation of effective feedback loop enabling continuous improvement of feed quality

This paper aims to provide a brief, thematic and non-comprehensive review of some of resources and programs available to feed manufacturers.

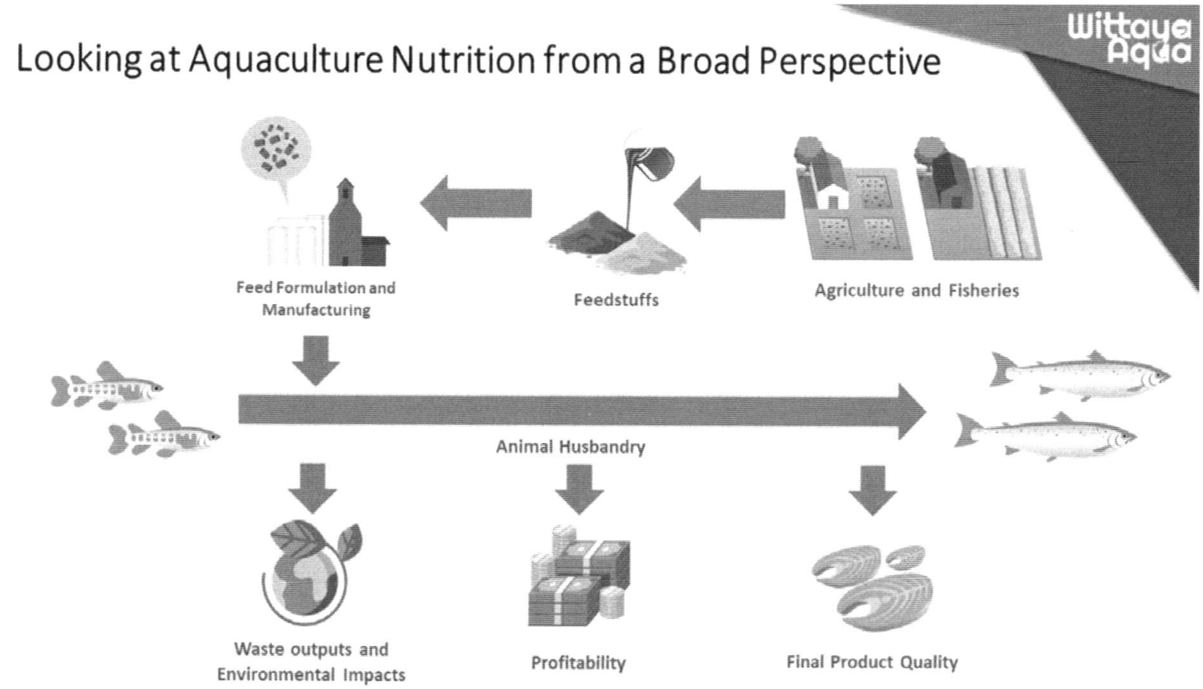

Figure 1. Animal production, such as aquaculture, rely on a complex nutritional supply chain.

Online Nutritional Information Databases

Feed formulation relies on meeting the nutritional requirements of animals using a variety of feed ingredients. Nutritionists need access to detailed and accurate information on the nutritional composition of feed ingredients in order to formulate feed that meet the nutritional requirements of the animals and the targeted nutritional specifications and other important characteristics (e.g. pellatability, palatability, sustainability, etc.) of the feeds.

There is an increasing variety of feed ingredients available to feed manufacturers. Many ingredients come in various declinations (e.g. origin, crude protein content, etc.). There is a large collective body of analytical work and research efforts on feed ingredients generated each year by different organizations. Compiling, analyzing, summarizing and making this information available to potential users has been the focus of a limited number of efforts.

Feedipedia (https://www.feedipedia.org/) is amongst the most comprehensive of those efforts. It is a joint project of the *Institut national de recherche pour l'agriculture, l'alimentation et l'environnement* (INRAE), the *Centre de Coopération Internationale en Recherche Agronomique pour le Développement* (CIRAD) and the *Association Française de Zootechnie* (AFZ) and the Food and Agriculture Organization of the United Nations (FAO).

The main objective of Feedipedia is to provide animal agriculture industry stakeholders with scientific information on the origin, characteristics and chemical composition of a multitude of conventional and unconventional feed resources. It is an open access online resources that provides information for nearly 1400 feed resources. The emphasis is on emerging countries where animal production is intensifying and information on the nutritive value of local feed resources is important.

The INRAE-CIRAD-AFZ feed tables (https://www.feedtables.com/) are important complementary resources to Feedipedia. These feed tables contain data on the chemical and nutritional composition of more than 300 feed ingredients and mineral supplements. Data are made available in the form of tables, on as fed (as is) and dry matter basis, and include statistics of the reported variability for these ingredients. The INRAE-CIRAD-AFZ feed tables are populated with data from chemical analyses performed by a number of laboratories using standard methods and then adjusted to standardized chemical profiles. Nutritional value of the feed ingredients is computed using equations derived from *in vivo* trials with different livestock species. The feed tables also provide environmental impact values that originate from the ECO-ALIM project (https://www6.inrae.fr/ecoalim_eng/Project). Efforts are currently invested in making the data available in the INRAE-CIRAD-AFZ feed tables more dynamic using RM-Link, a software developed by A-Systems (https://www.a-systems.fr).

The IAFFD: A Canadian Success Story

The aquaculture industry is one of the fastest food production industries globally. This industry is supported by a multitude of feed manufacturers with very different scientific and technical capabilities. The large of number (500+) of species cultivated, the diversity of the production systems and conditions encountered, the very large number of feed ingredients used and the advanced manufacturing techniques utilized are factors that make the production of successful aquaculture feed especially challenging.

The International Aquaculture Feed Formulation Database (IAFFD) (http://iaffd.com) is a free online resource that was developed and supported by an informal partnership that included the University of Guelph's Fish Nutrition Research Laboratory (UG-FNRL), the United States Soybean Export Council (USSEC), Wittaya Aqua International (http://wittaya-aqua.ca), MITACS and a few other organizations.

The IAFFD is a "Made in Canada" success story deserving some attention. It addresses challenges that aquaculture feed manufacturers are facing in an innovative and comprehensive manner via two major modules: the Feed Ingredient Composition Database (FICD) and the Aquaculture Species Nutritional Specifications Database (ASNS).

The FICD is comparable to other feed tables presented above but differentiates itself by the very large number of ingredients (650+) and the many nutrients and nutritional specifications (230+)

it covers. The focus is on ingredients used in aquaculture feeds. As such it includes many ingredients of marine origins (e.g. fish meals, fish oils, etc.) and several novel ingredients (e.g. insect meals, algal biomass, fermentation products, etc.) without overlooking all the important agricultural commodities, such grains and oilseeds and their by-products, processed animal proteins, amino acid, vitamin and mineral supplements, feed additives, etc.

The ASNS contains nutritional specifications for over 30 commercially important aquaculture species. Defining optimal nutritional specifications for aquaculture feed is especially challenging given the large number of species cultivated (500+), wide variety of culture environment (freshwater, marine, brackish) and production systems (ponds, tanks, net-pens) used, degree of intensiveness (semi-intensive, intensive, super-intensive) of production, feed types (pelleted vs. extruded, regular vs. high energy feeds) and wide difference in socio-economical background of aquaculture producers (e.g. small-scale farmers with very modest means vs. large integrated multinational corporations).

The ASNS provides complete sets of nutritional specifications for commercial species at different life stages (or liveweight ranges) reared under different production conditions. The nutritional specifications are obtained from a series of nutritional models developed at the University of Guelph over the past three (3) decades that are calibrated for the different species, feed types and production conditions. These nutritional specifications are theoretical values. However, efforts are made to compare and adjust these theoretical model predictions against commercially relevant benchmarks and the results from controlled trials.

The IAFFD was originally developed in 2014 as the Asian Aquaculture Feed Formulation Database (AAFFD). It was the first standardized database of nutritional information for aquaculture species. Interest in this resource grew rapidly and a more international outlook was adopted in 2016. As opposed to many online resources, the IAFFD is a relatively long-lived (7+ years) effort and it is continuously updated and improved. Updates are released approximately every six months. Version 7.1 was recently released and version 8.0 will be available in September 2022.

The IAFFD was developed in close collaboration with Adifo BestMix (Adifo.com). Efforts were invested in ensuring the data is available in a format that is highly compatible with popular least-cost feed formulation software (e.g. BestMix 4, WebAllix, Brill, Format, etc.). The IAFFD is made available under a Creative Common (CC) license and users can download, use, modify and share the data as they wish.

Both the FICD and ASNS have proven very useful to aquaculture feed manufacturers around the world. A survey carried out by USSEC indicated that 70% of the aquaculture feed manufacturers in Latin America made regular use of the IAFFD databases. Several feed manufacturers in Southeast Asia have adopted the FICD and ASNS as their reference databases for formulations of their feeds. The IAFFD met significant needs of the aquaculture feed manufacturers in those two regions. Some feed manufacturers in the Middle East and North Africa (MENA) region have also adopted the IAFFD as their reference database.

The IAFFD has also emerged as a crucial tool for different educational and training activities. Over the past eight (8) years, USSEC organized over 40 feed formulation short courses with participant from around the world and all these short courses relied on the use of the IAFFD. The IAFFD has also been used for teaching activities at the University of Guelph and elsewhere.

The IAFFD may be highly relevant for other animal feed manufacturers, notably petfood manufacturers. There are several commonalities between the aquaculture feed and the petfood industries in terms of feed ingredients used (e.g. processed animal proteins), nutrients of interest (e.g. taurine, DHA, etc.) and feed manufacturing techniques (e.g. extrusion). Consequently, the FICD could potentially be expanded to accommodate the needs of the petfood industry. Efforts could also be invested to making the IAFFD better adapted to the needs and interests of terrestrial livestock feed manufacturers.

Raw Materials Quality Management and Tracking Tools

Feed manufacturers invest considerable efforts in analyzing raw materials and finished feeds as part of their Quality Assurance and Quality Control (QA/QC) program. These programs typically generate an inordinate amount of data using tedious and idiosyncratic collection of data (record sheets, spreadsheets, etc.). These data are most often underutilized because of the limited capacity to collate, analyze and use the data in a meaningful and timely fashion.

Sustained interactions with feed industry stakeholders worldwide suggest that many are relying on relatively static, potentially outdated, and error-prone information about the chemical composition and nutritive value of feed ingredients. Nutrition and QA/QC teams also often spend several hours or even days tracking information about feed ingredients and feed quality in preparation for monthly or quarterly management meetings. Large feed manufacturers can afford tailor-made ERP systems that can enable tracking of this kind of information. However, for many medium and small-size manufacturers, such systems are overly expensive and complicated. Many feed manufacturers need flexible, user-friendly and affordable solutions.

Innovative platforms addressing this challenge are now available to feed manufacturers of any size. These include BESTMIX® Quality Control (https://www.adifo.com/en/brands/bestmix-quality-control), KAllix from A-Systems (https://www.a-systems.fr/index.php/en/software-en/quality-security) and the Raw Material Map (RMM) which is part of the AquaOp Feed platform offered by Wittaya Aqua (https://wittaya-aqua.ca/feed.html). These solutions can help streamline the QA/QC data collection and handling processes. They allow feed manufacturers to save time, avoid errors, update feed ingredient composition databases used in the feed formulation software on a more regular basis, improve the monitoring and benchmarking of the quality of raw materials sourced from different suppliers and support decisions made by the nutrition and QA/QC teams.

AquaOp Feed's Raw Material Mapping (RMM) Tool is being developed using a cloud-based approach. The advantage of using this type of approach is that there is no redundant or inconsistent data and users can dynamically change and edit data without any back-and-forth communication between different users or providers of information anywhere and instant synchronization of data between users around the world.

Integrated Systems Linking Field Performance and Feed Composition

While there is no doubt that the analyses carried out as part of QA/QC programs are of crucial importance to ensuring stable quality, the ultimate measure of quality is in the performances of the animals fed these feeds. Many of the tests and measurements carried out as part of QA/QC programs are surrogates for true measures of the nutritive value of ingredients and feeds. For example, crude protein is not an indication of the true protein content of an ingredient. Nutrient analyses, such as amino acid analysis, does not provide any information about the digestibility and bioavailability of the nutrients.

There would be great value for feed manufacturers in effectively linking animal performance with information about the formulation and chemical composition of their feeds. Feed manufacturers cannot manage and afford to test every batch of raw materials and finished feeds through controlled laboratory or pilot-scale growth trials. Consequently, they are highly dependent on monitoring of performance of their feeds under commercial conditions at their clients' operations. However, obtaining accurate and detailed information about animal performance from clients can be challenging as it is often difficult to share information. In addition, farm performance are affected by a multitude of factors and it difficult to attribute effects to the feeds or to other factors.

Adoption by feed manufacturers and animal agriculture enterprises of systems enabling the efficient capture, compilation, integration and vetting of data would be an important step towards establishing an appropriate linkage between feed formulation and composition information and commercial field performance. Integrating information from the entire nutrition supply chain could help better characterize the effects of various factors (e.g. ingredient quality, nutritional specifications, physical characteristics of the feeds, etc.) on different production parameters under field conditions and potentially enable fine-tuning of feed formulation and composition to improve sustainability and profitability of the sector.

Wittaya Aqua is a Canadian-based software and service company that endeavors to support global aquaculture with industry leading tools, knowledge, and solutions to empower people to make informed decisions and improve the provision of nutritious food in an environmentally, economically, and socially sustainable way. The company is developing and commercializing AquaOp Farm and AquaOp Feed, two software platforms that are uniquely integrated to create a robust feedback loop between field performance on commercial aquaculture operations and feed ingredient composition and nutritional characteristics of the feed (Figure 2).

These two platforms are powered by a series of proprietary, cutting-edge hybrid knowledge and data-driven mathematical models. These models enable the AquaOp Farm and AquaOp Feed platforms to provide true predictive and prescriptive analytics, not simply descriptive analytics. The platforms are designed to foster confidential, seamless and non-interfering cross-cooperation amongst stakeholders (aquaculture producers, feed manufacturers, hatcheries/breeders, processing plants, etc.). Moreover, the platforms can also be used as objective reporting tools to third parties such as regulatory agencies, certification bodies, financial institutions, and insurance brokers.

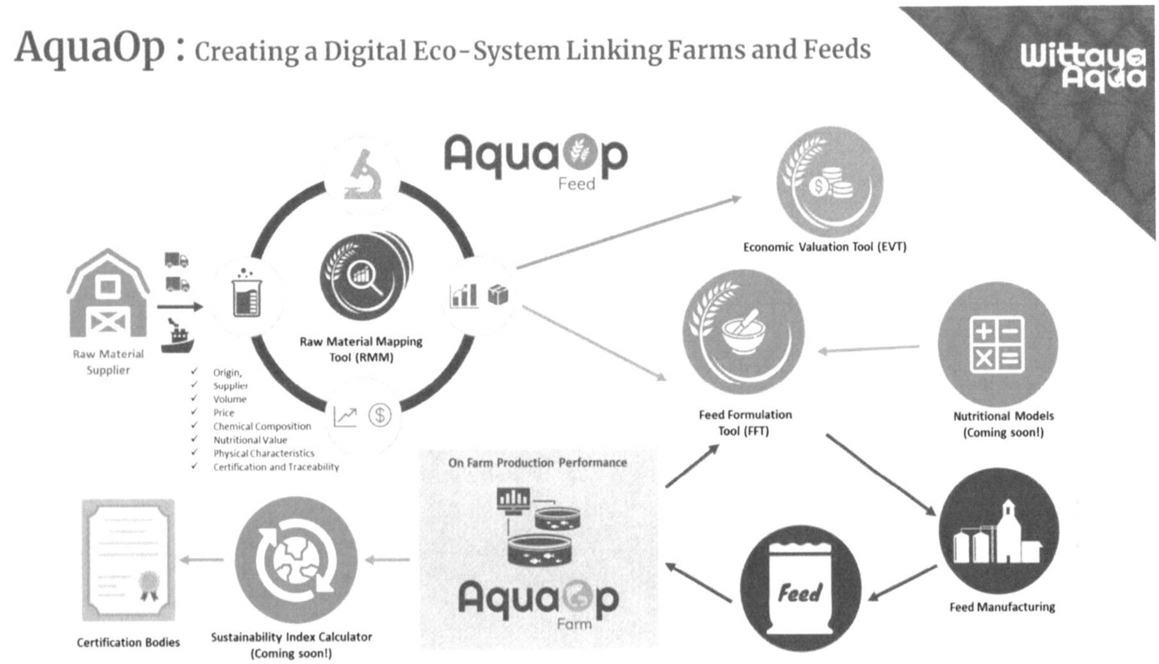

Figure 2. The development of a digital ecosystem linking feed composition and performance achieved on aquaculture operations.

Conclusion

The production of animal feeds is a complex process that relies in large part on having detailed and accurate information about the composition of feed ingredients, defining optimal nutritional specifications, adjusting feed formulations to changing market conditions, tracking a complex nutritional supply chain and monitoring performance of animals under commercial conditions.

Several online resources and programs have been developed and are accessible to feed manufacturers. The sustained development and implementation robust systems capable of tracking the nutritional supply chain and linking animal performance on farms and environmental impacts to feed formulation and composition could be a highly effective solution to improving the production efficiency and ultimately, the environmental and economic sustainability of animal productions.

Feed Processing for a Profitable and Sustainable Feed Business

La transformation des aliments dans un contexte d'activité commerciale rentable et durable

Menno Thomas[1,2] and Guido Bosch[2]
[1] Zetadec, Wageningen, The Netherlands
[2] Animal Nutrition Group, Wageningen University & Research, Wageningen, The Netherlands
menno.thomas@zetadec.com

Sponsored by / Commandité par :

Abstract

The expected increase in world population and in animal-derived food protein consumption will further impose pressure on our food system, which will stimulate innovations that make the system more effective and productive. To accommodate the demand in animal-derived food proteins, the feed business will also be forced to become more effective and productive using feedstuffs that fit with a more circular food system. For feed manufacturing operations either a mitigating strategy can be employed, in which further refinements of existing feed manufacturing technologies is aspired, or adaptations to the system of feed production for animals is sought. In the latter case alternative processing technologies are used to increase the value of existing and new raw materials, including co-products, to prepare animal feed. To adapt the current feed manufacturing technologies (mixing, milling, conditioning, pelleting and cooling) advancements in knowledge of currently used unit operations is desired. Here, existing concepts applied in food processing technology, like state phase diagrams, may further enhance the efficiency of existing feed manufacturing operations. For adaptive technology, a range of alternative technologies can be made available which may act upstream of existing feed manufacturing schemes; e.g. microbial or fungal treatment of e.g. fibrous by-products before incorporating in pelleted feed. Or, alternative technologies may develop in parallel to pelleting operations in which case an animal on-farm is given a diet consisting of pelleted feed supplemented with a blend of treated raw materials of various origin. In this contribution, innovations in feed technology are discussed and a brief overview is given on possible adaptive technologies which may be integrated in the animal feeding system.

Key words: feed manufacturing operations, new raw materials, mitigative and adaptive technology development, future feed manufacturing requirements

Résumé

L'augmentation prévue de la population mondiale et de la consommation de protéines alimentaires d'origine animale imposera une pression supplémentaire sur notre système alimentaire, ce qui stimulera les innovations visant à en améliorer l'efficacité et la productivité. Pour répondre à la demande de protéines alimentaires d'origine animale, le secteur de l'alimentation animale sera également contraint de devenir plus efficace et plus productif en utilisant des ingrédients qui s'inscrivent dans un système alimentaire plus circulaire. Pour les opérations de fabrication d'aliments pour animaux, il est possible soit d'adopter une stratégie d'atténuation, dans laquelle on aspire à un perfectionnement des technologies de production existantes, soit de chercher à adapter le système de production. Dans ce dernier cas, des technologies de transformation de substitution sont utilisées pour augmenter la valeur des matières premières existantes et nouvelles, y compris les coproduits, pour préparer les aliments pour animaux. Pour adapter les technologies actuelles de fabrication des aliments pour animaux (mélange, mouture, conditionnement, granulation et refroidissement), il est nécessaire d'améliorer les connaissances sur les opérations unitaires actuellement utilisées. Dans ce cas, les concepts existants appliqués à la technologie de la transformation des aliments, comme les diagrammes de phase, peuvent améliorer l'efficacité des opérations de fabrication actuelles. En ce qui concerne les technologies adaptatives, diverses technologies de substitution peuvent être mises à disposition afin d'agir en amont des schémas de fabrication existants, notamment le traitement antimicrobien ou antifongique des sous-produits fibreux (par exemple) avant leur incorporation aux aliments granulés. Ou encore, des technologies autres peuvent être développées parallèlement aux opérations de granulation; dans un tel cas, un animal à la ferme recevrait, par exemple, un régime composé d'aliments granulés complété par un mélange de matières premières de diverses origines traitées. Cet exposé porte sur les nouvelles technologies de production d'aliments pour animaux, et un bref aperçu est donné sur les technologies adaptatives possibles qui peuvent être intégrées dans le système d'alimentation animale.

Introduction

The animal feed manufacturing industry's main purpose is to recombine different flows of nutrients into safe, nutritious, attractive and handleable feeds which are efficiently converted in animal-sourced food products (meat, eggs, milk). The raw materials harboring these nutrients vary greatly in composition and origin. Raw materials may be co-products sourced more locally from food or biofuel industries whereas others may travel large distances on the globe, depleting nutrients in one region, and leaving excreted components like nitrogen and phosphorus in animal-dense regions resulting in profound challenges for, and negative impact on the environment at a global scale (IPC, 2022). The capacity of the feed manufacturing industry to transform the great diversity of raw materials into quality animal feed is crucial for the efficiency of our food system. For this, manufacturers have implemented various processing technologies to assure feed quality (Thomas and Van der Poel., 1996; Thomas *et al.*, 1997; Van der Poel *et al.*,

2018). Each unit-operation entails a processing step that requires energy and resources, and it is of interest to improve processes, where possible, to minimize costs and environmental impact while also assuring feed quality. Piecemeal engineering brought such improvements in feed manufacturing technology in terms of enhancing process performance and feed quality. Examples are the appearance of expanders in the '90s, Post Pelleting Applications of liquid additives, the appearance of new types of mixers (twin- and single shaft paddle mixers) which slowly replace the ribbon mixer as the mixing work horse in the industry, and improvements/enhancements/additions to this equipment to make production of animal feed better suited to the requirements of the feed manufacturer. To further enhance and optimize future feed production Van der Poel *et al.* (2020) indicated several relevant areas such as accurate and fast testing technologies to determine variability in ingredients and use of (big) data to understand and optimize feed production. The objective of this paper is to expand on this and more specifically investigate on a technical level where progress in the main feed unit-operations (milling, mixing and pelleting) may lead to reductions in operational costs (energy) while maintaining the possibility to manufacture feed with its great diversity in raw materials and co-products. We will also look into technologies that may rework co-products or ingredients into components of an animal's ration without the use of standard feed manufacturing unit-operations. To do so it classifies technology into 'mitigating' and 'adaptive' technology. Mitigating technology in feed manufacturing, in this paper's context, is related to improvements in our current understanding of feed manufacturing unit operations (milling, grinding, pelleting) with the aim of improving one or more aspects of the production process such as reduction in energy consumption per tonne of feed produced, optimizing physical or nutritional quality or increasing the capacity of a production line. Adaptive technology takes into account novel ingredients and novel processing methods. In case of co-products derived from human food production, food waste, or fodder by-products, this may require additional processing steps to make nutrients more available or to assure food safety.

Mitigating technology in feed manufacturing

Milling

Milling, or grinding, together with mixing and the pelleting process, is one of the key unit-operations in feed manufacturing. Milling requires energy for breaking raw materials into smaller particles. Reasons for milling are rooted in facilitating nutrient digestion and supporting gastrointestinal functioning and health in animals whilst at the same time preparing the ingredients to obtain a more homogeneous feed mix and improving the pelleting process. The optimum particle size of feeds in terms of processing costs and animal performance and health has been evaluated for pigs (Patience 2012) and reviewed for pigs and poultry (Kiari and Mills 2019). The high relevance of particle size with respect to animal performance and feed manufacture operations warrants the question how current milling operations can be improved and what research is required to support these operations.

Within feed manufacture there are two strategies to grind raw materials, i.e. pre-milling and post-milling. In pre-milling, raw materials are ground upon entry into the factory and stored as ground materials. In principle, this permits more stringent particle size specifications according to nutritionist or feed mill requirements as raw materials can be ground separately with specific and controllable operational settings. Post-milling is the most common adopted strategy in grinding, due to lower capital costs for additional storage bins and transporting equipment compared to pre-milling. Raw materials are weighed and dosed according to recipe after which the entire batch is transported to the milling bin, up-stream of the grinding system. At this point, the mill starts working on a segregated set of raw materials entering the mill depending on dosing order. Although, in principle, the order of raw materials arriving at the mill could be estimated by the dosing order, this is not common practice. Controlling a hammer mill or roller mill consists of decreasing or increasing the flow of raw material to a pre-set motor load. With many ingredients regularly making up an animal feed, it is not uncommon in the Netherlands to have between 10 and 20 different raw materials in one batch, it means the mill is continuously adapting to the flow of the new raw material entering the mill in which case very rarely a steady state grinding situation is obtained. The fluctuation in mass flow and constant adaptation to the desired load of the motor will affect the degree of fill of the hammer mill, the flow pattern through the mill and therefore affect particle-size and nutrient distribution in ways that hitherto have not been investigated in the feed manufacturing industry.

As reducing particle size requires energy and the particle size distribution impacts the pelleting process, it is important to have a deeper understanding of how particles break under set conditions. The machine parameters (design), process parameters (operational settings) and the characteristics of the raw materials ultimately define how particle size is reduced into a specific distribution. A comparison between the amount of energy required to comminute wheat, maize and soybeans by three different mill types (roller mill, hammer mill and multicracker) using various operational settings relevant to the specific mill type, showed that energy requirements are the lowest for the crushing type of mills (e.g. roller mill and multicracker) as compared to the impact mill (hammer mill) to obtain the same mean particle size (Table 1). The total Specific Mechanical Energy load (SME) is the total amount of energy consumed including air exhaust for the hammer mill and idle-load of the motors. Effective SME is corrected for both idle running and air exhaust and is on average 35% lower than Total SME with the largest reduction in SME for the hammer mill (43.8%), multicracker (32.8%) and roller mill (27.7%), respectively. This shows that changes in mill type or comminution principle can reduce energy requirements in grinding. Thomas *et al.* (2018) also showed Kick's constant values for these three different mill types and three different feed raw materials. The development of breakage equations for single feed raw materials would be instrumental for tailoring particle size requirement to animals and optimisation on lowering the energy input in milling. This would also be of benefit for post-mill grinding applications as the raw materials arrive as individual components to the mill. For a limited number of feed ingredients breakage equations have been established. For the roller milling of wheat such a breakage equation has been established by Fang and Campbell (2003[a]), where the breakage equation included shape, size, moisture, and hardness values of the wheat kernels.

Table 1. Energy requirements (Kick's constants: kJ/kg) for three different mill types (hammer mill, roller mill and multicracker) in grinding of maize, soybeans and wheat (Adapted from Thomas *et al.,* 2018).

Parameter	Mill type		
	Hammer mill	Roller mill	Multicracker
Total Specific Mechanical Energy			
Maize	9.76	6.62	7.6
Soybean	30.28	11.05	14.26
Wheat	30.42	12.38	20.41
Effective Specific Mechanical Energy			
Maize	5.11	4.57	5.15
Soybean	20.48	8.42	10.22
Wheat	14.80	8.89	12.67
Average percentual reduction[1]	43.8%	27.7%	32.8%

[1]Calculated as (Total SME – Eff. SME) / Total SME * 100% per SME class, corrected for raw material.

Mass Population Balance Models (MPBM) predict the particle size distribution of the ground material as a function of the incoming raw materials particle size distribution and a breakage equation. This breakage equation estimates the redistribution of comminuted particles over the smaller size classes (Lyu, 2021). Most preferable these breakage equations contain information on breakage behaviour of the raw materials and the equipment being used. The use of MPBM in milling is well established in pharmaceuticals (Diemer, 2021) and chemical industry (Vogel and Peukert, 2005) to predict the particle size distribution of the ground material as a function of the incoming raw materials particle size and in many cases to obtain an estimate on the amount of grinding energy involved. The use of MPBM in milling operations in feed manufacturing on mixtures or batches of feed ingredients has not been done. However, grinding of individual raw materials which are used in the feed industry have been analysed. MPBM have been described for maize in hammer mills (Cotabarren *et al.* 2020), for wheat in roller mills (Fang and Campbell, 2002, 2003[a], 2003[b]; Fistes and Tanovic, 2006; Galindeze-Najera, 2014; Galindeze-Najera *et al.*, 2016), the latter because of the large interest in the topic from the flour milling industry. Cotabarren *et al.* (2020) uses a more general approach in which required parameters in the breakage function are fitted from experimental data to predict breakage of maize in a hammer mill.

Finally, multistage grinding in which a combination of classification and (re-)grinding is used may further enhance required particle size distribution in the feed mash whilst reducing energy consumption (Lucht, 2011).

Mixing

Mixing, together with milling is the main unit-operation in a feed plant. Mixing as a unit-operation has been well described, usually in association with the problem of demixing/segregation occurring after mixing (Poux *et al.*, 1991; Ottino and Khakhar, 2000; Mosby *et al.*, 1996). In terms of energy use, mixing requires less energy compared to milling and pelleting, and it is of interest to minimize mixing time to increase the capacity of a production line. Furthermore, it is evident that mixing is crucial as heterogenous and unbalanced feeds can depress animal performance and health, reduce nutrient utilization efficiency and, consequently, increase spillage of nutrients to the environment. Physical factors that affect blending or segregation of feed particles have been described by Axe (1995) and Tang and Puri (2004) and contain particle associated factors and dynamic factors. Examples of dynamic factors include rise or fall of particles due to vibration, elutriation of small particles due to upward air displacement, and tribocharging.

Issues in dry mixing of feed materials are usually associated with inhomogeneous distribution of particles or nutrients during mixing. Even though a homogeneous mix can be achieved in the mixing stage, subsequent transport and storage operations will lead to segregation of the dry mix. Root causes responsible for demixing are, in order of relevance – most relevant factor first: i.) differences in particle size, ii.) density differences between particles, iii.) interactions between specific components or iv.) degree of agglomeration (Deveswaran *et al.*, 2009). In the feed industry a large variety of particle size ratios, density between particles, chemical or physical interactions, and agglomeration between feed mash particles is present, which originates from both the large number of feed raw materials used and the wide range of particle size characteristics (size and shape) emanating from the milling process. Reducing problems in mixing, therefore, requires similar particle size, a high particle number, similar shape and density, a minimal amount of dust, and good flowability (Axe, 1995). Based on the large differences in these traits in both heterogenous raw materials and premixes, these conditions will never be met in feed manufacturing. Free flowing mixtures are easy to mix but suffer from post mixing processing, where these same properties promote segregation. Cohesive mixtures are more difficult to mix, but once mixed have a lower tendency towards segregation (Deveswaran *et al.*, 2009). Ordered mixing is the process in which mechanical, adhesion, electrostatic or coating forces are used to prepare ordered units (Saharan *et al.*, 2008) which after mixing show a lower tendency to segregate. The process involves breaking up of the finer or smaller component and binding these fine components to the coarser carrier (Figure 1).

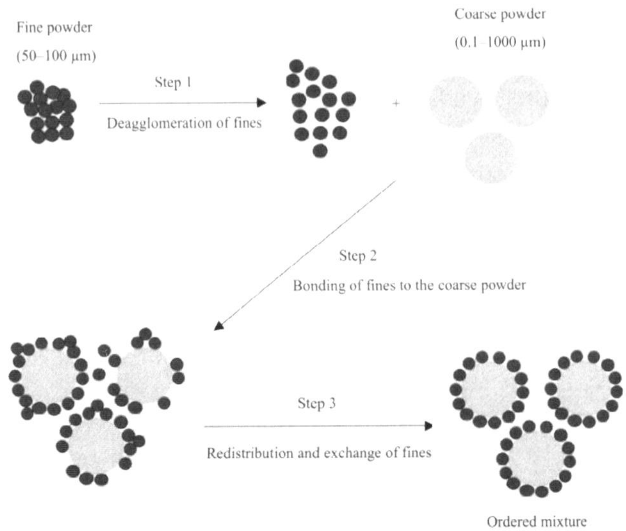

Figure 1. Various steps in the process of ordered mixing (Saharan *et al.,* 2008).

In addition, the order of mixing may influence the mixing process and ease of particle dispersion as a function of particle size and cohesive properties of the particles. This has been shown for a pharmaceutical three component system containing Avicel, aspirin and povidone 90, where required mixing time was shorter if povidone was added before the aspirin, as measured using a Near Infra-Red (NIR) system (Bellamy, 2008). A simulation of two discrete sized particle types showed that dosing the larger particles *after* the smaller particles led to reduced mixing times and better mixing (Xiao *et al.* 2015). To what extent these results are transferable to formulations with multiple raw materials in a feed manufacturing system is currently not known.

We, the feed industry, need to improve our understanding of obtaining a homogeneous mix in terms of nutrients whilst reducing energy use in the mixing process and reduce processing times in the factory. In addition, this need becomes more pertinent when future feed formulations containing more fibrous co-products with difficult handling properties are used. With respect to mixing in the feed industry (both premix and complete feeds) the following directions of research may provide us with information on how to better mix our materials: i) Order of mixing, where the dosing order in the mixer is determined by its particle characteristic(s) e.g. size and distribution. This would require a further integration with the milling operation as well. ii) Adding liquids (hydrophilic: water, molasses, sugar syrups; and hydrophobic: fats and oils) with respect to order and time in the mixing process to make maximum use of these liquids as binders of fine particle material. As such the concept of 'ordered mixing' as used in the pharmaceutical industry may be adopted. Ultimately, the outcome of such research is the manufacture of non-segregating feed mashes, which may make the use of pelleting as a fixation process for the structure of particles, obsolete.

Pelleting

The pelleting process typically entails the conditioning of a mash, followed by compaction through a die and subsequent cooling of the pellets obtained. Concepts related to conditioning, pelleting and cooling have been discussed in Thomas and Van der Poel (2020). Relevant contributing factors to a deeper understanding of the pelleting process and resulting pellet quality traits are the interaction between moisture, temperature and residence time given certain physico-chemical properties of the original feed mash (e.g. chemical composition, particle size distribution). In addition, the dynamics in moisture and temperature alter the physico-chemical properties in the raw material constituents, which affect the energy required to push pellets through the die and the overall observable quality trait characteristics of the formed pellets. The connecting element between processing conditions and physico-chemical changes is the state-phase diagram for relevant raw material constituents. With use of state-phase diagrams interactions between processing conditions and subsequent changes in raw material components can be understood and used as guiding framework in understanding resultant quality traits. For wheat flour and its components gluten and starch, examples on the use of such a state-phase diagram are given by Cuq *et al.* (2003). Taking native starch as an example, increasing the moisture and/or temperature will result at some point in a change from the glassy state to the rubbery state for those polymers which are in an amorphic state. Typically, these changes occur already at room temperature with moisture levels in the range of the base moisture level of many raw materials used (e.g. between 8 – 12%). Increasing the moisture level will lower the glass transition temperature. Despite different sources in raw materials for starch, it follows that many of the changes of state for starch can be described according to the state-phase diagram given in

Figure 2 (Van der Sman and Meinders, 2010). The heterogeneous structure in proteins as for instance in pulses or animal proteins, leads to a broader range of glass-transition phenomena as compared to starch rich raw materials, e.g. rice or corn. Starch at the outer part of a particle will be the first to change states and depending on temperature, moisture and time, starch in the inner parts of the particle may change states given sufficient diffusion of moisture to the particle interior within the permitted (residence) time of the equipment (Thomas and van der Poel, 2018).

Starch in a rubbery state (above the glass transition) can deform during compaction, allowing particles to stack more densely and flow differently through the die, compared to starch below the glass transition which behaves as hard and brittle material. These changes translate to a certain amount of (mechanical) energy required to produce pellets. Furthermore, when particles are stacked more densely, there is more opportunity for interparticle bonds to form. During cooling, temperature and moisture drop, leading to a change from rubbery back to glass state. Both a combination of deformed particles and solid bonds appearing at the contact planes between particles affect the final physical pellet quality. To advance our understanding how to effectively pellet feed particles varying in composition, more information is required to build similar diagrams for the different proteins, fibers and other polymeric components present in raw materials. A better insight in the dynamic sorption or diffusion of heat and water in feed mash particles is also required to properly align with information provided in the (static) state-phase diagrams of feed raw materials. Moreover, for single raw materials such diagrams may have limited application in practical feed manufacturing yet. A framework needs to be established for mixtures of raw materials and how physico-chemical changes in individual raw materials contribute to overall behavior of the mix of interacting raw materials in conditioning, pelleting, cooling, and resultant pellet quality.

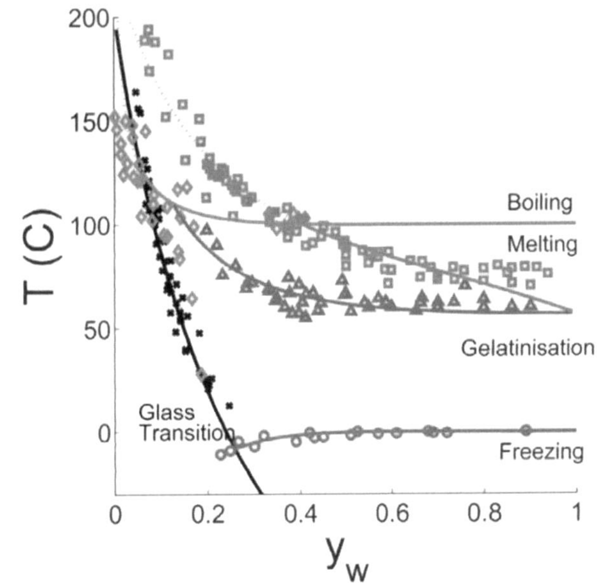

Figure 2. State-phase diagram of starch showing various moisture and temperature conditions at which changes of state occur. Included in the diagram are various changes of state relevant in compound feed manufacture as e.g. the glass-transition, melt-transition line and gelatinisation line (Van der Sman and Meinders, 2010).

Adaptive technology in feed manufacturing

Where mitigating technology investigates and refines current feed manufacturing practices, adaptive technology is focusing on new methods and procedures in feed manufacturing. Such processing technologies can include grinding, toasting, baking, infrared radiation, flaking, steam cooking, expander processing of single ingredients in combination with pelleting and extrusion cooking to enhance their nutritional value. For example, for fibrous feedstuffs these have been investigated by De Vries *et al.* (2012). The individual treatment of feedstuffs warrants the question to what extent energy (electric, thermal) increases the digestibility of the by-product and if this increased (fossil fuel) energy consumption outweighs the enhanced nutritional performance. New methods and procedures in feed manufacturing are also required to accommodate the ever-changing availability of feedstuffs. In the future, primary products like cereal grains will be exclusively used for human food production. Higher levels of co-products derived from these ingredients become available as feed raw material (van Zanten *et al.* 2019). These encompass products such as hulls, middlings and pulps made from whole raw materials soy, wheat or citrus products. These co-products are typically rich in protein (meals) and/or fiber (brans, pulps) and low in starch and oil (FEFAC, 2019). Co-products from such processed human food ingredients become depleted in terms of digestible nutrients, richer in fiber and, usually, have poor handling characteristics. The supply and quality of these co-products will be more diverse, and most likely with different functional properties in unpredictable quantities, depending on processing origin. Manufacturers will have to be more flexible and able to handle more heterogeneous ingredient mixtures, both in nutritional and in handling properties. This makes detailed knowledge of the physico-chemical processes important for effective use of the co-products in quality pelleted animal feeds. Currently in the Netherlands, co-products are derived from a.o. the potato industry (potato steam peelings, potato-starch, off-spec fries), flour milling industry (wheat bran, wheat middlings, wheat gluten feed, liquid wheat starch), bakery products (bread meal, off-spec cake and chocolate-cake meal), fermentation industry (wet corn or wet wheat separation processing (DDGS), yeast concentrates), dairy industry (permeates, milk concentrates, milk powders), and from the brewing industry) (yeast, brewery spent grains). In addition, trimmings and off-spec foodstuffs not suitable for human consumption and non-sold food products from e.g. supermarkets enter the feed chain. Finally, and though currently prohibited in Europe as a feed ingredient due to food safety measures, kitchen leftovers (swill) from e.g. restaurant or canteens which have a high feeding value especially for pigs and poultry (Westendorf, 2000) may re-enter the feed chain. If the quality and safety of these product can be guaranteed, these products might become, again, available for feed manufacturers. These energy-rich, low fiber products contain high levels of starch, sugars, proteins, and/or fats as compared to high fibrous co-products from processed human food ingredients. Table 2 shows a small portion of co-products generally included in feed rations where a large part of the wet products is not included in pelleted feed.

Table 1. Examples of the chemical composition of wet by-products originating from different human food industries and used in animal feed production.

Origin	Raw material	Dry matter[1]	Protein[2]	Fat[2,3]	Fibre[2]	Ash[2]	Starch[2,4]	Sugar	Remark	Ref.
Potato industry	Potato peelings	138	137	11	56	69	416	19	350 < Starch[2,4] < 475	a
	Potato pulp	150	109	7	194	62	225	11	pressed	a
	Potato starch	451	22	4	14	19	863	2	Untreated, solid	a
Wheat (wet milling)	DDGS	916	324	68	71	46	11	49		a
Bakery	Bread meal	897	124	54	11	27	499	68	remains	a
	Chocolate-cake mix	907	111	165	39	39	255	258		b
Brewery	Beer spent grains	265	247	104	179	43	20		low moisture	a
	Brewers yeast	150	233	18	14	38	30			a
Maize	DDGS	903	268	129	67	44	29	17		a
	Steep liquor	476	429	6	5	177	3	80		a
Fermentation industry	Wheat yeast concentrate	256	296	63	29	62	11	124	275 < CP[2] < 325	a
Dairy	Whey permeate	350	92	1	0	326	0	396	Sugar is lactose	b
	Lactose concentrate	256	24	4	0	96	0	829	Sugar is lactose	b

[1]g/kg, [2]g/kg dry matter, [3]after acid hydrolysis, [4]amyloglucosidase when ref is 'a', otherwise unknown.
a CVB (2021), b Looop (2022)

Many of the available dry (>88% dry matter) co-products are easily used in combination with the current feed manufacturing practice. This is, after all, at the core of existence of feed manufacture, increasing the value of edible waste streams and co-products by standardization and turning this mix in a handleable form (pellets). Wet streams (< 75% dry matter) as such, however, are generally not included in pelleted feeds for a number of reasons: i.) blending high moisture feed raw materials with dry ingredients leads to lumping and caking of feed material on the inside of mixers and transport elements leading to problems in transport and sanitary status of the feed(-mash) due to microbial and mold formation. ii.) a high level of free moisture in feed mash leads to blocking of the pelletizer. The base moisture level of feed ingredients plus added moisture from condensing steam, in practice, does not exceed 18% moisture level after which blockage of the pellet press may occur. This limits the use of moist co-products in dry feed mash for pelleting. The use of an expander permits higher moisture levels, since expanders tend to run more stable at elevated moisture levels. Expanders permit to run to moisture levels of the feed mash up till 30% (Mościcki, 2011). However, if the expander is used in combination with a pelletizer, still the upper moisture limit should be respected in order to prevent blockages of the pelletizer. iii.) The whole concept of feed manufacture rests on being able to reduce the moisture and heat-content of the hot pellets after manufacture by using the cooling process, without the addition of drying energy to actively reduce the moisture content of the pellets. The addition of heat by condensing steam and friction in the die of the pelletizer and/or heat dissipation by an expander motor introduces enough energy that pellet moisture content comes back to approximately base moisture level of the ingredients with pellet temperature within 5°C of the ambient air temperature. If more water is added, not enough energy is stored in the pellets to dry to base moisture levels. The installation of drying equipment with its high energy requirement makes the process of standard feed manufacture uneconomical and prevents the inclusion of more free water possibly derived from wet co-products.

Adaptive technologies are therefore required to (further) upcycle these resources as feedstuffs in the food chain. Such new methods, procedures and techniques aim to facilitate the use and/or to increase the nutritional value of certain (novel) ingredients or co-products as originating from human food industry. Figure 3 shows technology possibilities for high fibrous co-products when moving from wet to dry state and the possible technologies that can act upon these co-products during transition from wet to dry. In addition, it shows technology options when leaving such co-products in the wet stage. These include, amongst others, enzymatic, fermentation, ensiling and fungi treatment of such products. For example, the application of enzymes to modify raw materials in the processing chain of feed manufacture is hampered by the high required moisture content and residence time to bring about the desired effect. Wet milling processes like starch production do incorporate separation stages at elevated moisture contents in which, in principle, additional enzyme modification steps could be introduced. Enzymes can also be added to (spontaneous) fermentation process to increase efficiency of this process and target specific ANFs or nutrients (Goodarzi Boroojeni et al., 2017, Zentek and Boroojeni, 2020). Treating large volumes of wet, fermented co-products is traditionally not incorporated in the current industrial wet separation processes. Further processing of these co-products on farm level seems to be most logical step in further increasing its feeding value.

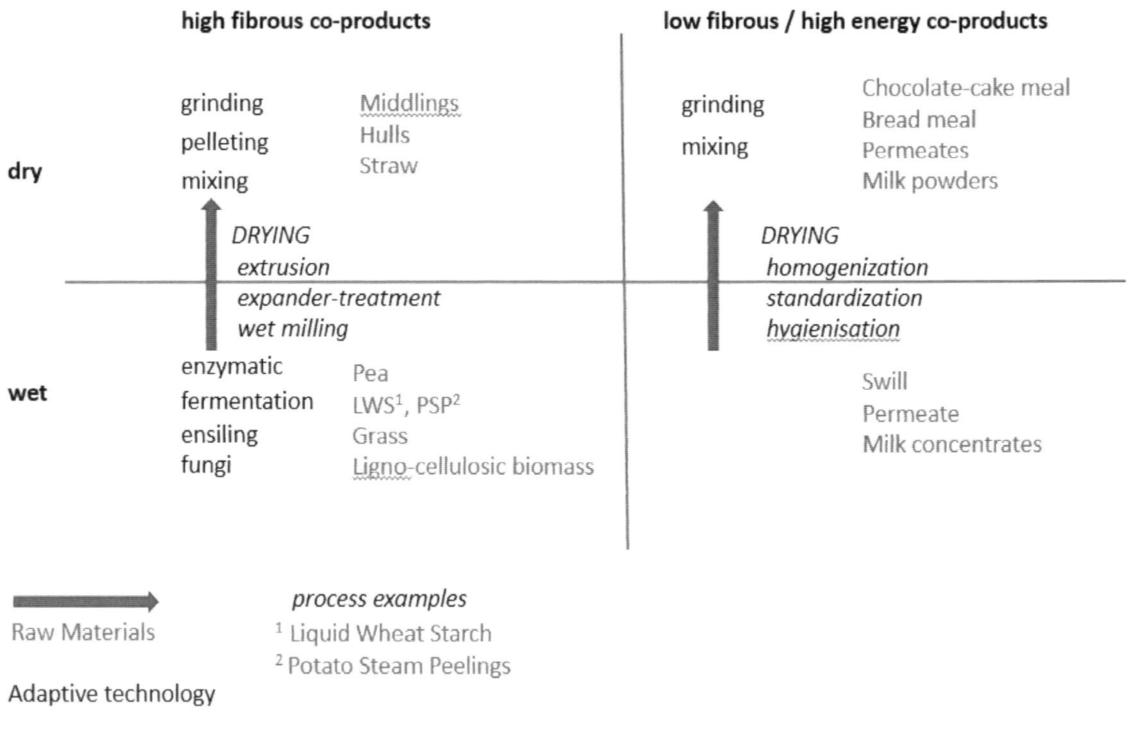

Figure 3. Principles of operation for adaptive technologies (see text).

White rot fungi can be used to degrade lignin in lignocellulosic biomass to improve the availability of the cellulose and hemi-cellulose components for either ruminant nutrition or biofuel production (van Kuijk, 2016). Although the technology is low cost, simple and environmentally friendly, loss of carbohydrates during the process of fungal incubation requires further study and optimization of the technology to be used in large scale feed applications (van Kuijk, 2016). Prior to the drying stage but after enzymic or fermentative stages, wet co-products may be subjected to other thermal technologies such as extrusion, expander treatment, or wet-milling to modify these co-products into higher nutritional value products.

The economic value for high-energy, low fibrous wet co-products and food waste permits drying (including homogenization, standardization and hygienization) of these products. This makes these products suitable for inclusion in standard feed manufacturing operations. For example, Japan and South Korea collect and heat-treat swill and upcycle in total 36% and 43% of the food leftovers (Sugiura et al., 2009; Zu Ermgassen et al., 2016). In many cases, such co-products are included in sensitive stages of animal production like young animal feeds, since often such feeds include high levels of sugars and fats which are desirable to enhance feed intake and energy level.

Much of the focus in animal nutrition is on replacing soy as protein source for alternative protein rich feed ingredients. Tallentire *et al. (2018)* investigated micro- and macro-algae, duckweed, yeast protein concentrate, bacterial protein meal, leaf protein concentrates and insects as alternative novel feed ingredients. Using a modelling approach, they found that soybean products could be replaced by the novel feed ingredients while at the same time reducing greenhouse gas emissions and arable land requirements, pending sufficient upscaling of the production processes. To reduce drying costs, it is of interest to either build processing plants at places where residual heat from industrial processes can be transferred and used to dry these novel feedstuffs or further integration of the production of such novel feed ingredient with energy production (e.g. biodiesel or bio ethanol for drying). In addition, improving the drying process may additionally lead to claimed reductions of up to 65% in net energy for drying (Geelen, 2022).

Feed manufacturing is one of the most viable and efficient industries to upcycle co-products and incorporate novel feed ingredients into valuable animal products. The increases in variability in future feed raw materials with reduced nutrient content and poor handling properties, combined with societies demand for reduced greenhouse gas emission and energy consumption pose new challenges for the feed manufacturer. Part of the solution lies in further upstream integration with suppliers of (novel) feed ingredients and further refinement and understanding of the current unit operations.

Take home points

- Post mill grinding as currently executed, limits control over the resulting particle size of the ground raw materials. As a result, changes in dosing order to the mill may affect particle size distribution of a raw material to an unknown extent.

- The effects of dosing order according to particle size or cohesive properties on mixing time and homogeneity have not been assessed for animal feed.

- Addition of liquids during the mixing stage may be combined with the concept of ordered mixing, to reduce segregation and improve homogeneity of the feed. Investigations on how to best dose, in terms of time, order of hydrophilic and hydrophobic liquids in addition to equipment design (mixer and liquid dosing system) should be set up.

- Combining process conditions with state-phase diagrams enhances our understanding of the interactions between conditioning, pelleting, cooling and functional properties of feed raw material components. Currently, information is lacking to include dynamic behavior e.g. sorption or diffusion of heat and water in feed mash particles to properly align with information provided in the (static) state-phase diagrams of feed raw materials.

- There's an upper limit in moisture content in standard feed manufacturing practice of about 18% moisture. Higher moisture levels lead to blocking of the pelletizer. In addition, residual heat in the hot pellets is not high enough to reduce the moisture content to the original raw materials base moisture level (usually between 8 – 12% moisture). This limits the inclusion of wet co-products.

- Increasing the nutritional value of high-fiber co-products in the wet stage by processes such as wet milling or enzymic treatment is paramount for upcycling. If, as part of the production, energy carriers like bio-ethanol or bio-diesel are produced, an economical combination with drying is possible.

References

Axe D.E. 1995. Factors affecting uniformity of a mix. Animal Feed Science and Technology, 53, 211-220.

Bellamy L.J., Nordon A. & Littlejohn D. 2008. Effects of particle size and cohesive properties on mixing studied by non-contact NIR. International Journal of Pharmaceutics, 361, 87-91

Boroojeni F.G., Senz M., Kozłowski K., Boros D., Wisniewska M., Rose D., Männer K. & Zentek J. 2017. The effects of fermentation and enzymatic treatment of pea on nutrient digestibility and growth performance of broilers. Animal, 11, 1698-1707.

Cotabarren I., Fernández M.P., Di Battista A. & Piña J. 2020. Modeling of maize breakage in hammer mills of different scales through a population balance approach. Powder Technology, 375, 433-444.

Cuq B., Abecassis J. & Guilbert S. 2003. State diagrams to help describe wheat bread processing International Journal of Food Science & Technology, 38, 759-766.

CVB. 2021. CVB Veevoedertabel 2021. Stichting CVB.

Deveswaran R., Bharath S., Basavaraj B., Abraham S., Furtado S. & Madhavan V. 2009. Concepts and techniques of pharmaceutical powder mixing process: A current update. Research Journal of Pharmacy and Technology, 2, 245-249.

De Vries S., Pustjens A., Schols H., Hendriks W.H & Gerrits W.J.J. 2012. Improving digestive utilization of fiber-rich feedstuffs in pigs and poultry by processing and enzyme technologies: A review. Animal Feed Science and Technology, 178, 123-138.

Fang C. & Campbell G. 2002. Effect of roll fluting disposition and roll gap on breakage of wheat kernels during first-break roller milling. Cereal Chemistry. 79, 518-522.

Fang C. & Campbell G. 2003[a]. On predicting roller milling performance IV: Effect of roll disposition on the particle size distribution from first break milling of wheat. Journal of Cereal Science, 37, 21-29.

Fang C. & Campbell G. 2003[b]. On predicting roller milling performance V: Effect of moisture content on the particle size distribution from first break milling of wheat. Journal of Cereal Science, 37, 31-41.

FEFAC. 2019. Co-products, an essential part of animal nutrition. Bruxelles, Belgium: FEFAC

Fistes A. & Tanovic G. 2006. Predicting the size and compositional distributions of wheat flour stocks following the first break roller milling using the breakage matrix approach. Journal of Food Engineering, 75, 527-534.

Galindez-Najera S. 2014. Modelling first break milling of debranned wheat. In: Galindez-Najera S. (Ed.), A compositional breakage equation for first break roller milling of wheat. University of Manchester. UK, pp. 81-109.

Galindez-Najera S., Choomjaihan P., Barron C. & Lullien-Pellerin V. 2016. A compositional breakage equation for wheat milling. Journal of Food Engineering, 182, 46-64.

Geelen S. 2022. Electrification of dryers for petfood. International petfood. February.

Looop. 2022. Derived from website, 20 April. Available at: https://www.looop.company/nl/bijproducten.

Lucht, T., GmbH, A.K., KG, Co. 2011. Stage grinding with hammer mill and crushing roller mill. Feed Compounder, 31, 22-26.

Lyu F. 2021. Hammer-milling maize and soybean meal. Physical and nutritional characteristics of particles. PhD-thesis Wageningen University and Research.

Mosby J., de Silva S.R. & Enstad G.G. 1996. Segregation of particulate materials – Mechanisms and testers. KONA Powder and Particle Journal, 14, 31-43.

Mościcki L. 2011. Extrusion-cooking techniques: applications, theory and sustainability. Wiley-VCH Verlag, Chapter 11.

Ottino J.M. & Khakhar D.V. 2000. Mixing and segregation of granular materials. Annual Review of Fluid Mechanics, 32, 55-91.

Poux M., Fayolle P., Bertrand J., Bridoux D. & Bousquet J. 1991. Powder mixing: Some practical rules applied to agitated systems. Powder Technology, 68, 213-234.

Saharan V.A., Kukkar V., Kataria M., Kharb V. & Choudhury P. 2008. Ordered mixing: mechanism, process and applications in pharmaceutical formulations. Asian Journal of Pharmaceutical Sciences, 3, 240-259.

Van der Sman R. & Meinders M. 2011. Prediction of the state diagram of starch water mixtures using the Flory-Huggins free volume theory. Soft Matter, Royal Society of Chemistry, 7, 429-442.

Tallentire C., Mackenzie S. & Kyriazakis I. 2018. Can novel ingredients replace soybeans and reduce the environmental burdens of European livestock systems in the future? Journal of Cleaner Production, 187, 338-347.

Tang P. & Puri V.M. 2004. Methods for minimizing segregation: A review. Particulate Science and Technology, 22, 321-337.

Thomas M. & Van der Poel A.F.B. 1996. Physical quality of pelleted animal feed. 1. Criteria for pellet quality. Animal Feed Science and Technology, 61, 89-112.

Thomas M., van Zuilichem D.J. & Van der Poel A.F.B. 1997. Physical criteria of pelleted animal feed. 2. Contribution of processes and its conditions. Animal Feed Science and Technology, 64, 173-192.

Thomas M., Hendriks W.H. & Van der Poel A.F.B. 2018. Size distribution analysis of wheat, maize and soybeans and energy efficiency using different methods for coarse grinding. Animal Feed Science and Technology, 240, 11-21.

Thomas M. & Van der Poel A.F.B. 2020. Fundamental factors in feed manufacturing: Towards a unifying conditioning/pelleting framework. Animal Feed Science and Technology, 268, 114612.

Sugiura K., Yamatani S., Watahara M. & Onodera T. 2009. Ecofeed, animal feed produced from recycled food waste. Veterinaria Italiana, 45, 397-404.

Van der Poel A.F.B., De Vries S. & Bosch G. 2018. Feed processing. In: Feed evaluation science, Moughan P.J. & Hendriks W.H. (Eds). Wageningen, The Netherlands: Wageningen Academic Publishers, pp. 295-336.

Van Kuijk, S.J. 2016. Fungal treatment of lignocellulosic biomass. PhD-thesis Wageningen University and Research.

Van Zanten H.H., Van Ittersum M.K. & De Boer I.J. 2019. The role of farm animals in a circular food system. Global Food Security, 21, 18-22.

Westendorf, M. L. (Ed.) 2000. Food waste to animal feed (1st ed.). Ames: Iowa State University Press.

Xiao X., Tan Y., Zhang H., Jiang S., Wang J., Deng R., Cao G. & Wu B. 2015. Numerical investigation on the effect of the particle feeding order on the degree of mixing using DEM. Procedia Engineering, 102, 1850-1856.

Zentek J. & Boroojeni F.G. 2020. (Bio) Technological processing of poultry and pig feed: Impact on the composition, digestibility, anti-nutritional factors and hygiene. Animal Feed Science and Technology, 268, 114576.

Zu Ermgassen E.K., Phalan B., Green R.E. & Balmford A. 2016. Reducing the land use of EU pork production: where there's swill, there's a way. Food Policy, 58, 35-48.

The Influence of Transition Diet Energy and Protein Content on Colostrum and Early Lactation Milk Composition and Bioactive Compound Concentrations in Holstein Dairy Cattle

Influence de la teneur en énergie et en protéines de la ration de transition sur la composition et les concentrations en composés bioactifs du colostrum et du lait de début de lactation chez les vaches laitières Holstein

Amanda J. Fischer-Tlustos[1], M. Victoria Sanz-Fernandez[2], Dave J. Seymour[2], Koki Nishihara[1], Justin B. Renaud[3], John P. Cant[1], and Michael A. Steele[1]
[1]Department of Animal Biosciences, University of Guelph, Guelph, ON
[2]Trouw Nutrition Research and Development, Amersfoort, the Netherlands
[3]London Research and Development Centre, Agriculture and Agri-Food Canada, London, ON
amandaf@uoguelph.ca

ANAC Scholarship Recipient / Récipiendaire de la bourse d'études de l'ANAC

Animal Nutrition Association of Canada
ANAC
Association de nutrition animale du Canada

Abstract

Colostrum and transition milk (**TM**) contain elevated levels of energy substrates and bioactive factors that are crucial to calf development; however, dam factors that control their concentrations are not well understood. The objective of this study was to evaluate how prepartum dietary energy density affects colostrum composition and how pre- and postpartum dietary energy and protein content, respectively, affect TM and mature milk composition. Multiparous (**MP**; n = 28) and primiparous (**PP**; n = 20) Holstein cows were randomly assigned within block to a close-up diet (**CUD**) containing low (**LED**; 1.10 Mcal NE_L/kg DM) or high (**HED**; 1.52 Mcal NE_L/kg DM) levels of energy from 19 ± 4.0 d prior to expected calving date, and to a high protein (**HPD**; 18.5% crude protein (**CP**), 1.73 Mcal NE_L/kg DM) or average protein (**APD**; 15.5% CP, 1.68 Mcal NE_L/kg DM) postpartum diet (**PPD**) after calving. Fat, CP, lactose, milk urea nitrogen (**MUN**) and total solids (**TS**) concentrations were determined by infrared spectroscopy, and IgG was quantified by radial immunodiffusion in colostrum (milking 1), TM (milkings 2 to 7) and mature milk (16 ± 1.9 d postpartum). Data were analyzed using a linear mixed model considering the fixed effects of parity, milking, CUD, PPD and their interactions, and the random effects of cow and block. The CUD had a greater effect on MP cows than PP cows; HED MP cows had greater ($P < 0.0001$)

DMI from wk -3 relative to calving and increases in yields of milk (24.8%; $P = 0.02$), fat (37.7%; $P = 0.007$), CP (14.0%; $P = 0.04$), and TS (35.3%; $P = 0.02$) throughout the entire sampling period compared to LED MP cows. There were no differences between CUD within specific milkings. In contrast to the CUD, the PPD did not differentially affect MP and PP cows. On average, APD cows tended ($P = 0.06$) to have 200 g greater TS yield and had 85.4 g greater ($P = 0.04$) CP yield and 3.4 mg/dL lower ($P = 0.04$) MUN than HPD cows. No differences were observed for IgG concentrations; however, LED-APD cows produced an additional 31.2 g ($P = 0.02$) of IgG compared to LED-HPD cows over the sampling period. The results suggest that increasing close-up diet energy density may be a strategy to improve component yields, with the exception of IgG, in multiparous cows in early lactation.

Keywords: colostrum, transition milk, immunoglobulin G

Résumé

Le colostrum et le lait de transition (TM) sont riches en substrats énergétiques et facteurs bioactifs indispensables au développement du veau; cependant, les facteurs maternels qui en modifient les concentrations ne sont pas bien compris. L'objectif de cette étude était d'évaluer comment la densité énergétique de la ration servie avant le vêlage influence la composition du colostrum et comment la teneur et énergie et en protéines des rations servies avant et après le vêlage, respectivement, influencent la composition du lait de transition et du lait mature. Des vaches Holstein multipares (MP; n = 28) et primipares (PP; n = 20) ont été assignées de manière aléatoire dans un même bloc à une ration de préparation au vêlage (CUD) pauvre (LED; 1,10 Mcal ÉNL/kg de MS) ou riche (HED; 1,52 Mcal ÉNL/kg de MS) en énergie, de 19 □ 4,0 j avant la date prévue de vêlage, ainsi qu'à une ration de début de lactation (PPD) riche en protéines (HPD; 18,5 % de protéines brutes [CP], 1,73 Mcal ÉNL/kg de MS) ou à teneur moyenne en protéines (APD; 15,5 % CP, 1,68 Mcal ÉNL/kg de MS), après le vêlage. Les concentrations en matières grasses, protéines brutes, lactose, azote uréique du lait (MUN) et solides totaux (TS) ont été mesurées par spectroscopie à infrarouge, et les IgG ont été quantifiées par immunodiffusion radiale dans le colostrum (traite 1), dans le lait de transition (traites 2 à 7) et dans le lait mature (16 ± 1,9 j post-partum). Les données ont été analysées à l'aide d'un modèle mixte linéaire considérant les effets fixes de la parité, de la traite, de la CUD, de la PPD et de leurs interactions ainsi que les effets aléatoires de la vache et du bloc. La CUD a eu un effet plus marqué sur les vaches multipares que sur les primipares; les vaches multipares qui ont reçu la ration HED ont présenté une consommation volontaire de matière sèche plus élevée ($P < 0,0001$) à compter de la semaine -3 avant le vêlage et des augmentations des rendements en lait (24,8 %; $P = 0,02$), en matières grasses (37,7% ; $P = 0,007$), en protéines brutes (14,0 %; $P = 0,04$) et en solides totaux (35,3 %; $P = 0,02$) pendant toute la période d'échantillonnage, par rapport aux vaches multipares qui ont consommé la ration LED. Aucune différence n'a été observée entre les rations de préparation au vêlage pour les traites spécifiques. Contrairement aux rations pré-partum, les rations post-partum n'ont pas entraîné de différences entre les vaches multipares et primipares. En moyenne, les vaches APD ont eu tendance à produire 200 g de solides totaux de plus ($P = 0,06$), 85,4 g de protéines brutes de plus ($P = 0,04$) et 3,4 mg/dL d'azote uréique de moins ($P = 0,04$) que les vaches HPD. Aucune différence n'a été observée pour les concentrations d'IgG; cependant, les vaches LED-APD ont produit 31,2 g ($P = 0,02$) d'IgG supplémentaires par rapport aux vaches LED-HPD au cours de la période d'échantillonnage. Les résultats suggèrent qu'augmenter la densité énergétique de la ration de préparation au vêlage pourrait améliorer les rendements en composants du lait, à l'exception des IgG, chez les vaches multipares en début de lactation.

Introduction

Bovine colostrum is largely known for its crucial role in providing the newborn dairy calf with immunoglobulin G (**IgG**) to establish passive immunity. However, IgG is only one of numerous colostral bioactive compounds – including hormones, fatty acids, and sialylated oligosaccharides (**OS**) – that have an unrealized potential to positively stimulate calf development (Fischer-Tlustos et al., 2021). In contrast to IgG, many of these bioactive factors remain elevated in transition milk (**TM**; milkings 2-6; Fischer-Tlustos et al., 2021) and may be responsible for observed benefits on calf intestinal development (Pyo et al., 2020), growth (Van Soest et al., 2020), and health (Conneeley et al., 2014) when TM is fed after the initial colostrum meals. However, concentrations of bioactive compounds in colostrum and TM vary greatly (Cabral et al., 2016; Fischer-Tlustos et al., 2020) and prepartum strategies to maximize their concentrations remain to be elucidated.

It is well known that altering prepartum dietary energy density influences prepartum dry matter intake (**DMI**), metabolism and energy balance (Janovick and Drackley, 2010; Mann et al., 2015; Haisan et al., 2021). It is hypothesized that these prepartum alterations in dam metabolism may influence the process of colostrogenesis in primiparous (**PP**) and multiparous (**MP**) cows; however, research regarding this concept is scarce. In addition, it is important to evaluate how prepartum energy density may interact with differing fresh cow diets, such as altered crude protein (**CP**) content, to influence TM composition. During the fresh period, cows experience a negative energy and protein balance (Bell et al., 2000) and protein derived from both the diet and body reserves is crucial in supplying amino acids and glucose to the mammary gland for milk synthesis (Bell et al., 2000). Thus, it is hypothesized that prepartum dietary energy density and postpartum CP content may interact to alter dam metabolism, which may affect TM composition and yield. Therefore, the objective of this study was to evaluate how prepartum dietary energy density affects colostrum composition and how pre- and postpartum dietary energy and CP content, respectively, affect TM and mature milk composition.

Materials and Methods

The animal experiment was a 2 × 2 factorial, randomized complete block design that involved MP (average parity = 2.4 ± 0.50, n = 28) and PP (n = 20) Holstein cows housed at Trouw Nutrition AgResearch Dairy Facility (Burford, ON, Canada). From -57 ± 5.8 d prior to expected calving, MP cows were dried off, and both MP and PP cows began a low energy diet (**LED**; 1.10 Mcal NE_L/kg DM; 94% NE_L requirements). From d -19 ± 4.0, animals were randomly assigned within block to a close-up diet (**CUD**) to either remain on the LED or to a high energy diet (**HED**; 1.52 Mcal NE_L/kg DM; 129% NE_L requirements). During the dry period, animals were housed in group pens and provided with treatment diets in an individual automated feed bunk (Calan Broadbent Feeding System). After calving, cows were moved to individual tie stalls and assigned to a postpartum diet (**PPD**) containing high protein (**HPD**; 18.5% CP, 1.73 Mcal NE_L/kg DM) or average protein (**APD**; 15.5% CP, 1.68 Mcal NE_L/kg DM) content. Cows were offered all diets once daily at 0900 h and refusals were collected and weighed daily to calculate DMI. A 100 mL colostrum sample was collected as soon as possible after calving, after which cows were milked twice daily at 0500 and 1600 h, and 50 mL samples of TM (milkings 2 to 7) and mature milk (16 ± 1.9 d postpartum) were collected. In all samples, fat, CP, lactose, milk urea nitrogen (**MUN**) and total solids (**TS**) concentrations were determined by mid infrared spectroscopy, IgG was quantified by radial immunodiffusion, and sialylated OS (3'sialyllactose (**3'SL**), 6'sialyllactose, 6'sialyllactosamine, and disialyllactose) were semi-quantified by LC-MS/MS using HILIC chromatography. Statistical analysis was conducted using SAS Studio (version 9.4, SAS Institute Inc., Cary, NC), with cow considered the experimental unit and as a random effect. Using PROC GLIMMIX, two separate models were used to determine 1) the effect of CUD on colostrum yield and composition, with the fixed effects of CUD, parity and their interactions and the random effect of block and CUD × block; and 2) the effect of CUD and PPD on TM and milk yield and composition, with the fixed effects of CUD, PPD, parity and their interactions, random effect of block, and repeated effect of milking. All values reported are least squares means and significance was declared at $P < 0.05$ and tendencies at $0.05 \leq P < 0.10$.

Results

There was no effect of CUD on colostrum yield ($P = 0.625$), colostrum IgG concentrations ($P = 0.673$) or other milk components ($P > 0.251$). However, colostrum SCC increased ($P = 0.036$) by 68.9% while colostrum 3'SL ($P = 0.091$) and total sialylated OS ($P = 0.065$) concentrations tended to decrease by 14.8% and 16.5%, respectively, in HED cows compared to LED cows. Aside from colostrum yield and composition, the HED also decreased the average concentrations of 3'SL (22.9%; $P = 0.0079$) and total sialylated OS (22.8%; $P = 0.0027$) concentrations compared to the LED over the entire sampling period. In regard to additional TM and milk components, it was clear that the prepartum diet had a greater influence on MP cows than PP cows; specifically, HED-MP cows had greater DMI at each week during the close-up period ($P < 0.001$) and during the 4 week fresh period ($P = 0.049$) compared to LED-MP cows, while HED-PP and LED-PP cows did not differ ($P > 0.912$) in DMI at any specific timepoints. These differences in pre- and postpartum DMI in MP cows may explain the observed increases in energy-corrected milk yield (29.4%; $P = 0.008$), fat yield (37.7%; $P = 0.007$), protein yield (14.0%; $P = 0.036$), total solids yield (35.3%; $P = 0.020$), and lactose yield (27.7%; $P = 0.009$) over the entire sampling period in HED-MP cows compared to LED-MP cows. No differences ($P > 0.999$) were observed on colostrum, TM, and mature milk yield and composition within CUD between PP cows.

The PPD did not affect postpartum DMI ($P = 0.996$), TM and mature milk yield ($P = 0.182$) and OS concentrations ($P > 0.728$). In contrast, cows fed the HPD had a 200, 86, and 15 g decrease in TS yield ($P =0.062$), protein yield ($P = 0.040$) and IgG yield ($P = 0.046$), respectively, compared to cows fed the APD diet over the sampling period. The HPD cows also had 3.4 mg/dL greater ($P = 0.042$) average MUN than HPD cows during the postpartum period. In contrast to the CUD, no differences were observed within parity between the PPD. The results revealed that there was largely no effect of CUD × PPD on TM and milk yield and composition ($P > 0.271$). Yet, HED-HPD and LED-HPD cows had 59.1 and 43.2% higher ($P < 0.014$) MUN, respectively, than HED-

APD cows. Interestingly, LED-APD cows produced an additional 31.2 g (P = 0.025) of IgG compared to LED-HPD cows over the sampling period.

Implications and Conclusions

In contrast to previous studies (Mann et al., 2016; Fischer-Tlustos et al., 2021), altering prepartum dietary energy density did not influence colostrum IgG concentrations. These inconsistencies between studies may be explained by differences in energy level, as well as additional factors, such as management and production level. Interestingly, colostrum SCC increased in response to a high energy CUD; however, it is clear that prepartum udder health in HED cows was not compromised to such an extent that colostrum composition was negatively affected. Furthermore, colostrum OS concentrations decreased in HED cows, suggesting that further investigation regarding how dietary energy density may influence OS synthesis at the mammary gland level is required. Aside from colostrum, the results suggest that increases in pre- and postpartum DMI in response to increasing prepartum dietary energy density, as observed only in MP cows in the present study, may be a crucial factor responsible for the alteration of early lactation milk production and component yields.

Increasing CP content postpartum had negative effects on TS, protein, and IgG yield during the sampling period, and the observed higher MUN concentrations indicated inefficient protein utilization. The phenomena by which increasing postpartum CP content decreased IgG yield over the sampling period, especially in LED cows, requires further investigation given the low concentrations and yields of IgG in TM and mature milk compared to colostrum. In conclusion, these results suggest that increasing prepartum dietary energy density in MP cows may be a feasible strategy to increase TM component yields to improve young calf development and health.

Acknowledgements

The authors would like to thank the staff of the Trouw Nutrition AgResearch Dairy Facility (Burford, ON) for their assistance in animal feeding and sampling. This project was financially supported by Trouw Nutrition Canada and a Natural Sciences and Engineering Research Council of Canada (NSERC, Ottawa, ON) Alliance Grant in collaboration with the Saskatoon Colostrum Company Ltd. (Saksatoon, SK), Elanco Animal Health (Greenfield, IN, USA), and Land O' Lakes Animal Milk Solutions Inc. (Arden Hills, MI, USA).

References

Bell, A.W., W.S. Burhans, and T.R. Overton. 2000. Protein nutrition in late pregnancy, maternal protein reserves and lactation performance in dairy cows. Proc. Nutr. Society. 59:119-126.

Cabral, R.G., C.E. Chapman, K.M. Aragona, E. Clark, M. Lunak, and P.S. Erickson. 2016. Predicting colostrum quality from performance in the previous lactation and environmental changes. J. Dairy Sci. 99:4048-4055.

Conneely, M., D.P. Berry, J.P. Murphy, I. Lorenz, M.L. Doherty, and E. Kennedy. 2014. Effect of feeding colostrum at different volumes and subsequent number of transition milk feed on the serum immunoglobulin G concentration and health status on dairy calves. J. Dairy Sci. 97:6991–7000.

Fischer, A.J., A. Lopez, K.S. Hare, K.M. Wood, and M.A. Steele. 2021. Effects of colostrum management on transfer of passive immunity and the potential role of colostral bioactive components on neonatal calf development and metabolism. Can. J. Anim. Sci. 101:405-426.

Fischer-Tlustos, A.J., K. Hare, J. Haisan, W. Shi, J.P. Cant, M. Oba, and M.A. Steele. 2021. Transition diet starch content impacts colostrum and transition milk composition, immunoglobulin G and insulin concentrations in Holstein dairy cattle. J Dairy Sci. 104(Suppl. 1):142.

Fischer-Tlustos, A.J., K. Hertogs, J.K. Van Niekerk, M. Nagorske, D.M. Haines, and M.A. Steele. 2020a. Oligosaccharide concentrations in colostrum, transition milk, and mature milk of primi- and multi-parous Holstein cows during the first week of lactation. J. Dairy Sci. 103:3683-3695.

Haisan, J., Y Inabu, W. Shi, and M. Oba. 2011. Effects of pre- and postpartum dietary starch content on productivity, plasma energy metabolites, and serum inflammation indicators of dairy cows. J. Dairy Sci. 104:4362-4374.

Janovick, N.A. and J.K. Drackley. 2010. Prepartum dietary management of energy intake affects postpartum intake and lactation performance by primiparous and multiparous Holstein cows. J. Dairy Sci. 93:3086-3102.

Mann, S., F.A. Leal Yepes, T.R. Overton, J.J. Wakshlag, A.L. Lock, C.M. Ryan, and D.V. Nydam. Dry period plane of energy: Effects on feed intake, energy balance, milk production, and composition in transition dairy cows. J. Dairy Sci. 98:3366-3382.

Mann, S., F.A. Leal Yepes, T.R. Overton, A.L. Lock, S.V. Lamb, J.J. Wakshlag, and D.V. Nydam. 2016. Effect of dry period dietary energy level in dairy cattle on volume, concentrations of immunoglobulin G, insulin, and fatty acid composition of colostrum. J. Dairy Sci. 99:1515-1526.

Pyo, J., K. Hare, S. Pletts, Y. Inabu, D. Haines, T. Sugino, L. L. Guan, and M. Steele. 2020. Feeding colostrum or a 1:1 colostrum:milk mixture for 3 days postnatal increases small intestinal development and minimally influences plasma glucagon-like peptide-2 and serum insulin-like growth factor-1 concentrations in Holstein bull calves. J. Dairy Sci. 103:4236-4251

Van Soest, B., F. Cullens, M.J. VandeHaar, and M. Weber Nielsen. 2020. Short communication: Effects of transition milk and milk replacer supplemented with colostrum replacer on growth and health of dairy calves. J. Dairy Sci. 103:12104-12108.

ADVANCING THE SCIENCE OF NUTRITION
World Leader in Nutrient Expertise
Focused on Our Customers
Delivering innovation to the industry
Biomin® is now part of DSM
Hy•D® Stand Strong for Lifetime Productivity
VevoVitall® Improve Nursery Performance
HiPhos™ The Stable Phytase
Biomin® II Absolute Protection
If not us, who? If not now, when?
WE MAKE IT POSSIBLE
NUTRITION · HEALTH · SUSTAINABLE LIVING
DSM
BRIGHT SCIENCE. BRIGHTER LIVING.

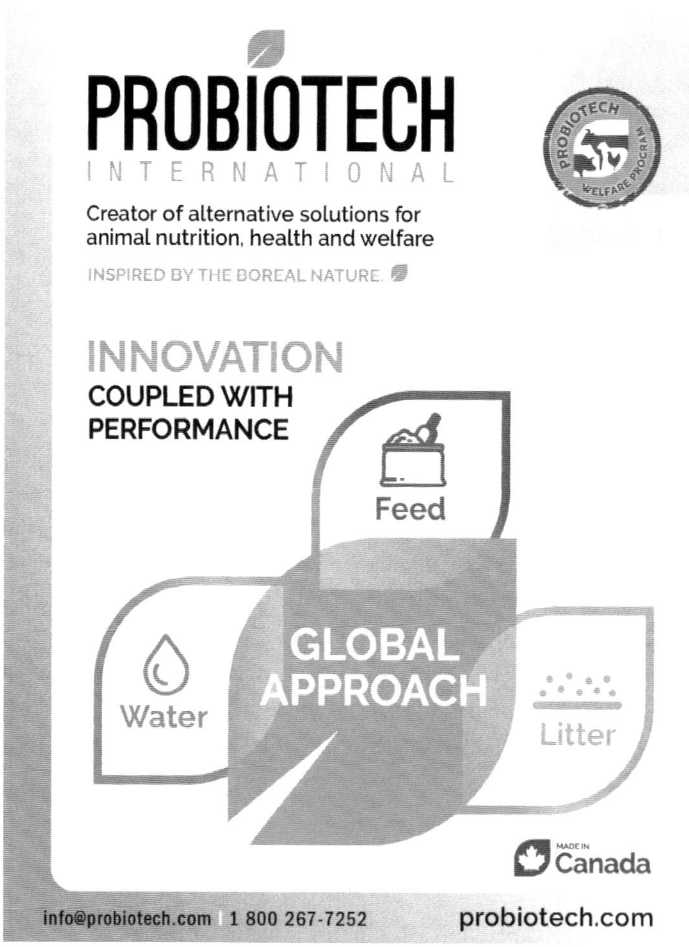
PROBIOTECH
INTERNATIONAL
PROBIOTECH WELFARE PROGRAM
Creator of alternative solutions for animal nutrition, health and welfare
INSPIRED BY THE BOREAL NATURE.
INNOVATION
COUPLED WITH PERFORMANCE
Feed
GLOBAL APPROACH
Water
Litter
MADE IN Canada
info@probiotech.com | 1 800 267-7252
probiotech.com

dcl
Naturally
NEW
electro +
DON'T INVEST YOUR MONEY ANYWHERE
« STEP UP NOW TO electro + »
BETTER QUALITY - BETTER SOLUBILITY - BETTER PRICE
electro +
WATER ADDITIVE
HELP TO IMPROVE OSMOTIC REGULATION AND WATER BALANCE.
Formulated to reduce oxidative stress and maintain healthy immune functions.
Provides electrolytes to help maintain hydration.
Use during periods of heat stress, as well as other periods of stress such as water consumption failures, handling and vaccination.
ELECTROLYTE
WWW.DCL.AG

100% PURE BOVINE PLASMA
SOURCED IN CANADA
SWINE
COMPANION
POULTRY
AQUA
Help support health and immunity with functional proteins.
Learn more about plasma and red cell ingredients at www.APCproteins.com
WATCH THEM thrive
APC

Research proven nutrition
to combat rising ingredient costs
www.halchemix.ca
halchemix@halchemixcanada.com
800-540-4756
HALCHEMIX

Intersection of the Nutritive Value and Bioactive Potential of Two Emerging Classes of Novel Feed Ingredients: Insect and Algae Products

Intersection de la valeur nutritive et du potentiel bioactif de deux classes émergentes d'ingrédients alimentaires nouveaux : les produits à base d'insectes et les produits à base d'algues

Stephanie A. Collins
Department of Animal Science and Aquaculture, Faculty of Agriculture,
Dalhousie University, Bible Hill, NS
s.collins@dal.ca

Sponsored by / Commandité par :

PROBIOTECH INTERNATIONAL

Abstract

Insect and algae products are two emerging classes of novel feed ingredients that are celebrated for their nutritive value and/or bioactive capacity. Insect and algae products and extracts are being developed to replace traditionally used nutrient sources, such as fish meal and soy, and in the case of bioactive capacity, in-feed antibiotics. The fact that production of these feed ingredients may redirect waste streams, can require low-level inputs and/or can be achieved vertically adds to their allure and the impetus to normalize their use in the diets of multiple commercial animal species. Replacing plants and animals with insects and algaes cannot be accomplished without considering all their constituent components. This consideration must include bioactive or other chemo-protective components of the ingredients that could also act as antinutrients, and their influence on feed consumption and digestion, gut microbiome, and immunity. Although insects and seaweeds are naturally consumed in the wild by species that relate closely to animals used in human food production (fish, poultry, swine), feeding rates and applications of use in a commercial setting may differ from self-selective levels of consumption exhibited by wild animals. As we move in the direction of small- (*in-ovo* injection) and large- (use of fat and protein products in dietary formulations) scale use of insects and algaes in animal diets, a measured approach must be taken to fully assess and appreciate these up-and-coming feed ingredients and utilize them to the fullest of their potential and limitations, supporting food security for present and future generations.

Key words: novel feed ingredients, seaweed, microalgae, insect meal; bioactives; antinutrients

Résumé

Les produits à base d'insectes et d'algues sont deux classes émergentes d'ingrédients alimentaires nouveaux vantés pour leur valeur nutritive ou leur pouvoir bioactif. Les produits et extraits à base d'insectes ou à base d'algues sont développés pour remplacer les sources de nutriments traditionnellement utilisées, comme la farine de poisson et le soya et, dans le cas du pouvoir bioactif, les antibiotiques alimentaires. Le fait que la production de ces ingrédients alimentaires puisse rediriger des circuits de déchets, nécessiter des intrants de faible valeur ou être réalisée verticalement ajoute à l'attrait de ces ingrédients et à l'élan en faveur de la normalisation de leur utilisation dans les régimes alimentaires de plusieurs espèces animales commerciales. Remplacer les plantes et les animaux par des insectes et des algues ne peut se faire sans prendre en compte tous les éléments qui les composent, dont les composants bioactifs ou autres composants chimio-protecteurs qui pourraient également agir comme facteurs antinutritionnels, et leur influence sur la consommation et la digestion des aliments, le microbiome intestinal et l'immunité. Bien que les insectes et les algues soient naturellement consommés à l'état sauvage par des espèces proches des animaux utilisés dans la production alimentaire humaine (poissons, volailles, porcs), les concentrations et les applications dans un cadre commercial peuvent différer des niveaux de consommation instinctifs affichés par les animaux sauvages. Considérant l'intérêt marqué pour une utilisation à petite (injection in ovo) et grande échelles (formulations avec des matières grasses et des protéines) des insectes et des algues dans les aliments des animaux, une approche mesurée doit être adoptée pour évaluer et apprécier pleinement ces ingrédients alimentaires en développement et pour les utiliser au maximum de leur potentiel, sans en dépasser les limites, dans un contexte de promotion de la sécurité alimentaire pour les générations actuelles et futures.

Introduction

Monogastric feeds are reliant on plant-based protein sources, like corn and soybean. However, increasing demands on these plant-based materials for non-food industries, such as biofuels, is also associated with an increase in the price of these feed ingredients, placing an impetus on the demand for these alternative protein sources (Leiber et al., 2017). Emerging areas of monogastric animal nutrition research include the use of insect meal, microalgae, single cell proteins and fermentation products as alternative feed sources (Aas et al., 2022).

The use of these novel feed ingredients originating from insects and algae as protein and lipid sources in monogastric diets continues to gain interest. Many of these feed ingredients display an efficient nutritional profile for use in monogastric feeds, in addition to possessing additional bioactive properties and promise in terms of sustainability and feed security. Many of these products have yet to be approved as regulated feed ingredients, or have only recently been approved by regulating bodies, such as the Canadian Food Inspection Agency (CFIA). As an example, at present, only one dried microalgae product, three black soldier fly larvae product (one meal and two oils) and a single earthworm lysate extract are the only monogastric-approved feed ingredients of note in these two categories listed on CFIA's Schedule IV and V List of Approved Ingredients in the Feed Regulations. (CFIA, 2021). Additionally, the dietary inclusion level of whole black soldier fly larvae meal is currently limited to a maximum dietary inclusion level of 7.5 and 10% in tilapia and salmonids, respectively, the defatted larvae meal is limited to 10% in these same fish and the oil is limited to a maximum dietary inclusion level of 5% in poultry and fish (CFIA, 2021).

For companies to produce and sell similar feed sources for use in Canada, adequate data is required for their regulatory approval, which includes production data, with focus on nutrient digestibility and performance (growth, egg production) of animals fed these feed ingredients. Secondary research in novel feed ingredients involves research into their sustainability and impacts of additional bioactive compounds (or extracts) on animal production and health. The findings from this research are helping to contribute to our understanding of the potential use of these products in animal feeds. It may be detrimental to treat these feed ingredients like plant or animal products, leading to confusing results when the additional properties present in these novel feed ingredients are not considered.

Literature review and discussion

Insect product use in monogastric animal feeds

Many insects (black soldier fly, cricket, mealworm) possess an acceptable nutritional profile for use in monogastric feeds including adequate amino acid profiles in addition to other potential nutrients and bioactive compounds (Biasato et al., 2017; Leiber et al., 2017; Bovera et al., 2016; Cullere et al., 2016; De Marco et al., 2015; Khusro et al., 2012; Velten et al., 2016; NRC, 1993). The insect kingdom is vast, thus nutritional variation among species, as well as for life stages within species, should be expected. Some of the more commonly explored species for use in animal feeds have a nutrient profile that does not replicate, but can be considered comparable to that of corn and soybeans (Khusro et al., 2012; Liu et al., 2017).

Additional information on individual insect species and their nutritional impact when included in animal feeds will be valuable, as the current depth of knowledge is limited, but promising. Much of the nutritional evidence involving these feed ingredients suggest minimal disturbances in production performance at reasonable dietary inclusion levels (Fisher et al., 2020; Bovera et al., 2016; Cullere et al., 2016). The presence of antimicrobial peptides in insect products are also interesting, particularly in antibiotic-free animal production, as they may protect against pathogenic microbial organisms responsible for intestinal disturbances that impair health, immune function and production performance (Biasato et al., 2017; Józefiak et al., 2016; Chernysh et al., 2015; Yi et al., 2014).

Use of algaes in monogastric animal feeds

Numerous opportunities exist to include algaes (macro and micro) in monogastric diets. The benefits of microalgae is attractive to many industries beyond the feed industry, including biofuel applications, flavor enhancers/feed attractants and sources of nutrients in human nutrition (proteins/amino acids, vitamins, polysaccharides and lipids/long-chain fatty acids) (Pignolet et al., 2013; Becker, 2007; Spolaore et al., 2006; Brown, 1991).

In the animal feed and human food industries, functional compounds, such as bioactive peptides derived from microalgae are also attracting attention due to evidence that these peptides exhibit positive impacts on immune function and physiological regulation (Suetsuna et al., 2004; Morris et al., 2007; Sheih et al., 2009). Macroalgae contains additional bioactive properties, including antioxidant and antimicrobial compounds with the potential to maintain or improve the function of the intestinal barrier (Tresserra-Rimbau et al., 2018; Abdel-Moneim et al., 2020; Ford, et al., 2020; Zhong et al., 2014). Microalgae-derived, sustainable sources of *n-3* HUFAs also show promise for future-focused sustainability goals (Tocher, 2015), as this source of fatty acids will provide necessary nutrients to the industry without impacting wild fish stocks.

Allelochemicals in feed ingredients

Many non-animal-based feed ingredients produce and possess chemical compounds responsible for physiological functions and protective defense that will affect the growth, survival and/or reproduction of other organisms, which is referred to as allelopathy. The effect of these allelochemicals on the organism involved may be beneficial (positive allelopathy) or harmful (negative allelopathy) (Zhao et al., 2022; Mendes and Vermelho, 2013). Some of these compounds carry through feed processing and remain in the "as fed" animal feed.

In traditional monogastric nutrition, allelochemicals are commonly observed in plant-based feed ingredients as antinutrients (negative allelochemicals) and bioactives (positive allelochemicals), some of which are reduced, neutralized or removed through formulation and processing techniques. A short list of antinutrients in plant-based feed ingredients includes tannins, alkaloids, terpenes (such as saponins), glucosinolates, cyanogens, fibres, mucilage, isoflavones lectins and haemagluttanins (Collins et al., 2013; Htoo et al., 2008; Francis et al., 2001). The level of presence of each antinutrient, degree of severity of the effect of feeding these ingredients varies by plant species, as well as the species to which they are fed. For example, a carnivorous fish, such as a salmonid will be more sensitive to some fibrous components of the diet than an herbivorous / selectively omnivorous animal, such as a chicken or pig.

Allelochemicals in insects

The most-commonly referenced allelochemical in insects is located in their exoskeleton. Insect exoskeleton contains chitin, which can be subdivided further to a compound with antimicrobial properties: chitosan. Chitosan is also present in shrimp shells and fungi and is suggested to play a role in food allergies, but in the case of insect-based feed ingredients, has not been definitively tested and confirmed. Because of the antimicrobial properties of chitin / chitosan, research is currently underway on this compound in the hopes of large-scale application of this compound in antibiotic-free poultry production (Marono et al., 2017; Józefiak et al., 2016).

In feeding studies involving the inclusion of insects and insect products in animal feeds, reduced production performance observations in animals fed a test diet containing insect products vs animals fed a control diet is often attributed to chitin (Mwaniki et al., 2020; Kawasaki et al., 2019; Bovera et al., 2016; Makkar et al., 2014). However, in these studies, chitin and/or chitosan is rarely measured in order to compare feeding dose with animal response. Due to a dearth of quantified data, there is not enough scientific evidence available in the literature to definitively attribute reduced production performance in animals fed insect products to chitin or at the least, solely to chitin.

In addition to chitin, insects contain a number of self-protective compounds with allelopathic properties, including toxic cyanides and deterrent alkaloids, ketones, aldehydes and terpenes

(Boevé and Giot, 2021; Zagrobelny et al., 2018), many of which are known to impair nutrient digestion, reduce feed intake and/or animal growth when present as antinutrients in plant-based feed ingredients (Collins et al., 2013; Francis et al., 2001). Although evidence of these allelochemicals in insects have been identified, the effects of the presence of these compounds in insect-based feed ingredients used in animal diets and the conditions required to reduce, remove or neutralize these allelochemicals have yet to be determined.

Allelochemicals in algaes

Marine microalgaes maintain protective defensive functions against environmental pressure through allelopathy. Allopathic functions in marine microalgae may involve inhibition of predatory protozoans through the inhibition of ciliate population growth and density (Zhou et al., 2022) to the use of phagotrophy and osmotrophy to ingest prey and organic molecules, respectively, in microalgae present in zooplankton (Mendes and Vermelho, 2013), and defensive responses to antibiotics, such as tetracycline and microalgae-bacterial granules (Wang et al., 2020).

Major compounds of interest in macroalgaes include non-starch polysaccharides, such as phlorotannins (tannins) and polyphenolic compounds, such as alginates and fucoidans (Naiel et al., 2021; Leyton et al., 2016; Fleming, 1995). Depending on dose and application, the positive and negative allelopathic properties of these compounds may be considered interchangeably.

In industries unrelated to animal agriculture, the allelochemicals produced by marine algaes also have application as biological herbicides and pesticides, which should be considered when analyzing production-based data in animals fed marine microalgaes in their diets. Many of the allelochemicals produced by microalgaes as a protective mechanism are similar to those found in plants that have know negative impacts on the nutritive value of animal feeds, such as lactones, aldehydes, phenolic compounds, alkaloids, oligopeptides and cyclic peptides. Additional allelochemicals produced by marine microalgaes are enzyme inhibitors, which may impair nutrient digestion, including glucosidase, glycosidase, peptidase and alpha-amylase inhibitors (Mendes and Vermelho, 2013).

Conclusion

Insect meals and algae protein products are up-and-coming ingredients in the monogastric feed industry. These innovative feed production approaches have the capability to utilize minimal space, redirect waste nutrients and lead the way for regenerative agricultural approaches in animal agriculture. Their many beneficial properties have the potential to play a role in ensuring an affordable, sustainable food supply for future generations.

When assessing the nutritional value of insects and algaes, limitations on maximum ingredient inclusion levels are too often held at face value, attributing reduced animal production performance in animals fed insects to chitin and performance reductions in animals fed algae to the rigid hemicellulose-supported structure of its cell wall (Becker, 2007). Although these components may provide barriers, algaes in particular have had a generous deal of research and development devoted to disrupting the cell wall, including cell lysis, ultrasound, thermal and osmotic shock (Ursu et al., 2014; Sari et al., 2013; Doucha and Lívanský, 2008; Middelberg, 1995; Hopkins,

1991), overcoming any challenges it may pose to an animal's ability to access the nutrients within. To truly understand a feed ingredient, one must thoroughly see all aspects of its composition, capabilities and obstacles to overcome.

Novel feed ingredients such as insect and algae oil and protein products, as well as additionally processed feed additives, including lysates, offer numerous benefits beyond the plant and animal-based feed ingredients commonly included in monogastric diets. As with plant-based feed ingredients, optimal formulation and processing techniques will add nutritive power to insect and algae-based feed ingredients, by evaluating these ingredients for their nutritional and allelochemical profiles and utilizing this knowledge to maximize their benefits and minimize detrimental impacts.

Adopting the approaches used in developing value-added plant-based feed ingredients, but not duplicating these processes will allow nutritionists to treat algaes like algaes and insects like insects, rather than seeing them as plant- and animal-adjacent feed ingredients. This strategy will inform future directions in processing, formulation and maximum dietary inclusion in this emerging era of the feed industry.

References

Aas, T.S., T. Ytrestøyl, T. and E. Åsgård. 2022. Utilization of feed resources in Norwegian farming of Atlantic salmon and rainbow trout in 2020. Professional report. Nofima AS. https://hdl.handle.net/11250/2977260.

Abdel-Moneim, A.E., A.M. Shehata, S.O. Alzahrani, M.E. Shafi, N.M. Mesalam, A.E. Taha, A. Swelum, M. Arif, M. Fayyaz and M.E. Abd El-Hack. 2020. The role of polyphenols in poultry nutrition. J. Anim. Physiol. Anim. Nutr. 104(6) 1851-866.

Becker, E.W. 2007. Micro-algae as a source of protein. Biotechnol. Adv. 25, 207-210.

Biasato, I., E. Biasibetti, L. Spuria, A. Schiavone, L. Gasco, C. Dall'Aglio and M.T. Capucchio. 2017. Histological, Morphometric and Histochemical Findings in Broiler Chickens Fed Diets Containing Insect Meal. J. Comp. Pathol. 156(1) 81.

Boevé, J. and R. Giot. 2021. Chemical composition: Hearing insect defensive volatiles. Patterns 2(11) 100352.

Bovera, F. 2016. Use of *Tenebrio molitor* larvae meal as protein source in broiler diet: Effect on growth performance, nutrient digestibility, and carcass and meat trait. J. Anim. Sci. 94(2) 639-647.

Brown, M.R. 1991. The amino acid and sugar composition of sixteen species of microalgae used in mariculture. J. Exp. Mar. Biol. Ecol. 145, 79-99.

Canadian Food Inspection Agency (CFIA). 2021. Feed Regulations. Schedules IV and V. List of Approved Ingredients. Version: 30 August 2021.

Chernysh, S., N. Gordya and T. Suborova. 2015. Insect Antimicrobial Peptide Complexes Prevent Resistance Development in Bacteria. PLoS ONE 10(7) E0130788.

Collins, S.A., G.S. Mansfield, A.R. Desai, A.G. Van Kessel, J.E. Hill and M.D. Drew. 2013. Structural equation modeling of antinutrients in rainbow trout diets and their impact on feed intake and growth. Aquaculture 416-417 219-227.

Cullere, M., G. Tasoniero, V. Giaccone, R. Miotti-Scapin, E. Claeys, S. De Smet and A. Dalle Zotte. 2016. Black soldier fly as dietary protein source for broiler quails: Apparent digestibility, excreta microbial load, feed choice, performance, carcass and meat traits. Animal 10(12) 1923-1930.

De Marco, M., S. Martínez, F. Hernandez, J. Madrid, F. Gai, L. Rotolo, M. Belforti, D. Bergero, H. Katz, S. Dabbou, A. Kovitvadhi, Z. Ivo, L. Gasco and A. Schiavone. 2015. Nutritional value of two insect larval meals (*Tenebrio molitor* and *Hermetia illucens*) for broiler chickens: Apparent nutrient digestibility, apparent ileal amino acid digestibility and apparent metabolizable energy. Anim. Feed Sci. and Technol. 209 211-218.

Doucha, J. and K. Lívanský. 2008. Influence of processing parameters on disintegration of Chlorella cells in various types of homogenizers. Appl. Microbiol, Biotechnol. 81 431-440.

Fleming, A.E. 1995. Growth, intake, feed conversion efficiency and chemosensory preference of the Australian abalone, *Haliotis rubra*. Aquaculture 132 297-311.

Fisher, H., S.A. Collins, C. Hanson, B. Mason, S. Colombo and D. Anderson. 2020. Black soldier fly larvae meal as a protein source in low fish meal diets for Atlantic salmon (*Salmo salar*). Aquaculture 521 734978.

Ford, L., A.C. Stratakos, K. Theodoridou, J.T.A. Dick, G.N. Sheldrake, M. Linton, N. Corcionivoschi and P.J. Walsh. 2020. Polyphenols from brown seaweeds as a potential antimicrobial agent in animal feeds. ACS Omega 5(16) 9093-9103.

Francis, G., H.P.S. Makkar and K. Becker. 2001. Antinutritional factors present in plant-derived alternate fish feed ingredients and their effects in fish. Aquaculture 199(3-4) 197-227.

Htoo, J.K., X. Meng, J.F. Patience, M.E.R. Dugan and R.T. Zijlstra. 2008. Effects of coextrusion of flaxseed and field pea on the digestibility of energy, ether extract, fatty acids, protein, and amino acids in grower-finisher pigs. J. Anim. Sci. 86(11) 2942-2951.

Hopkins, T.R. 1991. Physical and chemical cell disruption for the recovery of intracellular proteins. Bioprocess Technol. 12 57-83.

Józefiak, D., A. Józefiak, B. Kieronczyk, M. Rawski, Mateusz, S. Swiatkiewicz, J. Dlugosz and R.M. Engberg. 2016. Insects - A Natural Nutrient Source for Poultry - A Review. Ann. Anim. Sci. 16(2) 297-313.

Kawasaki, K., Y. Hashimoto, A. Hori, T. Kawasaki, H. Hirayasu, S. Iwase, A. Hashizume, A. Ido, C. Miura, T. Miura, S. Nakamura, T. Seyama, Y. Matsumoto, K. Kasai and Y. Fujitani. 2019. Evaluation of Black Soldier Fly (*Hermetia illucens*) Larvae and Pre-Pupae Raised on Household Organic Waste, as Potential Ingredients for Poultry Feed. Animals 9(3) 98.

Khusro, M., N. Andrew and A. Nicholas. 2012. Insects as poultry feed: A scoping study for poultry production systems in Australia. World's Poult. Sci. J. 68(3) 435-446.

Leiber, F., T. Gelencsér, A. Stamer, Z. Amsler, J. Wohlfahrt, B. Früh, B. and V. Maurer. 2017. Insect and legume-based protein sources to replace soybean cake in an organic broiler diet: Effects on growth performance and physical meat quality. Renew. Agric. Food Syst. 32(1) 21-27.

Leyton, A. R. Pezoa-Conte, A. Barriga, A.H. Buschmann, P. Maki-Arvela, J.P. Mikkola and M.E. Lienqueo. 2016. Identification and efficient extraction method of phlorotannins from the brown seaweed *Macrocystis pyrifera* using an orthogonal experimental design. Algal Res. 16 201–208.

Liu, X., X. Chen, H. Wang, Q. Yang, K. Ur Rehman, W. Li and L. Zheng. 2017. Dynamic changes of nutrient composition throughout the entire life cycle of black soldier fly. PLoS ONE 12(8).

Makkar, H., G. Tran, V. Heuzé and P. Ankers. 2014. State-of-the-art on use of insects as animal feed. Anim. Feed Sci. Technol. 197I 1-33.

Marono, S., R. Loponte, P. Lombardi, G. Vassalotti, M. Pero, F. Russo, L. Gasco, G. Parisi and A.P.J. Middelberg. 1995. Process-scale disruption of microorganisms. Biotechnol. Adv. 13 491-551.

Morris, H.J., O. Farnés, A. Almarales, R. Bermúdez, Y. Lebeque, R. Fontaine, G. Llauradó and Y. Beltrán. 2007. Immunostimulant activity of an enzymatic protein hydrolysate from green microalga *Chlorella vulgaris* on undernourished mice. Enzyme Microb. Technol. 40 456-460.

Piccolo, G., S. Nizza, C. Meo, Y. Attia, Y. and F. Bovera. 2017. Productive performance and blood profiles of laying hens fed *Hermetia illucens* larvae meal as total replacement of soybean meal from 24 to 45 weeks of age. Poult. Sci. J. 96(6) 1783-1790.

Mendes, L.B.B and A.B. Vermelho. 2013. Allelopathy as a potential strategy to improve microalgae cultivation. Biotechnol. Biofuels 6 151.

Mwaniki, Z., M. Neijat and E. Kiarie. 2018. Egg production and quality responses of adding up to 7.5% defatted black soldier fly larvae meal in a corn–soybean meal diet fed to Shaver White Leghorns from wk 19 to 27 of age. Poult. Sci. J. 97(8) 2829-2.

Naiel, M.A.E., M. Alagawany, A.K. Patra, A.I. El-Kholy, M.S. Amer and M.E. Abd El-Hack. 2021. Beneficial impacts and health benefits of macroalgae phenolic molecules on fish production. Aquaculture 534 736186.

Pignolet, O., S. Jubeau, C. Vaca-Garcia and P. Michaud. 2013. Highly valuable microalgae: biochemical and topological aspects. J. Ind. Microbiol. Biotechnol. 40 781-796.

Sari, Y.W., M.E. Bruins and J.P.M. Sanders. 2013. Enzyme assisted protein extraction from rapeseed, soybean, and microalgae meals. Ind. Crops Prod. 43 78-83.

Sheih, I.C., T.J. Fang and T.K. Wu. 2009. Isolation and characterisation of a novel angiotensin I-converting enzyme (ACE) inhibitory peptide from the algae protein waste. Food Chem. 115 279-284.

Spolaore, P., C. Joannis-Cassan, E. Duran and A. Isambert. 2006. Commercial applications of microalgae. J. Biosci. Bioeng. 101 87-96.

Suetsuna, K. and J.R. Chen. 2001. Identification of antihypertensive peptides from peptic digest of two microalgae *Chlorella vulgaris* and *Spirulina platensis*. Mar. Biotechnol. 3 305-309.

Tresserra-Rimbau, A., R.M. Lamuela-Raventos and J.J. Moreno. 2018. Polyphenols, food and pharma. Current knowledge and directions for future research. Biochem. Pharmacol. 156 186–195.

Ursu, A.-V., A. Marcati, T. Sayd, V. Sante-Lhoutellier, G. Djelveh and Michaud, P. 2014. Extraction, fractionation and functional properties of proteins from the microalgae *Chlorella vulgaris*. Bioresour. Technol. 157 134-139.

S.R.N. Velten, C. Neumann and F Liebert. 2016. Evaluation of partly defatted insect meal from *Hermetia illucens* as a substitute for soybean meal in broiler chicken diets. 10.13140/RG.2.2.24796.39042.

Wang, S., B. Ji, M. Zhang, Y. Ma, J. Gu and Y. Liu. 2020. Defensive responses of microalgal-bacterial granules to tetracycline in municipal wastewater treatment. Bioresour. Technol. 312 123605.

Yi, H., M. Chowdhury, Y. Huang and X. Yu. 2014. Insect antimicrobial peptides and their applications. Appl. Microbiol. Biotechnol. 98(13) 5807-22.

Zagrobelny, M., É.C.P. de Castro, B.L. Møller and S. Bak. 2018. Cyanogenesis in arthropods: from chemical warfare to nuptial gifts. Insects 9(2) 51.

Zhao, L., X. Geng, Y. Zhang, X. Hu, X. Zhang, H. Xu, G. Yang, K. Pan and Y. Jiang. 2022. How do microalgae in response to biological pollution treat in cultivation? A case study investigating microalgal defense against ciliate predator *Euplotes vannus*. Environ. Sci. Pollut. Res. 29 32171-32179.

Zhong, X., Y. Shi, J. Chen, J. Xu, L. Wang, R.C. Beier, X. Hou and F. Liu. 2014. Polyphenol extracts from *Punica granatum* and *Terminalia chebula* are anti-inflammatory and increase the survival rate of chickens challenged with *Escherichia coli*. Biol. Pharm. Bull. 37(10) 1575–1582.

Viral Transmission Through Feed

Transmission virale par l'intermédiaire des aliments

Chad B. Paulk
Department of Grain Science & Industry, Kansas State University, Manhattan, KS, USA
cpaulk@ksu.edu

Sponsored by / Commandité par :

WLT Distributors Inc.

Abstract

The ability of African Swine Fever Virus (ASFV) to move easily throughout a country due to movement of animals and contaminated fomites showcases the ability of ASFV to cause disastrous consequences on a naïve pig population. An introduction of ASFV into North America is a significant threat not only to the health and wellbeing of the swine population, but also to our significant trade relationships with countries that have endemic ASFV. Regulatory control of live animals and pork-containing foods substantially reduces this risk ASFV introductory into North America. However, because ASFV can survive in feed during shipping, there is concern that contaminated feed or ingredients will introduce ASFV into the North American swine population. Regardless of its method of entry, there is concern that infection of US or Canadian pigs may result in contamination of the feed supply chain, leading to rapid and widespread distribution of the virus like what was seen with Porcine Epidemic Diarrhea Virus (PEDV). Field evidence suggests that ASFV can be distributed throughout the feed supply chain. Recent research confirmed that the distribution of ASFV into the feed manufacturing environment is widespread and persists even after manufacturing additional feed batches initially free of ASFV. This is similar to what is observed with PEDV and indicates that it is extremely important to prevent the entry of ASFV into feed mills. Therefore, reducing the risk of ASFV or other biological hazards in feed manufacturing facilities is an important part of the complete biosecurity plan for pig producers.

Key words: feed mill biosecurity, viral transmission

Résumé

La capacité du virus de la peste porcine africaine (VPPA) à circuler facilement dans un pays par le biais des déplacements d'animaux et des matières contaminées montre à quel point le VPPA peut avoir des conséquences désastreuses sur une population porcine jamais exposée.

L'introduction du VPPA en Amérique du Nord constitue une menace sérieuse non seulement pour la santé et le bien-être de la population porcine, mais aussi pour nos importantes relations commerciales avec les pays où le virus est endémique. Le contrôle réglementaire des animaux vivants et des aliments contenant du porc réduit considérablement ce risque d'introduction du VPPA en Amérique du Nord. Cependant, parce que le VPPA peut survivre dans les aliments pour animaux pendant le transport, on redoute que des aliments ou des ingrédients contaminés n'introduisent le VPPA dans la population porcine nord-américaine. Quelle que soit la porte d'entrée, on craint que l'infection des porcs américains ou canadiens n'entraîne une contamination de la chaîne d'approvisionnement en aliments pour animaux, ce qui conduirait à la propagation rapide et étendue du virus, comme cela a été le cas avec le virus de la diarrhée épidémique porcine (VDEP). Les données recueillies sur le terrain suggèrent que le VPPA pourrait se propager tout au long de la chaîne d'approvisionnement des aliments pour animaux. Des travaux récents ont confirmé que le VPPA est très répandu dans le milieu de la fabrication des aliments pour animaux et qu'il persiste même après la fabrication de lots supplémentaires initialement exempts du virus. Ce phénomène, similaire à celui observé avec le VDEP, indique qu'il est extrêmement important d'empêcher l'entrée du VPPA dans les usines de fabrication d'aliments pour animaux. Par conséquent, diminuer les risques de présence du VPPA ou d'autres dangers biologiques dans les installations de fabrication d'aliments pour animaux est un élément important du protocole de biosécurité complet des producteurs de porcs.

Introduction

On August 28, 2021, the United States Department of Agriculture (USDA) confirmed that African swine fever virus (ASFV) had been diagnosed in the Dominican Republic (https://www.aphis.usda.gov/aphis/newsroom/news/sa_by_date/sa-2021/asf-confirm). This poses a tremendous threat to the swine industry of the United States (U.S.) given the proximity to the mainland. When ASFV emerged in China, the virus was able to move rapidly and easily throughout the country due to movement of animals and contaminated fomites, showcasing the ability of ASFV to cause disastrous consequences on a naïve pig population. An introduction of ASFV into North America is a serious threat because of significant trade relationships with countries that have endemic ASFV. Regulatory control of live animals and pork-containing foods substantially reduces this risk, but there is evidence that feed and/or ingredients may be potential vectors of ASFV introduction.

Previous research has been conducted to determine the minimum infectious dose of ASFV in water and feed (Niederwerder et al. 2019). The authors concluded that ASFV can be easily transmitted orally through natural consumption of both liquid and feed. Recent research has also demonstrated that ASFV can survive in various feed ingredients during transboundary, transatlantic shipping (Dee et al., 2018). Because ASFV can survive in feed during shipping, the US is rightfully concerned that a contaminated feed or ingredient will introduce ASFV into the US swine population. Regardless of its method of entry, there is concern that infection of US pigs may result in contamination of the feed supply chain, and rapid and widespread distribution of the virus like what was seen with PEDV. Therefore, the purpose of this paper will be to briefly review the risk of feed as a vector for ASFV and key feed mill biosecurity practices to reduce the risk of spreading ASFV via the feed.

Distribution of African Swine Fever Virus Into the Feed Manufacturing Environment

Regardless of potential ASFV method of entry into North America, there is valid concern that once in North America there will be rapid and widespread distribution of the virus like what was observed with PEDV (Schumacher et al., 2017). There are multiple known ASFV transmission routes that have been previously described including direct pig to pig contact, consumption of contaminated food products (swill feeding), or other fomites such as vehicles, workers, or other equipment. (Bellini et al., 2016; Guinat et al., 2016; Olesen et al., 2020). Recent surveillance research conducted in Vietnam by Kansas State University and collaborators investigated the epidemiology of ASFV transmission, particularly focusing on the movement of ASFV through the supply of swine feed and transportation. Our goal was to understand the potential areas within a swine production, feed manufacturing, and distribution system where detection of ASFV DNA could be found. It was discovered that a common source of ASFV DNA detection was in the truck cabs (Gebhardt et al., 2021). Identifying this gap led the production system to alter their biosecurity practices and incorporate truck cab decontamination procedures and restrictions on employee traffic associated with feed delivery. These changes led to major improvements in biosecurity. Although these biosecurity practices are important to consider for reducing the spread of ASFV, they would also be beneficial to reduce the spread of current endemic diseases. It has also been demonstrated that truck foot pedals can be a common area with contamination of PEDV or porcine delta coronavirus (Greiner, 2016).

For biosecurity purposes, it is important to reduce the risk of bringing a virus from the farm back to the feed mill. Elijah et al. (2021) investigated the risk of ASFV mill and subsequent batches of feed contamination when using ASFV contaminated ingredients in the feed manufacturing process. The authors determined that the distribution of ASFV into the feed manufacturing environment is widespread and persists even after manufacturing additional feed batches initially free of ASFV (Table 1). It was also demonstrated that transient surfaces play an important role in the spread of the virus through the feed mill. This is similar to what is observed with PEDV (Schumacher et al., 2017) and indicates that it is extremely important to prevent the entry of ASFV or other endemic viruses into feed mills. Once these viruses enter a feed mill, they have the potential to remain in the environment for an extended period of time. This information becomes vary pertinent because feed manufacturing facilities are not designed to easily be cleaned and disinfected. If a feed manufacturing facility becomes contaminated with a virus, there are no current recommendations for best practices to clean and disinfect these facilities. Research is being conducted to determine optimal methods for disinfecting feed manufacturing facilities, especially equipment that is not designed to be disinfected. Therefore, moving objects like people, PPE, and trucks should be taken in account when designing feed biosecurity protocols and feed/feed mill surveillance could be pivotal in maintaining appropriate feed biosecurity.

Table 1. Main effect of feed batch and zone on detection of African swine fever virus (ASFV) during manufacture of virus inoculated feed (Elijah et al., 2021)

Main effect	Detectable DNA/Total[3]	Cycle threshold[4]	Genomic copy number/mL[5]
Batch			
Negative	0/36	45.0	0
Positive	26/36	37.4 ± 0.70[a]	16,580 ± 7,581
After sequence 1	21/36	39.4 ± 0.70[a,b]	4,437 ± 996
After sequence 2	19/36	40.0 ± 0.70[a,b]	2,479 ± 288
After sequence 3	16/36	40.4 ± 0.70[b]	11,057 ± 8,941
After sequence 4	17/36	39.9 ± 0.70[a,b]	5,423 ± 1,735
Zone			
Feed contact	31/60	40.5 ± 0.63[b]	8,422 ± 7,148[a,b]
Non-feed contact, < 1 m	21/50	41.7 ± 0.62[b]	914 ± 252[a]
Non-feed contact, > 1 m	17/40	42.4 ± 0.66[b]	321 ± 121[a]
Transient surface	30/30	33.1 ± 0.59[a]	22,325 ± 6,276[b]

Strategies to Reduce Risk of Viral Transmission Through Feed

Keeping feed mills clean from contamination of viruses, such as PEDV and ASFV, is important to reduce the risk of disease spread. Multiple factors must be considered to determine what steps need to be considered or implemented to the reduce the risk of virus contamination. Traditionally, feed mills were not designed with biosecurity in mind and not all feed mills are designed the same. Therefore, acceptable degree of risk and the practical implication of biosecurity practices must be considered on a case-by-case basis. Below are key actions to consider to reduce the risk contaminating feed mills.

Ingredient Sourcing

Ingredient sourcing is an important step to prevent biological hazards from entering the feed manufacturing facility via ingredients. Ingredients from areas experiencing disease outbreaks have a greater potential for being high-risk. To accurately identify ingredient risk, knowledge of the ingredient supply chain should extend from the point of ingredient manufacture through transportation to the feed mill, including any intermediaries or blending locations. Supplier identification is important to maintain transparency across the feed supply chain. It is also important to identify biosecurity practices that may be implemented throughout this supply chain. Some of these practices include heat treating the ingredients or feed or holding the ingredient outside of the mill long enough that the virus or bacteria becomes inactive. Regardless of where an ingredient is being sourced from, each step of the shipping and transportation process from manufacturing to the feed mill should be tracked. Identified biological hazards will depend on the species in which feed is being manufactured. However, the entire feed mill needs to be considered as opposed to species specific diets. To prevent mill contamination via ingredients, a balance required for quality, feed safety, and procurement areas of the mill to all work together on ingredient sourcing.

Receiving and Loadout Practices

Risk of ingredients previously discussed the movement of trucks and people through receiving poses a significant risk in terms of biosecurity. In terms of biosecurity, the main goal of feed mill should be to prevent contaminated material for entering the receiving pig. Ensuring the underside of the truck is clean before entering the receiving bay also helps reduce the chance of debris entering the pit. Instituting a tire and undercarriage wash prior to trucks entering the mill will help to minimize the chance of contaminants being brought into the receiving bay. In addition, using pit covers while trucks are entering and leaving the receiving bay helps to prevent any debris from falling into the pit and contaminating ingredients. Sweeping spilled ingredients into the pit is another thing to be avoided. Due to the nature of trucks entering and leaving, the ground in the receiving area can not be considered "clean" or safe for biosecurity and sweeping spilled ingredients into the pit is a risk of introducing disease into the mill.

In addition to the vehicles themselves, truck drivers pose another entry point of contamination. Drivers delivering ingredients or picking up finished feed should not enter the mill beyond the receiving bay to prevent tracking potential contaminants to other areas. Implementing a degree of separation between feed delivery drivers and employees in the mill also prevents contamination from trucks that have been on farms from entering the rest of the mill. Having drivers receive paperwork through a window or from the other side of a door at the office can create this separation. Also having drivers observe potential biosecurity risks at the farms they deliver to can help determine better delivery route options. Farms where exhaust fans are directed toward the feed bins or carcass disposal is near the driveway may need to receive feed later in the day to reduce the chance of bringing disease back to the mill. Having a biosecurity kit in feed delivery trucks helps reduce the risk of contamination. This kit should contain disposable plastic booties for whenever the driver needs to exit their truck, disinfectant to wipe or spray down their cabs, and a method to dispose of these items. In winter months, multiple sets of rubber overshoes could be used instead of the plastic booties, but each pair would need to be disinfected after use.

Zones in the Mill

Feed mill managers should focus on ways to separate the mill into specific zones and limit employee crossover between zones. This helps prevent employees from carrying contaminants from their shoes or clothing through the mill. For instance, an employee may help in receiving and carry contaminated feed dust on their clothes as they walk through the mill to change a die in pelleting. That contamination is no longer limited to the receiving area and now can come in contact with finished feed. Having one zone designated for receiving, another for mixing and processing, and another at load-out helps to prevent this sort of cross-over by preventing employees from moving between areas without going through a decontamination process. This zoning of the mill also needs to cover tools and equipment that might be shared between areas. To have this process be effective, each zone needs its own sets of tools, brooms, ladders, etc. to prevent these items from carrying contamination through the mill as well.

Feed Handling

Preventing finished feed from contacting surfaces that also handle raw ingredients adds another layer of feed security. Disease mitigation measures like heat treatment (pelleting or extrusion) are effective at killing or inactivating bacteria and viruses in the feed but do not prevent recontamination. As a result, sharing equipment between finished feed and raw ingredients should be avoided. When complete separation is not possible, equipment should be thoroughly disinfected before finished feed is run through it. Dust and screenings are also known to contain higher levels of pathogens than the bulk of the ingredient load. These residues should be discarded instead of being added back to ingredients or feed to prevent the accumulation of pathogens.

Employee Training

Employees need to receive proper training for any biosecurity plan to succeed. Ensuring that they are aware of the mill's biosecurity plan, FSMA regulations, and pathogen mitigation and prevention strategies helps to encourage compliance. Proper disinfecting procedures should not be overlooked. Many disinfectants require a certain amount of time in contact with a surface in order to kill pathogens. The presence of dust, grain, or other materials can also limit the effectiveness of disinfectants. All of this should be communicated to employees and biosecurity refreshers should be implemented at intervals throughout the year to keep the plan up-to-date and fresh in employee's minds.

Culture

The most important aspect of a mill biosecurity plan is the culture of the workplace. A strong employee culture encourages everyone to implement the procedures and follow them. If management does not take their policies seriously, there is little incentive for the rest of the employees to follow the policies as well. Encouraging employees to follow a biosecurity protocol starts with management and is strengthened with knowledge of why the plan matters.

Conclusion

In conclusion, data has demonstrated the ability of viruses, such as PEDV and ASFV, can be distributed throughout the feed supply chain. It has also been confirmed that the distribution of ASFV into the feed manufacturing environment is widespread and persists even after manufacturing additional feed batches initially free of ASFV. This is similar to what is observed with PEDV and indicates that it is extremely important to prevent the entry of ASFV into feed mills. Therefore, reducing the risk of ASFV or other biological hazards in feed manufacturing facilities is an important part of the complete biosecurity plan for pig producers.

References

Bellini, S., D. Rutili, & V. Guberti. 2016. Preventive measures aimed at minimizing the risk of African swine fever virus spread in pig farming systems. *Acta Veterinaria Scandinavica*, 58, 82. https://doi.org/10.1186/s13028-016-0264-x

Dee, S.A., F.V. Bauermann, M.C. Niederwerder, A. Singrey, T. Clement, M. de Lima, C. Long, G. Patterson, M.A. Sheahan, A.M.M. Stoian, V. Petrovan, C.K. Jones, J. De Jong, G.D. Spronk, L. Minion, J. Christopher-Hennings, J.J. Zimmerman, R.R.R. Rowland, E. Nelson, P. Sundberg, D.G. Diel. 2018. Survival of viral pathogens in animal feed ingredients under transbounday shipping models, *PLoS One*, 13, e0194509. https://doi.org/10.1371/journal.pone.0194509

Elijah, C.G., J.D. Trujillo, C.K. Jones, N.N. Gaudreault, C.R. Stark, K.R. Cool, C.B. Paulk, T. Kwon, J.C. Woodworth, I. Morozov, C. Gallardo, J.T. Gebhardt, and J.A. Richt. 2021. Evaluating the distribution of African swine fever virus within a feed mill environment following manufacture of inoculated feed. *PLoS ONE*, 12, e0187309. https://doi.org/10.1371/journal.pone.0256138

Gebhardt, J.T., S.S. Dritz, C.G. Elijah, C.K. Jones, C.B. Paulk, and J.C. Woodworth. 2021. Sampling and detection of African swine fever virus within a feed manufacturing and swine production system. *Transboundary and Emerging Diseases*, 69, 103-114. **https://doi.org/10.1111/tbed.14335**

Greiner, L.L. 2016. Evaluation of the likelihood of detection of porcine epidemic diarrhea virus or porcine deltacoronavirus ribonucleic acid in areas within feed mills. *Journal of Swine Health and Production*, 24, 198– 204.

Guinat, C., A. Gogin,S. Blome, G. Keil, R. Pollin, D.U. Pfeiffer, and L. Dixon. 2016. Transmission routes of African swine fever virus to domestic pigs: Current knowledge and future research directions. *Veterinary Record*, 178, 262– 267. https://doi.org/10.1136/vr.103593

Niederwerder, M.C., A. Stoian, R. Rowland, S.S. Dritz, V. Petrovan, L.A. Constance, J.T. Gebhardt, M. Olcha, C.K. Jones, J.C. Woodworth, Y. Fang, J. Liang, T.J. Hefley. 2019. Infectious Dose of African Swine Fever Virus When Consumed Naturally in Liquid or Feed. *Emerging Infectious Diseases*, 25(5), 891-897. https://doi.org/10.3201/eid2505.181495.

Olesen, A. S., G.J. Belsham, T. Bruun Rasmussen, L. Lohse, R. Bødker, T. Halasa, A. Boklund, and A. Bøtner. 2020. Potential routes for indirect transmission of African swine fever into domestic pig herds. *Transboundary and Emerging Diseases*, 67, 1472– 1484. https://doi.org/10.1111/tbed.13538

Schumacher, L. L., A.R. Huss, R.A. Cochrane, C.R. Stark, J.C. Woodworth, J. Bai, E.G. Poulsen, Q. Chen, R.G. Main, J. Zhang, P.C. Gauger, A. Ramirez, R.J. Derscheid, D.M. Magstadt, S.S. Dritz, & C.K. Jones. 2017. Characterizing the rapid spread of porcine epidemic diarrhea virus (PEDV) through an animal food manufacturing facility. *PLoS ONE*, 12, e0187309. https://doi.org/10.1371/journal.pone.0187309

Appreciating the Dynamics of Pellet Quality Improvements to Nutrient Segregation in Poultry Houses

Appréciation de la dynamique de l'amélioration de la qualité des granulés dans un contexte de ségrégation des nutriments dans les poulaillers

John Boney
Department of Animal Science, Pennsylvania State University, State College, PA, USA
jxb2002@psu.edu

Sponsored by / Commandité par :

DISTRIBUTION | MANUFACTURING | WAREHOUSING

Abstract

Improving pellet quality (PQ) and the percentage of pellets in the feed pan may be impeded by throughput demands, diet formulations, and ambient conditions. The effects of PQ improvements on broiler feed efficiency are known, however a clearer understanding on the impacts of feed milling, nutrient segregation, flock uniformity, and economics must be available for consideration by feed mill managers. The effects of PQ on nutrient segregation in commercial broiler houses differing in feed line length was studied. Dawkins et al. reported a strong negative correlation between aging broilers and movement, so nutrients were analyzed from feed pans in eight regions of broiler houses. Four scenarios were studied where PQ (poor vs improved) and feed line length (76-m vs 152-m) varied. Nutrient segregation was greatest when poor PQ feed was augered through 152-m feed lines. Nutrient segregation was not apparent when improved PQ feed was augered through 76-m feed lines. Intermediate degrees of nutrient segregation were apparent in the two other scenarios. These data indicated that nutrient segregation was impacted by both PQ and feed line length, offering multiple strategies to reduce nutrient variability throughout broiler houses. Amino acid density is known to impact nutrient digestibility, feed efficiency, and processing yields in both Ross and Cobb broiler strains. Phytase activity is also known to impact Ca and P digestibility, amino acid digestibility, bone mineralization, and broiler performance metrics. Improvements in PQ can reduce the variability of nutrients from feed pan to feed pan for more efficient and uniform flock performance.

Résumé

Les contraintes liées au débit du système de distribution, à la formulation des rations et aux conditions ambiantes peuvent être autant de freins à l'amélioration de la qualité des granulés (QG) et à l'augmentation du pourcentage de granulés dans la mangeoire. Les effets de l'amélioration de la QG sur l'indice de conversion alimentaire des poulets à griller sont connus, mais les responsables des usines d'aliments doivent acquérir une meilleure compréhension de l'impact de la mouture des aliments, de la ségrégation des nutriments, de l'uniformité du troupeau et des considérations financières. Les effets de la QG sur la ségrégation des nutriments dans les poulaillers commerciaux ont été étudiés en fonction de la longueur des lignes d'alimentation. Dawkins et al. ont signalé une forte corrélation négative entre le vieillissement des poulets à griller et le mouvement, de sorte que les éléments nutritifs ont été analysés à partir des mangeoires situées dans huit secteurs des poulaillers. Quatre scénarios ont été étudiés dans lesquels la QG (pauvre ou améliorée) et la longueur de la ligne d'alimentation (76 m ou 152 m) variaient. Une ségrégation plus importante des nutriments a été constatée lorsque des aliments de pauvre qualité étaient distribués par les lignes d'alimentation de 152 mètres. Aucune ségrégation n'a été observée avec des aliments de QG améliorée distribués par les lignes de 76 m. Des degrés intermédiaires de ségrégation des nutriments ont été notés dans les deux autres scénarios. Ces données indiquent que la ségrégation des nutriments est influencée à la fois par la QG et la longueur des lignes d'alimentation, ce qui multiplie les possibilités de stratégies pour réduire la variabilité des nutriments dans les poulaillers de poulets à griller. On sait que la densité des acides aminés a un impact sur la digestibilité des nutriments, l'efficacité alimentaire et les rendements de transformation chez les souches de poulets de chair Ross et Cobb. L'activité de la phytase est également connue pour son impact sur la digestibilité du Ca, du P et des acides aminés, la minéralisation osseuse et les paramètres de performance des poulets à griller. Améliorer la QG peut contribuer à réduire la variabilité des nutriments d'une mangeoire à l'autre et permettre d'améliorer l'efficacité et l'uniformité de la performance du troupeau.

Introduction

Feed ingredients and manufacturing are expensive components of integrated poultry production, sometimes reaching 70% of total operational costs. Fortunately, the added costs of manufacturing poultry feed improve feed efficiency and poultry performance. The performance benefits from feeding pelleted diets results in a positive return on investment, making it a common practice in broiler and turkey production. Improving the quality of pellets further enhances poultry performance and may offer a more uniform feed as it is conveyed throughout the house. Many factors contribute to a feed mill operator's ability or willingness to improve pellet quality such as throughput demands and the lack of association between pellet quality, performance, and economics. However, a considerable amount of recent pellet quality research supports improved feed intake, body weight gain, and feed efficiency when birds are provided high quality pellets. New perspectives on pellet quality are needed to support investments in the manufacturing process. Sellers et al. (2020) studied nutrient segregation in 58-m broiler houses and reported varying levels of phytase enzyme activity in feed pans. Phytase enzyme activity depended on the ratio of pellets and fines in the feed pan, suggesting that phytase may be a suitable marker for nutrient segregation. Considering the precision of modern broiler diet formulations, it became apparent that segregation of other nutrients should be explored. Four experiments were conducted to represent four different scenarios where feed line length and pellet quality differed in each scenario. The objectives of this paper are to offer nutrient segregation mitigation strategies to feed mill operators and poultry producers.

Studying nutrient segregation in four scenarios

Poultry producers in integrated operations finance the houses built on their property and contribute to house design and equipment installation decisions. It is common for feed bins to be located at the end of barn, having feed lines running the length of the barn. It is also common for feed bins to be centrally located, having feed lines running from the center of the house to either end. The difference is the distance feed travels in each scenario. Mechanical forces acting on the feed during conveyance contributes to pellet breakdown. Improving pellet quality minimizes the breakdown of pellets during transport and conveyance. Nutrient segregation was studied in two commercial broiler houses depicted in Figure 1. The four scenarios were:

- Poor pellet quality in long (152-m) feed lines
- Improved pellet quality in long (152-m) feed lines
- Poor pellet quality in short (76-m) feed lines
- Improved pellet quality in short (76-m) feed lines

To determine if nutrient segregation was occurring in each scenario, feed samples were collected from defined regions of each feedline. Each scenario was studied in four replicate feed lines. This allowed for a simple analysis of variance between each region of the feedline. Samples were sieved to determine the ratio of pellets and fines. Pellet durability analysis was performed on pellets samples. Pellet and fines samples were analyzed for the concentration of 12 amino acids as well as phytase enzyme activity at commercial laboratories. No broilers were in the barns during feed augering or sampling. Therefore, nutrient segregation affects to broiler performance were not measured in these four scenarios.

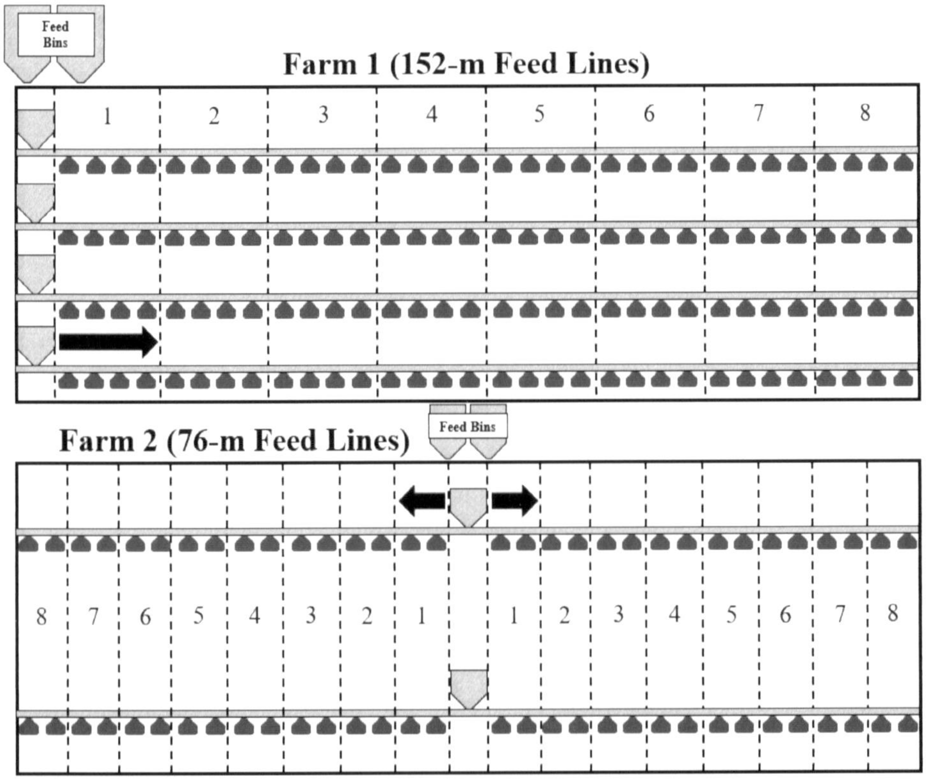

Figure 1. Schematic of replicate feed line regions at each farm. Arrows indicate feed flow direction at each farm. Commercial houses at farm 1 consisted of four 152-m-long feed lines equipped with 192 feed pans per line. Eight regions were created and consisted of 24 feed pans. In the diagram, a single feed pan represents six feed pans. The 152-m-long commercial house at farm 2 consisted of center of house fed feed lines, creating four 76-m-long feed lines with 95 feed pans per line. Feed was augered into centrally located feed hoppers and pulled to either end of the house. Each region consisted of 12 feed pans.

Nutrient segregation in broiler houses with long feed lines and poor-quality feed

In this scenario where poor-quality pellets were augered a long distance, nutrient segregation was apparent. Of the 12 measured amino acid concentrations, six of these amino acid concentrations varied across the 152-m feed line. Aspartate concentrations varied by 10.8%, glutamate by 12.8%, glycine by 7.8%, leucine by 7.4%, alanine by 6%, and lysine by 9.4% across the eight regions of the feed line (Table 1). Phytase segregation was also apparent with phytase activities varying by 50.3% across the eight regions of the feed line. In addition, pellet durability and the percentage of pellets in the feed pan varied by 5.6% and 10.6%, respectively.

Table 1. Nutrient segregation in long (152-m) feed lines conveying poor-quality pellets

	Region	Asp (%)	Glu (%)	Gly (%)	Ala (%)	Leu (%)	Lys[1] (%)	Phytase[2] (FTU/kg)	Pellet Durability[3] (%)	Pellets[4] (%)
Poor Pellet Quality	1	1.715^{cd}	3.110^{c}	0.746^{b}	0.967^{b}	1.598^{c}	1.027^{b}	487^{a}	72.6^{a}	52.8^{bcd}
	2	1.832^{abc}	3.361^{b}	0.788^{ab}	1.006^{ab}	1.683^{ab}	1.104^{ab}	392^{ab}	70.7^{ab}	55.7^{a}
	3	1.772^{bcd}	3.243^{bc}	0.766^{ab}	0.995^{ab}	1.657^{abc}	1.082^{ab}	263^{bc}	69.7^{bc}	54.8^{ab}
	4	1.854^{ab}	3.383^{ab}	0.797^{a}	1.021^{a}	1.710^{a}	1.121^{a}	366^{abc}	70.9^{ab}	53.7^{abc}
	5	1.707^{d}	3.146^{c}	0.751^{b}	0.979^{b}	1.613^{bc}	1.032^{b}	376^{abc}	70.3^{bc}	54.2^{ab}
	6	1.888^{ab}	3.412^{ab}	0.808^{a}	1.029^{a}	1.724^{a}	1.151^{a}	242^{c}	69.5^{bc}	50.5^{de}
	7	1.783^{bcd}	3.279^{bc}	0.768^{ab}	0.995^{ab}	1.664^{abc}	1.087^{ab}	311^{bc}	69.2^{bc}	51.4^{cde}
	8	1.914^{a}	3.541^{a}	0.809^{a}	1.026^{a}	1.726^{a}	1.134^{a}	272^{bc}	68.5^{c}	49.8^{e}
ANOVA P-Value		**0.019**	**0.002**	**0.038**	**0.047**	**0.032**	**0.035**	**0.019**	**0.015**	**0.001**
LSD		0.122	0.177	0.044	0.041	0.083	0.078	136	2.0	2.7
SEM		0.040	0.058	0.014	0.013	0.027	0.025	46	0.7	0.9

[1]Measured amino acids that did not segregate include: Thr, Pro, Cys, Val, Met, Ile
[2] Phytase activity was analyzed using the AOAC 2000.12 method
[3] Pellet durability was determined in a New Holmen Pellet Tester (NHPT100; TekPro Ltd., North Walsham, Norfolk, UK)
[4] Pellet percentage was determined by using a modified particle size separator fitted with a No. 5 American Society for Testing and Materials screen.

Nutrient segregation in broiler houses with long feed lines and improved-quality feed

In this scenario where improved-quality pellets were augered a long distance, only aspartate and glutamate concentrations varied (9.4% and 9.8%, respectively) across the eight regions of the feed line (Table 2). Pellet durability varied by 4.6% while phytase enzyme activity and the percentage of pellets in the feed pan were similar in this scenario. Improving pellet quality appears to decrease amino acid, phytase, and feed particle segregation when augering feed long distances.

Table 2. Nutrient segregation in long (152-m) feed lines conveying improved-quality pellets

	Region	Asp (%)	Glu[1] (%)	Phytase[2] (FTU/kg)	Pellet Durability[3] (%)	Pellets[4] (%)
Improved Pellet Quality	1	1.768^{c}	3.257^{c}	310	83.1^{ab}	62.7
	2	1.828^{bc}	3.337^{bc}	369	83.2^{a}	65.6
	3	1.782^{c}	3.258^{c}	273	82.7^{abc}	65.6
	4	1.850^{abc}	3.389^{bc}	341	82.4^{abc}	63.8
	5	1.913^{ab}	3.463^{ab}	251	81.3^{c}	63.0
	6	1.896^{ab}	3.454^{ab}	253	81.6^{abc}	63.4
	7	1.871^{abc}	3.410^{bc}	238	81.4^{bc}	63.8
	8	1.950^{a}	3.612^{a}	232	79.4^{d}	61.6
ANOVA P-Value		**0.046**	**0.012**	0.280	**0.002**	0.180
LSD		0.113	0.177	130	1.7	3.1
SEM		0.037	0.058	44	0.6	1.1

[1]Measured amino acids that did not segregate include: Thr, Pro, Gly, Ala, Cys, Val, Met, Ile, Lue, Lys
[2] Phytase activity was analyzed using the AOAC 2000.12 method
[3] Pellet durability was determined in a New Holmen Pellet Tester (NHPT100; TekPro Ltd., North Walsham, Norfolk, UK)
[4] Pellet percentage was determined by using a modified particle size separator fitted with a No. 5 American Society for Testing and Materials screen.

Nutrient segregation in broiler houses with short feed lines and poor-quality feed

In this scenario where poor-quality pellets were augered a short distance, threonine was the only amino acid concentration that varied (11.7%) across the eight regions of the 76-m feed line (Table 3). Phytase segregation was apparent in this experiment, where enzyme activity varied by 64.7% between region 2 and region 8. These data suggest that even when augering feed a short distance, applying phytase products to poor quality pellets results in enzyme activity variability from feed pan to feed pan.

Table 3. Nutrient segregation in short (76-m) feed lines conveying poor-quality pellets

	Region	Thr[1] (%)	Phytase[2] (FTU/kg)	Pellet Durability[3] (%)	Pellets[4] (%)
Poor Pellet Quality	1	0.740^{ab}	558^{abc}	71.8	44.3^{de}
	2	0.750^{ab}	782^{a}	71.0	42.2^{e}
	3	0.687^{c}	537^{abc}	69.6	45.9^{bcd}
	4	0.778^{a}	513^{bcd}	70.6	44.7^{cde}
	5	0.731^{b}	496^{bcd}	69.8	47.5^{abcd}
	6	0.759^{ab}	635^{ab}	69.6	48.3^{ab}
	7	0.731^{b}	359^{cd}	71.2	47.9^{abc}
	8	0.759^{ab}	276^{d}	70.1	50.9^{a}
ANOVA P-Value		**0.013**	**0.021**	0.429	**0.001**
LSD		0.041	261	2.3	3.6
SEM		0.013	89	0.8	1.2

[1]Measured amino acids that did not segregate include: Asp, Glu, Pro, Gly, Ala, Cys, Val, Met, Ile, Lue, Lys
[2] Phytase activity was analyzed using the AOAC 2000.12 method
[3] Pellet durability was determined in a New Holmen Pellet Tester (NHPT100; TekPro Ltd., North Walsham, Norfolk, UK)
[4] Pellet percentage was determined by using a modified particle size separator fitted with a No. 5 American Society for Testing and Materials screen.

Nutrient segregation in broiler houses with short feed lines and improved-quality feed

In this scenario where improved-quality pellets were augered a short distance, neither amino acid nor phytase segregation was apparent (Table 4). The percentage of pellets in the feed pan was the only measured variable that differed across the eight regions of the feed line.

Table 4. Nutrient segregation[1] in short (76-m) feed lines conveying improved-quality pellets

	Region	Phytase[2] (FTU/kg)	Pellet Durability[3] (%)	Pellets[4] (%)
Improved Pellet Quality	1	478	83.4	51.9[c]
	2	292	82.8	55.2[bc]
	3	443	83.0	56.2[abc]
	4	333	83.4	60.5[a]
	5	453	83.5	57.4[ab]
	6	346	83.2	58.0[ab]
	7	487	83.4	59.3[ab]
	8	264	82.5	60.5[a]
ANOVA P-Value		0.296	0.574	**0.028**
LSD		227	1.1	5.0
SEM		77	0.4	1.7

[1]Amino acid segregation was not apparent for the 12 amino acids measured: Asp, Thr, Glu, Pro, Gly, Ala, Cys, Val, Met, Ile, Lue, Lys

[2] Phytase activity was analyzed using the AOAC 2000.12 method

[3] Pellet durability was determined in a New Holmen Pellet Tester (NHPT100; TekPro Ltd., North Walsham, Norfolk, UK)

[4] Pellet percentage was determined by using a modified particle size separator fitted with a No. 5 American Society for Testing and Materials screen.

Impacts of on-farm nutrient segregation

Both the quality of the pellet and the length of feed conveyance affects the profile of nutrients present in feed pans. A summary of nutrient segregation in the four scenarios is found in Table 5. As broilers progress through their production cycle, they travel a shorter distance (Dawkins et al., 2012). Therefore, it is plausible that on-farm nutrient segregation affects broiler performance in later phases of production when feed intake is high, and distance traveled is low.

When studying the four scenarios in the current study, phytase segregation appeared to be affected largely by pellet quality. Walters et al. (2019) explained that higher doses of phytase improves performance, nutrient digestibility, and bone mineralization. Boney and Moritz (2017) explained that phytase dose affects may depend on ingredient composition. These papers support feeding similar levels of phytase enzymes to improve flock performance and uniformity. This may be most easily achieved by working with feed mill operators and nutritionists to create finished feeds with a greater percentage of pellets in the feed pan.

The four scenarios in the current study also indicated that amino acid segregation could be managed by shortening the distance that feed is conveyed or by improving pellet quality. Corzo et al. (2010) demonstrated that performance and yield were affected by amino acid density, especially in later phases of production. Furthermore, Kidd et al. (2005) explained how decreasing amino acid density reduced body weight, feed efficiency, and yields and suggested that optimizing amino acid nutrition in early production phases is critical. Once again, working with feed mill operators and nutritionists to create finished feeds with higher quality pellets is advised. If improving pellet quality is not possible, poultry producers are able to contribute to reduced nutrient segregation, flock performance, and flock uniformity by positioning feed bins near the middle of the barn so that feed is conveyed a shorter distance.

Table 5. Summary of nutrient segregation in commercial broiler barns

Feed Line Length	Pellet Quality	# of Amino Acids that Segregated[3]	Phytase Activity Segregation[4]
Long (152-m)	Poor[1]	6/12	Yes
	Improved[2]	2/12	No
Short (76-m)	Poor[1]	1/12	Yes
	Improved[2]	0/12	No

[1]Poor Pellet Quality = ~65% pellet
[2]Improved Pellet Quality = ~80% pellet
[3]Amino acids concentrations measured include aspartate, threonine, glutamine, proline, glycine, alanine, cystine, valine, methionine, isoleucine, leucine, lysine.
[4]Phytase activity was measured using the AOAC 2000.12 method at a commercial laboratory

References

Boney, J. W., and J. S. Moritz. 2017. Phytase dose effects in practically formulated diets that vary in ingredient composition on feed manufacturing and broiler performance. J. Appl. Poult. Res. 26:273-285.

Corzo, A., M. W. Schilling, R. E. Loar II, L. Mejia, L. C. G. S. Barbosa, and M. T. Kidd. 2010. Responses of Cobb x Cobb 500 broilers to dietary amino acid density regimens. J. Appl. Poult. Res. 19:227-236.

Dawkins, M. S., R. Cain, and S. J. Roberts. Optical flow, flock behaviour and chicken welfare. 2012. Anim. Behav. 84:219-223.

Kidd, M. T., A. Corzo, D. Hoehler, E. R. Miller, and W. A. Dozier III. 2005. Broiler responsiveness (Ross x 708) to diets varying in amino acid density. Poult. Sci. 84:1389-1396.

Poholsky, C. M., D. W. Hofstetter, D. Khezrimotlagh, and J. W. Boney. 2021. Effects of pellet quality to on-farm nutrient segregation in commercial broiler houses varying in feed line length. J. Appl. Poult. Res. 30:100157. https://doi.org/10.1016/j.japr.2021.100157.

Sellers, R. B., A. T. Brown, J. W. Boney, C. McDaniel, J. S. Moritz, and K. G. S. Wamsley. 2020. Impact of feed form, liquid application method, and feed augering on feed quality, nutrient segregation, and subsequent broiler performance. J. Appl. Poult. Res. 29:895-916. https://doi.org/10.1016/j.japr.2020.09.001.

Walters, H. G., M. Coelho, C. D. Coufal, and J. T. Lee. 2019. Effects of increasing phytase inclusion levels on broiler performance, nutrient digestibility, and bone mineralization in low-phosphorus diets. J. Appl. Poult. Res. 28:1210-1225.

The Potential for Feed Processing to Reduce Ergot Toxicity

Possibilité de réduire la toxicité de l'ergot par le traitement des aliments

Regina Halm, Rex Newkirk and Denise Beaulieu
Department of Animal and Poultry Science, University of Saskatchewan, Saskatoon, SK
denise.beaulieu@usask.ca

Sponsored by / Commandité par :

AGRIBUTION

Abstract

Ergot alkaloids (EA) are a group of compounds produced primarily by fungi of the *Claviceps spp.* The fungi are typically associated with grasses and rye, but wheat, barley, triticale, and oats can become contaminated when conditions are favourable. Symptoms of EA toxicity range from reduced feed intake and growth to sloughing of ears/tails and death. The EA content of heavily contaminated wheat screenings (>400 ppm EA) was reduced 50% by steam explosion (200 psi). Newly weaned pigs (7.4±1.3 kg BW) fed diets formulated to contain these screenings to achieve 0, 1.0, 2.0, or 4.0 ppm had comparable growth and feed intake throughout a 28 day trial. In the first week however, there was a linear reduction in growth with increasing EA, but only when the screenings were unprocessed (EA x processing, $P<0.01$). Extrusion of contaminated wheat screenings (240 ppm EA) had no effect on total EA content or individual alkaloids. Epimer analysis of the individual alkaloids, specifically ergocristine and ergocryptine, indicated a shift from the "ine" (R form) to the potentially less toxic "inine" (S) form. Growing pigs (65 to 125 kg BW) fed diets containing these screenings to achieve 4 ppm EA had reduced growth (1.07, 1.03 kg/d; 0 or 4 ppm, $P=0.02$) and feed intake (2.97, 2.78 kg/d, 0 or 4 ppm, $P=0.02$) regardless of whether the screenings were extruded (EA x extrusion, $P>0.50$). Further investigations are required to investigate appropriate and practical hydrothermal technologies to reduce EA toxicity.

Key words: ergot alkaloids, steam explosion, extrusion, swine

Résumé

Les alcaloïdes de l'ergot (AE) sont un groupe de composés fabriqués principalement par des champignons de l'espèce *Claviceps*. Les champignons sont généralement associés aux plantes herbacées et au seigle, mais le blé, l'orge, le triticale et l'avoine peuvent aussi être contaminés lorsque les conditions sont favorables. L'intoxication par les AE peut entraîner une diminution de la consommation alimentaire et de la croissance, la nécrose des tissus des oreilles et de la queue et même la mort des animaux. La teneur en AE de criblures de blé fortement contaminées (>400 ppm d'AE) a été réduite de 50 % à la suite d'un traitement de fragmentation par la vapeur (200 psi). Des porcs nouvellement sevrés (7,4±1,3 kg de poids corporel) nourris avec des régimes formulés avec ces criblures à 0, 1,0, 2,0 ou 4,0 ppm ont affiché une croissance et une consommation alimentaire comparables pendant un essai de 28 jours. Cependant, au cours de la première semaine, une réduction linéaire de la croissance a été constatée avec l'augmentation d'AE, mais seulement lorsque les criblures n'étaient pas traitées (AE x traitement, $P<0,01$). L'extrusion de criblures de blé contaminées (240 ppm d'AE) n'a eu aucun effet sur la teneur totale en AE ou sur les alcaloïdes individuels. L'analyse des épimères des alcaloïdes individuels, en particulier l'ergocristine et l'ergocryptine, a indiqué un passage de la forme « ine » (forme R) à la forme « inine » (forme S) potentiellement moins toxique. Des porcs en croissance (65 à 125 kg de poids corporel) nourris avec des régimes contenant ces criblures jusqu'à 4 ppm d'AE ont présenté une diminution de la croissance (1,07, 1,03 kg/j; 0 ou 4 ppm, P=0,02) et de la consommation alimentaire (2,97, 2,78 kg/j; 0 ou 4 ppm, $P=0,02$), que les criblures aient été extrudées ou non (AE x extrusion, $P>0,50$). D'autres travaux seront nécessaires pour étudier des processus hydrothermaux appropriés et pratiques permettant de réduire la toxicité des AE.

Introduction

Ergot alkaloids (EA) are a group of compounds produced primarily by fungi of the *Claviceps and Epichloë spp.* Ergot infection has traditionally been associated with grasses and rye, however when conditions are favourable, other important cereals such as wheat, barley, triticale and oats can become infected. The fungus prefers cool, moist conditions during the flowering phase of the host plant and there is evidence that the incidence is increasing. For example, in a survey of samples submitted to the Canadian Grain Commission harvest sample program from 2002 to 2013 it was shown that both the percent of samples down-graded due to ergot and the content of ergot in the samples had increased over the decade (Tittlemier et al. 2015).

Due to the pharmacological effects of the EAs they have been extensively studied and used in human medicine (Komarova and Tolkachev 2001) to treat a wide variety of clinical conditions such as post-partum hemorrhage, migraines, hyperprolactinemia, cerebral insufficiency and others (Beede 1980). Humans can be exposed to EA via ingestion of contaminated grains and there is historical evidence of outbreaks of ergotism with gangrene, vasoconstriction and neurotoxic symptoms reported (Mulac and Humpf 2011). However, strict biosecurity means that outbreaks are now relatively rare and ergotism is primarily a problem for livestock.

Unlike other mycotoxins prevalent in Western Canada such as deoxynivalenol or zearalenone which are a concern primarily to swine and poultry, ergot affects all classes of livestock, including ruminants. Consumption of the toxic alkaloids by livestock produces neurological and/or vasoconstrictive symptoms including gangrene resulting in losses of ear, tails and sloughing of hooves, especially in cattle (Klotz 2015). Reduced feed intake, growth, reproductive problems and agalactia are observed in both cattle and swine following EA ingestion. The negative effects on feed intake are observed at low levels and probably due to direct pharmacological effects of the

EA on receptors in the brain (Mulac et al. 2012). Negative effects of the EA on growth have been observed prior to decreased feed intake in swine, indicative of mechanisms independent of decreased feed intake (Oresanya et al. 2003) potentially intestinal and liver damage (Maurpo et al. 2017). Agalactia is due to the suppressive effect of the EA on prolactin, a hormone required for milk production. When nursery pigs were fed diets containing 0 to 1.0% ergot by weight, significant reductions in feed intake (13%) and growth (18%) were observed at dietary ergot concentrations of 0.10% and 0.25%, respectively. Prolactin, however, was reduced by 47% at the lowest level of ergot inclusion, 0.05% (Oresanya et al. 2003). This effect of EA on prolactin secretion has also been observed in lambs (Stanford et al. 2018) and cattle (Blaney et al. 2000) with dramatic effects on milk production and growth of offspring.

Decreased feed intake is the most sensitive indicator of the presence of EA in poultry diets, however reduced growth, diarrhea and increased mortality are also observed (Danicke 2017). There is some evidence that poultry may be less sensitive to EA than other classes of livestock (Mainka 2005) however results are variable. Danicke (2017) observed effects on feed intake with 2.5 mg/kg EA in the diet of young chicks while in a recent project conducted at the University of Saskatchewan we saw no effect on feed intake in chickens with 12 to 15 mg/kg EA in their diet (Beaulieu and Newkirk unpublished) and in work with laying hens, Danicke (2016) established a LOAEL (lowest observed adverse effect level) of 14.56 mg EA/kg diet and a NOAEL (no observed adverse effect level) of 3.72 mg/kg. Discrepancies in results may be partly due to differences in bird genetics, but are more likely a result of differences in the alkaloid composition of the ergot used in the studies, how the EA were quantified and the end-points used.

Chemistry of the Ergot Alkaloids

Although more than 50 different alkaloids are produced by *Claviceps purpurea* (Schummer et al. 2020) only 6 are consistently included in analytical methods. These can be divided into three related classes; clavine EA, lysergic acid derivatives and peptide EA with the latter being the most physiologically active (Hafner et al. 2008). The main EA consist of ergometrine (a simple lysergic derivative) and the ergopeptides; ergotamine, ergosine, ergocristine, α-ergocryptine and ergocornine. The amount and profile of the EA depend on fungal strain, host plant and climate (Hefner 2008). All EA have in common a tetracyclic ring system, ergoline (Figure 1), but vary in the structure of ring D, and the substitution around C8 (Komarova and Tolkachev 2001).

The physiological activity of the EA is related to the similarity of the ergoline ring to the noradrenaline transmitters, dopamine and serotonin (Figure 1; Komarova and Tolkachev 2001, Hafner 2008, Klotz 2015). The receptors associated with these amines are transmembrane G-proteins that consist of numerous families and sub-types (Klotz 2015). Ergot alkaloids can act at more than one receptor site, and the structural differences among the EAs means that binding may elicit a different response (Klotz 2015).

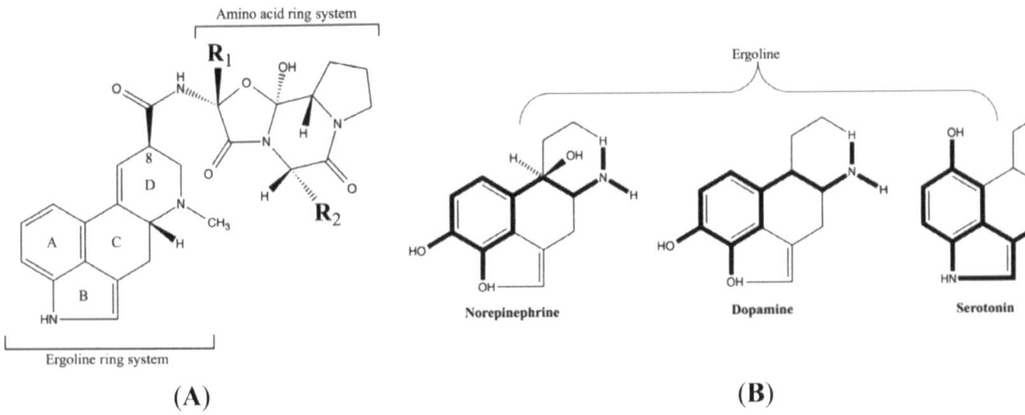

Figure 1. (A) Ergoline ring system common to all ergot alkaloids. Epimerization occurs at C8 and substitutions on R1 and R2 to create the various ergopeptine alkaloids. (B) Structural similarity to the catecholamines, norepinephrine, dopamine, and serotonin explains most of the physiological actions of the ergot alkaloids (adapted from Klotz, 2015).

The alkaloids have a stereocenter on C8 that exists in an "R" or "S" configuration. These are also distinguished by the final syllable, "ine" for the R-form and "inine" for the S form. Both epimers coexist in nature and epimerization is rapid, spontaneous and bidirectional (Komarova and Tolkachev 2001; Schummer et al. 2020). Extraction with buffers, especially alkaline buffers favours epimerization as does UV light (Komarova and Tolkachev 2001). Hefner et al. (2008) showed that epimerization depended on the solvent used for the extraction with methanol promoting the quickest epimerization and relative stability when the EA were stored in chloroform. Except for ergotametrine which was stable and did not epimerize to ergotametrinine, regardless of the solvent used, the ergopeptides, ergosine, ergotamine, ergocornine, α-ergocryptine and ergocristine behaved similarly with rapid epimerization (Hefner 2008). Following incubation of EA contaminated cookies in simulated salivary, gastric and duodenal juices, the percent of the "R" epimer of ergotamine and ergosine increased while ergocornine, α-ergocornine, ergocryptine and ergocristine shifted towards the "S" epimer (Merkel et al. 2012). Further studies indicated that the epimerization occurred primarily in the duodenal juice (pH 7.5) with little occurring in the saliva (pH 6.4) or gastric juices (pH 2.0; Merkel et al. 2012)

Despite their structural similarity, there is some evidence that EA vary in toxicity. For example, in a review of EA levels associated with gangrenous ergotism, intakes of 0.009 to 0.016 mg/kg BW ergovaline produced symptoms in cattle and sheep while 1.0 mg/kg ergotamine was required to produce symptoms in sheep (Klotz et al. 2010). In contrast, pelleting diets had no effect on total ergot content, but resulted in increased proportions of ergocristine and decreased ergotamine. Lambs fed the ergot contaminated, but pelleted diets had improved growth relative to those fed the unpelleted, contaminated diets indicating that the reduction in ergotamine may be responsible (Coufal-Majewski et al. 2017). Similarly, in an *in vitro* model of contractile response, ergovaline acted as an agonist in smooth muscle with maximal stimulation while ergocristine and ergocornine produced only 40 to 50% maximal stimulation (Klotz et al. 2010). Based on several parameters, ergocristine was shown to be the most toxic to cultured human renal cells, while ergometrine had no effect (Mulac and Humpf 2010). Similarly, in a model system of porcine brain endothelial cells it was demonstrated that the peptide EA could cross the blood brain barrier in high quantities within a few hours, while ergometrine (lysergic acid) required active transport (Mulac et al. 2011). The authors (Mulac et al 2011) noted that the peptide EAs are highly lipophilic compounds while the lysergic acid amides are slightly more soluble.

There are also consistent statements found in the literature such as ":..the alkaloids exist as different epimers, and there is evidence that one class of epimer ('in' or 'R' epimer) is more potent than the less active 'inine' (S) epimer" (Wolff et al. 1998) or "inine form is assumed to be inactive" (EFSA 2012). Although this observation is commonly stated (eg; Komarova and Tolkachev 2001; Merkel 2012; Danicke 2016; Schummer et al. 2020) it is difficult to find the experimental data supporting this claim. Several references lead back to a short article (less than 1 page in German) by Stadler and Sturmer (1970) and to Weber (1980). Weber (1980) describe a series of experiments examining the physical differences among the epimers and the potential for differences in biological activity but cites Sturmer (personal communication) as a reference for the data supporting his hypothesis of differential biological activity of the epimers.

Recent work has challenged this hypothesis. Ergocryptinine, ergocornine, ergocristinine and ergotaminine ("S" epimers) were all found to be vasoactive when tested in an *in vitro* system with bovine metatarsal arteries (Cherewyk et al. 2020) and had similar activity to that reported by others (ie. Klotz et al. 2010) for the "R" epimers. As mentioned above, using a model of the blood-brain barrier, while only the "R" epimers were able to cross the blood-brain barrier, the "S" epimer, ergocristinine was unable to cross; but this resulted in significant levels of cellular accumulation and disruption of the barrier integrity (Mulac et al. 2011).

Strategies to Mitigate Ergot Content of Feeds and Toxicity

Physical separation

Agronomists have several "tools" to limit contamination of crops with *Claviceps spp* and thus maintain grain quality (www.saskatchewan.ca). However, unpredictable weather and environmental conditions, means that periodic outbreaks still occur. The sclerotia or ergot bodies are black or purple and usually larger than the grain and thus easily separated from the grain by gravity separation or optical sorting. However, while this can reduce sclerotia content, these methods can be expensive and not always practical (Miedaner and Geiger 2015). Moreover, by-products from the milling process, which are often used in animal feeds, are a concentrated source of EAs (Tittlemier et al. 2019).

Hydrothermal processing

Hydrothermal processing refers to the use of heat, moisture and often pressure to change the physical and chemical characteristics of a feed, resulting in improved handling characteristics and nutrient digestibility. Steam explosion is a form of extreme hydrothermal processing. It consists of a steam explosion phase with temperatures of 170 to 210 °C and an explosion phase to convert the thermal energy into mechanical energy. It is commonly used in the bioprocessing industries to disrupt lignocellulosic materials to make the cellulose available to further enzymatic treatment (Ziegler-Devin et al. 2021). A similar process is "expansion", which is primarily used as a pre-treatment prior to pelleting. It is characterized by high temperature, and short time where steam-conditioned feed is forced through a barrel into an outlet when rapid expansion of the feed occurs (Danicke 2016). Expansion of diets contaminated with ergoty rye reduced total EA content by 11%. When the diets, containing up to 14 mg/kg EA, were fed to laying hens it was observed that while most performance parameters were compromised by the EA, this negative effect was mitigated when the diets were treated by expansion (Danicke 2016).

Steam explosion of heavily contaminated wheat and rye screenings reduced total EA content by 40 to 50% (Figure 2. Beaulieu and Newkirk, unpublished). Moreover, the response of the individual EA was variable with increases in proportions of ergocristine and ergotamine and decreases in ergocryptine and ergometrine following the SE (Figure 2). The processed or unprocessed wheat screenings, incorporated into diets to obtain final EA content of 0, 0.5, 1, 2 or 4 mg/kg were fed to newly weaned piglets for a 28 day trial. Overall, growth and feed intake were not significantly affected by EAs in the diet. The exception was during the first week, when growth rate was severely depressed by EA, but only when unprocessed (Figure 3; EA by processing interaction, $P < 0.01$). A similar response was observed with serum prolactin which was depressed at all levels of EA in the diet, but more so when the screenings were unprocessed (data not shown).

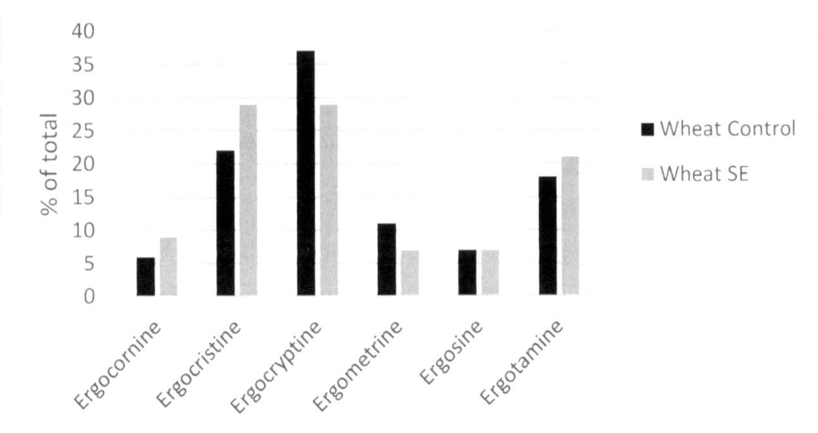

Figure 2. The change in proportions of wheat EA following steam explosion (SE). The total content of the EA was decreased by 40% with SE.

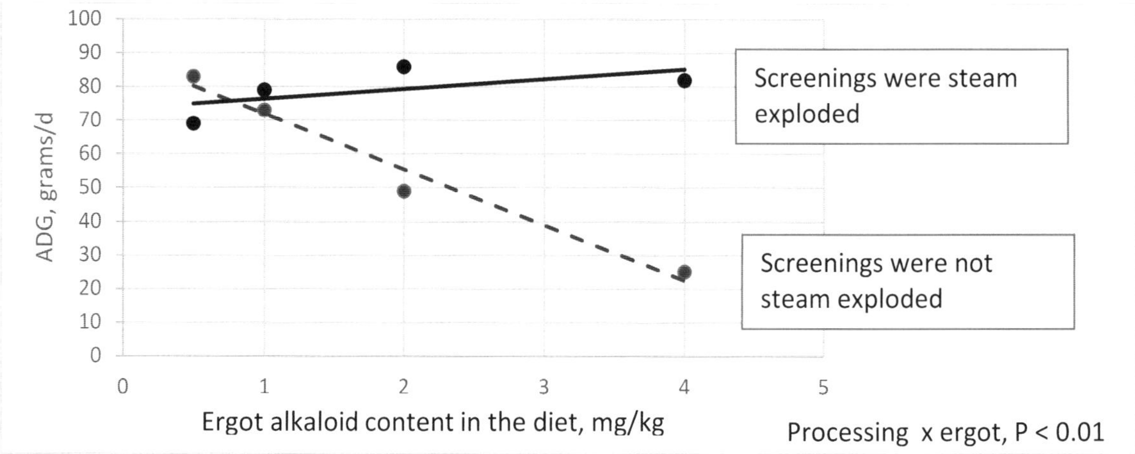

Figure 3. The response of weanling piglets (day 3 to 10 post weaning, initial BW 6.7 kg) to ergot alkaloids in the diet from contaminated wheat screenings processed by steam explosion.

The potential importance of the altered EA profile from hydrothermal treatment of the feed is also highlighted by the work of Coufal-Majewski et al. (2017). Pelleting diets for lambs had no effect on total EA content, but ergotamine and ergosine were 2-3 times greater in mash feeds, while ergocornine, ergocristine and ergometrine were 2-3 times greater in pelleted diets. Lambs fed the pelleted diets had 60 g/d greater ADG than those fed the mash diets. Moreover, the ergot induced depression in serum prolactin was less when the diets were pelleted (Coufal-Majewski et al. 2017.) As mentioned previously, there are numerous suggestions in the literature of a differential role of the "S" and "R" EA epimers, specifically referred to as the "biological inactivity of the 'S' epimer". Moreover, it has been demonstrated that processing affects this ratio. For example, baking cookies with ergot contaminated rye flour resulted in a 2-30% degradation of the EAs, plus a shift to the "S" epimer for all EAs (Merkel et al. 2012) and Tittlemier et al. (2019) reported a higher proportion of the "S" epimers resulting from the milling and cooking during the production of spaghetti.

We (Beaulieu and Newkirk) have tested out the hypothesis that "processing reduces EA toxicity by promoting a shift from the 'R' to the 'S' epimer" with experiments in swine and poultry. Diets formulated to contain 4 mg/kg EA using heavily contaminated wheat screenings, processed or not by extrusion (90 ºC; 80–100psi), or pelleting (90 ºC) were fed to growing pigs from 65 kg to 120 kg BW. Extrusion had no effect on total EA content or profile of the individual alkaloids (data not shown). However, there was a marked shift in the epimer ratio, resulting from both a decrease in the "R epimer" and an increase in the "S" epimer (Figure 4). Pigs fed the ergot contaminated diets were apparently healthy but had reduced growth and feed intake (Table 2). The lack of an effect on feed efficiency is evidence that reduced feed intake was the primary effect of the EAs. Extrusion, and the resulting shift in the epimer ratio, had no effect on the response of pigs to the EA contaminated diets (EA by extrusion interaction, $P > 0.10$).

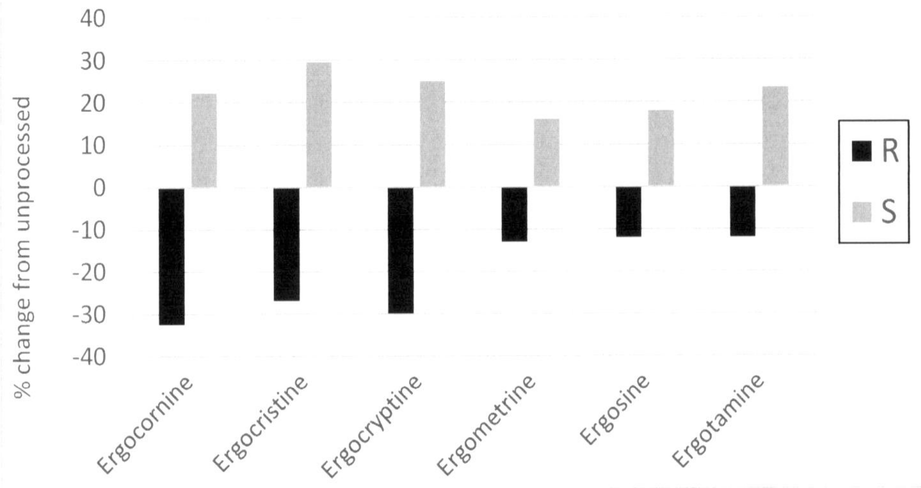

Figure 4. The percent change of the R and S epimers from the unprocessed control following extrusion of ergot contaminated wheat screenings.

Table 1. The response of growing pigs to diets contaminated with 4 ppm ergot alkaloids, unprocessed or extruded.

	Ergot, ppm		Extrusion		Pooled	P values*	
Parameter	0	4	No	Yes	SEM	Ergot	Extrusion
BW, kg							
Initial	65.94	65.82	65.87	65.9	1.54	0.84	0.97
d7	72.38	70.89	71.66	71.61	1.53	0.06	0.94
d14	79.5a	77.64	78.73	78.41	1.63	**0.03**	0.70
d21	88.10	85.89	87.30	86.6	1.18	**0.03**	0.55
d28	96.47	93.03	94.91	94.60	1.03	**0.002**	0.76
d42	112.56	109.55	111.33	110.77	1.15	**0.01**	0.62
ADG, kg/d							
d0-7	0.81	0.63	0.72	0.72	0.07	0.09	0.65
phase 1	1.07	0.95	1.02	0.99	0.02	**0.001**	0.34
phase 2	0.94	0.96	0.96	0.94	0.12	0.40	0.40
overall	1.07	1.03	1.05	1.04	0.01	**0.02**	0.31
ADFI, kg/d							
d0-7	2.10	1.92	1.98	2.03	0.07	0.09	0.65
phase 1	2.62	2.45	2.51	2.56	0.12	0.18	0.63
phase 2	3.22	3.02	3.15	3.11	0.05	**0.01**	0.54
overall	2.97	2.78	2.88	2.89	0.05	**0.02**	0.95
G:F							
d0-7	0.40	0.33	0.36	0.36	0.02	0.08	0.66
phase 1	0.41	0.40	0.41	0.39	0.01	0.39	0.12
phase 2	0.33	0.34	0.34	0.33	0.01	0.26	0.79
overall	0.36	0.37	0.37	0.36	0.01	0.19	0.37

In a similar experiment, diets were formulated for poultry to contain 0, 6, 12, 18 or 24 mg/kg EA (R + S epimer). The EA were obtained from heavily contaminated wheat screenings which were processed by either extrusion or pelleting prior to incorporation into the diet. Thus, there were 15 experimental treatments (5 EA levels x 3 processing treatments). These diets were fed to 480 Ross 308 chicks, housed 4 per cage. Temperature of a toe was taken individually using a thermocouple and a plastic clip attached to the middle toe of the right foot. Ergot alkaloid of the diets indicated they contained approximately 2.5 times the expected, formulated levels of EA. Birds were removed from the high EA treatments when toxicity symptoms were observed.

Neither pelleting nor extrusion resulted in the changes in the EA profile that has been reported by others (ie. Coufal-Majewski et al. 2017). However, processing consistently decreased the proportions of the "R" (ine) epimer and increased proportions of the "S" (inine) epimer, and this response was greater with extrusion than pelleting (Table 2). Overall, this resulted in a decreased R:S ratio with increased processing intensity.

Table 2. Changes in the ergot alkaloid epimer content and percentage when heavily contaminated wheat screenings were processed by pelleting or extrusion[1].

	Processing			% of total		
	None	Pelleted	Extruded	None	Pelleted	Extruded
Alkaloid, ppb						
Ergocornine[2]	928	911	712	7	6	5
Ergocorninine[2]	520	659	774	4	4	6
R:S	1.8	1.4	0.9			
Ergocristine	4,115	4,662	3,377	32	30	24
Ergocristinine	1,829	2,675	2,968	14	17	22
R:S	2.3	1.7	1.1			
Ergocryptine	1,598	1,581	1,199	13	10	9
Ergocryptinine	784	1,009	1,136	6	7	8
R:S	2.0	1.6	1.1			
Ergometrine	385	481	410	3	3	3
Ergometrinine	142	198	213	1	1	2
R:S	2.7	2.4	1.9			
Ergosine	576	686	552	5	4	4
Ergosinine	191	252	295	1	2	2
R:S	3.0	2.7	1.9			
Ergotamine	1,188	1,556	1,263	9	10	9
Ergotaminine	486	734	901	4	5	7
R:S	2.4	2.1	1.4			
Total	**12,742**	**15,406**	**13,802**			
Total R	8,791	9,878	7,514	69	64	54
Total S	3,951	5,528	6,288	31	36	46
R:S	2.2	1.8	1.2			

[1]Diets were prepared with the addition of heavily contaminated wheat screenings that had been processed with extrusion or pelleting prior to incorporation into the diet. The above data is from diets formulated to contain 6 mg/kg EA. Reasons for the discrepancy between formulated and analyzed EA content of the diet is unknown.
[2]Ine ("R" epimer), inine ("S" epimer).

Birds responded to increasing ergot content of the diet with reduced BW (Table 3a), feed intake (Table 3b) and external body temperature (Table 3c; P < 0.001). Feed intake and BW were comparable between birds consuming the 0 mg/kg diet and those consuming the diet formulated to contain 6 mg/kg. The analysis of this diet indicated 13 to 15 mg/kg EA, confirming our earlier observations that, compared to other livestock, poultry are relatively insensitive to EAs. The main effect of processing was significant for BW and feed intake. This is difficult to explain as only the screenings (ergot contaminated or not) were subject to processing prior to incorporation into the diet. Importantly, there were no EA level by processing interactions, indicating that even though processing changed the epimer ratio, this had no effect on EA toxicity. Further work is required to determine what hydrothermal processing conditions might cause a shift in EA profile and if this is indeed responsible for the observed effects.

Table 3a. Effect of ergot inclusion level and processing method on the body weight (g) of Ross 708 broilers[1,2].

	Ergot (E)*						Processing (P)			
Day of age	0	6	12	18	24	*P* value	None	Extruded	Pelleted	*P* value
1	46.6	47.2	47.1	46.9	47.3	0.661	47.1	47.1	47.0	0.918
8	205.5^a	197.4^a	166.7^b	137.4^c	123.4^d	**<0.001**	173.0^a	163.9^b	162.6^b	**0.001**
15	501.2^a	495.2^a	340.9^b	244.8^c	217.4^d	**<0.001**	379.5^a	355.8^b	348.5^b	**0.001**
22	985.1^a	962.7^a	601.5^b	.	.	**<0.001**	872.9^a	865.6^{ab}	821.7^b	**0.025**
29	1583.9^a	1549.7^a	1007.2^b	.	.	**<0.001**	1410.9	1406.4	1340.1	0.064

*Formulated EA content. Due to a mixing error, analyzed EA content 2-3 times the formulated amount (values represent R+S epimers.

Pooled SEM for the interaction, 0.14, 3.17, 11.51. 22.79 and 34.81 for day 1, 8, 15, 22 and 29 respectively.

$^{a-d}$ Means with different letters indicate a significant difference (P < 0.05)

[1]Ergot by processing, day 8, 15, 22 and 29 (P < 0.01). The difference between non-processed and processed increased with dietary EA content.

[2]Sex, day 15 (P < 0.01). Females heavier than males.

Table 3b. Effect of ergot inclusion level, processing method, on the average daily feed intake of Ross 708 broilers (g/bird/day).

	Ergot (E)*						Processing (P)			
Period	0	6	12	18	24	*P* value	None	Ext	Pellet	*P* value
d 1 to 8	25.0^a	23.7^a	20.6^b	18.1^c	15.5^d	**<0.001**	20.8	20.6	20.4	0.843
d 9 to 15	52.6^a	52.7^a	35.6^b	27.0^c	27.0^c	**<0.001**	42.2^a	38.1^b	37.0^b	**0.004**
d 16 to 22	88.4^a	86.7^a	53.0^b	.	.	**<0.001**	78.9^a	78.5^{ab}	73.0^b	**0.042**
d 23 to 29	120.7^a	119.5^a	80.9^b	.	.	**<0.001**	108.1^{ab}	112.2^a	102.2^b	**0.029**

*Formulated EA content. Due to a mixing error, analyzed EA content 2-3 times the formulated amount (values represent R+S epimers.

Pooled SEM for the interaction, 0.40, 1.27, 2.18 and 2.69 for day 1 to 8, 9 to 15, 16 to 22 and 23 to 29, respectively.

$^{a-d}$ Means with different letter indicate a significant difference (P < 0.05).

[1]Ergot by processing, day 9 to 15, 16 to 22 and 23 to 29 (P < 0.01). The difference between non-processed and processed increased with dietary EA content.

Table 3C. Effect of ergot inclusion level, processing method, and sex on the external temperature of Ross 708 broilers at 29 days of age (°C)[1].

Ergot (E)*				Processing (P)				Sex (S)		
0	6	12	*P* value	None	Extruded	Pelleted	*P* value	Male	Female	*P* value
31.49[a]	30.77[a]	29.20[b]	**0.0020**	31.06	29.93	30.60	0.1232	29.75	31.24	0.0045

P-values for interactions

E x P	E x S	P x S	E x P x S	SEM
0.2412	0.2157	0.1497	0.7118	0.303

*Formulated EA content. Due to a mixing error, analyzed EA content 2-3 times the formulated amount (values represent R+S epimers).

[1]Temperature of a toe was taken individually using a thermocouple and a plastic clip attached to the middle toe of the right foot

[a-b] Means with different letters indicate a significant difference ($P < 0.05$).

Summary and Conclusions

Severe hydrothermal processing (SE) of contaminated screenings reduced total EA content and the negative effects of the EA'S in a piglet model. However moderate processing; pelleting or extrusion, had no effect on total EA content or profile, but reduced the "R" epimer content and the overall R:S ratio. In contrast to our original hypothesis, this had no effect on toxicity of EA in a poultry or growing pig model. This implies that the "S" epimers of EAs have some or equivalent biological activity to the "R" epimers and need to be included in analysis and the consideration of amounts in livestock diets.

References

Beede, B. 1980. Ergot compounds. A synopsis. *in* Ergot Compounds and Brain Function: Neuroendocrine and Neuropsychiatric Aspects. ed. M. Goldstein et al. Raven Press, New York.

Coufal-Majewski, S., K. Stanford, T. McAllister, B. Blakley, J. Mckinnon, A.V. Chaves and Y. Wang. 2016. Impacts of cereal ergot in food animal production. Front. Vet. Sci. 3(15) doi: 10.3389.fvets.2016.00015.

EFSA (European Food Safety Authority), Arcella, D., J.A.G. Ruiz, M.L. Innocenti and R. Roldan. 2017. Human and animal dietary exposure to ergot alkaloids. EFSA J. 15(7): 4902, 53 p. doi: 10.2903/j.efsa.2017.4902

EFSA Panel on Contaminants in the Food Chain (CONTAM); Scientific opinion on ergot alkaloids in food and feed. EFSA J. 10(7):2798 [158 pp] doi:10.2903/j.efsa.2012.2798.

Hafner, M., M. Sulyok, R. Schuhmacher, C. Crews and R. Krska. 2008. Stability and epimerization behaviour of ergot alkaloids in various solvents. World Mycotoxin J. 1(1): 67-78.

Miedaner, T., and H.H. Geiger. 2015. Biology, genetics and management of ergot (*Claviceps* spp.) in rye, sorghum and pearl millet. Toxins. 7: 659-678. Doi: 10.3390/toxins7030659

Klotz, J.L. 2015. Activities and effects of ergot alkaloids on livestock physiology and production. Toxins. 7:2801-2821 doi: 10.3390/toxins7082801

Klotz, J.L., B.H. Kirch, G.E. Aiken, L.P. Bush and J.R. Strickland. 2010. Contractile response of fescue-naïve bovine lateral saphenous veins to increasing concentrations of tall fescue alkaloids. J. Anim. Sci 88:408-415 doi:10.24276/jas.2009-2243

Komarova, E.L., and O.N. Tolkachev. 2001. The chemistry of peptide ergot alkaloids. Part 1. Classification and chemistry of ergot peptides. Pharm. Chem. J. 35(9) 37-45

Liu, Y., J. Hubert, G. Yamdeu, Y.Y. Gong and C. Orfila. 2020. A review of postharvest approaches to reduce fungal and mycotoxin contamination of foods. Comp. Rev. Food Sci Food Safety 19:1521-1560 doi: 10.1111/1541-4337.12562

Mainka, S., S Dänicke, H. Böhem, J. Wolff, S. Mathes and G. Flachowsky. 2005. Comparative studies on the effect of ergot contaminated feed on performance and health of piglets and chickens. Arch. Anim. Nutr. 59(2):81-98 doi: 10.1080/174650390512331387909

Merkel, S., B. Dib, R. Maul, R. Köppen, M. Koch and I. Nehls. 2012. Degradation and epimerization of ergot alkaloids after baking and in vitro digestion. Ana. Bioanal Chem. 404: 2489-2497 doi: 10.1007/s00216-012-6386-8

Miedaner, T., and H.W. Geiger. 2015. Biology, genetics, and management of ergot (*Claviceps* spp) in rye, sorghum and pearl millet. Toxins 7: 659-678: doi:10.3390/toxins7030659

Mulac, D., H-U. Humpf. 2011. Cytotoxicity and accumulation of ergot alkaloids in human primary cells. Toxicology. 282: 112-121 doi: 10.1016/j.tox.2011.01.019

Mulac, D., S. Hüwel, H-J. Galla and H-U. Humpf. 2012. Permeability of ergot alkaloids across the blood-brain barrier in vitro and influence on the barrier integrity. Mol. Nutr. Food Res. 56: 475-485 doi: 10.1002/mnfr.201100431

Oresanya, T.F., J.F. Patience, R.T. Zijlstra, A.D. Beaulieu, D.M. Middleton, B.R. Blakley and D.A. Gillis. 2003. Defining the tolerable level of ergot in the diet of weaned pigs. Can. J. Anim. Sci. 83: 493-500

Pierri, L., I.H. Pitman, I.D. Rae, D.A. Winkler and P.R. Andrews. 1982. Conformational analysis of the ergot alkaloids ergotamine and ergotaminine. J. Med. Chem. 25: 937-942.

Smith, D.J., and N.W. Shappell. 2002. Technical note: Epimerization of ergopeptine alkaloids in organic and aqueous solvents. J. Anim. Sci. 80:1616-1622

Schummer, C., I. Zandonella, A. van Nieuwenhuyse and G. Moris. 2020. Epimerization of ergot alkaloids in feed. Heliyon 6(2020)e04336. doi: .org/10.1016/j.heliyon.2020.e04336

Stadler, P.A. and E. Stürmer. 1970. Comparative studies on the pharmacological properties of stereoisomers of ergotamine and dihydro-ergotamine. Naunyn Schmiedebergs Arch Pharmakol 266(4) 457-458

Weber, H.P. 1980. The molecular architecture of ergopeptines: a basis for biological interaction. *in* Ergot Compounds and Brain Function: Neuroendocrine and Neuropsychiatric Aspects. ed. M. Goldstein et al. Raven Press, New York

Wolff, J., Ch. Neudecker, Ch. Klug and R. Weber. 1988. Chemische und toxikologische untersuchungen über mutterkorn in nehl and Brot. Z. Ernährungswiss 27:1-22 (article in German, abstract in English)

Ziegler-Devin, I., L. Chrusciel and N. Brosse. 2021. Steam explosion pretreatment of lignocellulosic biomass: A mini-review of theorical and experimental approaches. Front. Chem. 9:705358 doi:10.3389/fehcm.2021.705358.

Growth Performance, Digesta pH and Organ Weight of Weaned Pigs Fed Barley Grain Differing in Fermentable Starch

Performances de croissance, pH du digesta et poids des organes chez des porcs sevrés nourris avec des grains d'orge de teneurs différentes en amidon et en fibres fermentescibles

J. Sanchez[1], L.F. Wang[1], E. Beltranena[1], A.D. Beattie[2], R.W. Newkirk[2], R.T. Zijlstra[1]
[1]University of Alberta, Edmonton, AB
[2]University of Saskatchewan, Saskatoon, SK
sanchezz@ualberta.ca

Sponsored by / Commandité par :

Cbs Bio Platforms

Abstract

Barley grain is underutilised as net energy (NE) source for weaned pigs. Whether fermentable starch and fibre in barley or reduced diet NE value would affect growth performance, digesta pH or organ weight in weaned pigs is unclear. Five diets contained 60–70% cereal grain: 1) low-fermentable wheat (LFW); 2) low-fermentable hulled barley (LFB); 3) LFB, without added oil, low NE (LFB-LNE); 4) high β-glucan, hull-less barley (HFBB); or 5) high amylose, hull-less barley (HFAB). Diets provided 2.45 and 2.33 Mcal NE/kg (2.25 and 2.19 for LFB-LNE), and 5.51 and 5.10 g standardised ileal digestible (SID) lysine/Mcal NE (6.01 and 5.45 for LFB-LNE) for phase 1 (day 1–14) and phase 2 (day 15–28), respectively. Forty pigs (8.6 kg) were fed 1 of 5 diets starting 1-week post-weaning. Overall, average daily feed intake (ADFI), average daily gain (ADG) and gain:feed (G:F) did not differ among diets formulated to equal NE and SID lysine/NE ratio. For day 15–21, ADG and G:F was greater ($P < 0.05$) for LFB than HFBB. Feces consistency, digesta pH and weights of kidneys, liver, spleen, carcass, non-carcass, and empty body weight (EBW) did not differ among pigs fed the 4 diets. The EBW (% of live BW) was 2.1% lower ($P < 0.05$) for pigs fed HFBB than pigs fed LFB and LFW. Empty colon and rectum weight (% of EBW) increased ($P < 0.05$) by 25% in pigs fed HFAB than pigs fed LFW and LFB. Empty cecum weight (% of EBW) increased ($P < 0.05$) by 27% in pigs fed HFBB than pigs fed LFB and LFW. Comparing LFB with LFB-LNE, low dietary NE value did not affect ADFI, ADG and G:F for the entire trial, but ADG was greater ($P < 0.05$) for LFB-LNE than LFB for day 1–7. Feces consistency, digesta pH, organ weights, carcass, non-carcass and EBW did not differ between pigs fed LFB and LFB-LNE. In conclusion, dietary fermentable starch and fibre did not affect growth, feces consistency and carcass weight in weaned pigs. High dietary fermentable carbohydrates may reduce EBW and increase large intestine weight. Reducing NE value of a hulled barley diet while maintaining SID AA content did not affect growth, EBW and organ weights. Dietary inclusion of barley grain and formulating low NE diets may reduce feed cost without affecting growth or carcass weight in weaned pigs.

Keywords: barley, fibre, pig, starch

Résumé

Le grain d'orge est sous-utilisé comme source d'énergie nette (NE) pour les porcs sevrés. Il n'est pas certain que l'amidon et les fibres fermentescibles de l'orge ou la réduction de la valeur de NE de la ration aient une incidence sur les performances de croissance, le pH du digesta ou le poids des organes des porcs sevrés. Cinq régimes alimentaires composés de 60 à 70 % de céréales : 1) blé peu fermentescible (LFW); 2) orge mondée peu fermentescible (LFB); 3) LFB sans huile ajoutée, faible NE (LFB-LNE); 4) orge nue à haute teneur en β-glucanes (HFBB); 5) orge nue à haute teneur en amylose (HFAB). Les régimes ont procuré 2,45 et 2,33 Mcal NE/kg (2,25 et 2,19 pour LFB-LNE) ainsi que 5,51 et 5,10 g de lysine digestible iléale standardisée (SID)/Mcal NE (6,01 et 5,45 pour LFB-LNE) pour la phase 1 (jours 1-14) et la phase 2 (jours 15-28), respectivement. Quarante porcs (8,6 kg) ont reçu l'un des cinq régimes à partir d'une semaine après le sevrage. Dans l'ensemble, la consommation moyenne par jour (ADFI), le gain moyen quotidien (ADG) et l'indice de conversion alimentaire (G:F) n'ont pas différé entre les régimes équivalents pour la NE et pour le rapport lysine SID/NE. Pour les jours 15 à 21, l'ADG et le G:F ont été plus élevés ($P < 0,05$) pour le régime LFB que pour le régime HFBB. Aucune différence dans la consistance des fèces, le pH du digesta ni le poids des reins, du foie, de la rate, de la carcasse et des abats ainsi que du poids corporel vide (EBW) n'a été observée entre les porcs nourris avec les 4 régimes. Une diminution du EBW (% du poids vif) de 2,1 % ($P < 0,05$) a été notée chez les porcs qui ont reçu le régime HFBB par rapport aux porcs nourris avec les régimes LFB et LFW. Le poids du côlon et du rectum vides (% du EBW) a augmenté ($P < 0,05$) de 25 % chez les porcs nourris avec le régime HFAB, comparativement aux porcs qui ont reçu les régimes LFW et LFB. Le poids du caecum vide (% de l'EBW) a augmenté ($P < 0,05$) de 27 % chez les porcs HFBB par rapport aux porcs LFB et LFW. En comparant les résultats obtenus avec les régimes LFB et LFB-LNE, on constate que la faible valeur de NE n'a pas influencé l'ADFI, l'ADG ni le G:F pour l'ensemble de l'essai et que l'ADG a été plus élevé ($P < 0,05$) avec le régime LFB-LNE qu'avec le régime LFB pour les jours 1 à 7. Aucune différence n'a été enregistrée quant à la consistance des fèces, au pH du digesta, au poids des organes et à celui la carcasse et des abats ainsi qu'au poids corporel vide entre les porcs LFB et les porcs LFB-LNE. En conclusion, l'amidon et les fibres fermentescibles n'ont eu aucune incidence sur la croissance, la consistance des fèces et le poids de carcasse des porcs sevrés. Une teneur élevée en glucides fermentescibles peut réduire l'EBW et augmenter le poids du gros intestin. Réduire la valeur de NE d'un régime à base d'orge mondée tout en maintenant la teneur en AA SID n'a pas eu d'effet sur la croissance, l'EBW ni le poids des organes. Les pratiques consistant à inclure des grains d'orge dans la ration et à formuler des régimes à faible teneur en énergie nette peuvent réduire le coût de l'alimentation sans nuire à la croissance ni au poids de carcasse des porcs sevrés.

Introduction

In Canada, barley is rarely fed to young pigs due to its relatively high fibre content that is associated with lower energy and nutrient digestibility (Che et al., 2012). Carbohydrates that are not digested in the small intestine may be fermented in the large intestine, and so provide energy and stimulate gut health (Fouhse et al., 2017). Barley cultivars with increased fermentable carbohydrates may enhance gut health in weaned pigs (Jha et al., 2019). Intestinal populations of *Bifidobacteria* spp. and *Lactobacillus* spp were increased in weaned pigs consuming 63% of dietary starch as amylose or barley-based diets (Fouhse et al., 2015). Moreover, fermentable carbohydrates may repress pathogen colonisation and enhance immune responses and gut function (Diao et al., 2019). However, fermentable carbohydrates may also alter the physicochemical properties of digesta, predisposing pigs to diarrhoea (Pluske et al., 2002). High-fibre diets may increase the mass of undigested residue flowing into the gut and increase mass of the gastrointestinal tract (GIT) (Agyekum and Nyachoti, 2017). The enlarged GIT would exert additional energy and nutrient demand to meet their increased maintenance costs (Nyachoti et al., 2000). The increase in energy expenditure is likely associated with increased protein synthesis and increased endogenous losses in the GIT when pigs are fed high-fibre diets (Jørgensen et al., 1996) and that may decrease amino acid availability for growth. Hence, research is required to validate the inclusion of barley high in fermentable starch and β-glucan in weaned pig diets and characterize its nutritive and functional value. Thus, the objective of this study was to evaluate growth performance, faeces consistency, digesta pH, and physical body composition of weaned pigs fed diets differing in fermentable carbohydrates content.

Methodology

Forty crossbred barrows and gilts (28 ± 1 days of age, 8.6 kg) were blocked based on gender and randomly assigned to individual metabolism pens (1.1 × 0.8 m). Pigs were allocated to 1 of 5 test diets in randomized complete blocks. Test diets contained 60–70% cereal grain: 1) low-fermentable wheat (LFW); 2) low-fermentable hulled barley (LFB); 3) LFB, without added oil, low NE (LFB-LNE); 4) high β-glucan, hull-less barley (HFBB); or 5) high amylose, hull-less barley (HFAB). Diets provided 2.45 and 2.33 Mcal NE/kg (2.25 and 2.19 for LFB-LNE), and 5.51 and 5.10 g standardised ileal digestible (SID) lysine/Mcal NE (6.01 and 5.45 for LFB-LNE) for phase 1 (day 0–14) and phase 2 (day 15–28), respectively. Test diets were fed ad libitum starting from 1 week after weaning. Pigs, added and remaining feed were weighed weekly to calculate average daily feed intake (ADFI), average daily gain (ADG) and gain-to-feed (G:F). Faeces consistency was recorded daily using an 8-grade score. Pigs were slaughtered at the conclusion of the study and emptied GIT, liver, spleen, kidneys, and blood were weighed. Digesta pH was measured in all GIT segments. Organs and blood were referred to as "non-carcass", whereas carcass including head and feet were referred to as "carcass." The empty BW (EBW) of the pig was the sum of carcass and non-carcass. Data were analysed using the GLIMMIX procedure (SAS Inst. Inc., Cary, NC). For growth performance and faeces consistency data the model included week and interaction of diet and week as fixed effects and block as random effect. Digesta pH and physical body composition data were analyzed with digesta temperature and initial BW as a covariate, respectively. A probability of $P < 0.05$ was considered significant.

Results and Discussion

For the 28-day trial, ADFI, ADG, G:F and faeces consistency did not differ among pigs fed diets formulated to equal NE and SID lysine/NE ratio. For individual weeks, ADG and G:F was greater ($P < 0.05$) for pigs fed LFB than for pigs fed HFBB for day 25–21. Fibre and fermentable carbohydrates content are negatively correlated with nutrient digestibility and energy value of feed (Le Goff et al., 2002) and may reduce pig growth (Che et al., 2012). Despite greater dietary fibre and fermentable carbohydrates content, pigs fed the barley diets had equal growth performance than pigs fed wheat diet in the present study. The sustained growth may be partly attributed to diets formulated to equal NE and SID AA, which reduces risks associated with feeding ingredients high in fibre on pig performance. Feeding increasing fibre and fermentable carbohydrates may increase visceral organ weight and gut fill which contributes to live BW gain (Agyekum et al., 2012). Additionally, fibre and fermentable carbohydrates in barley may maintain gut health and growth performance (Molist et al., 2014). Comparing LFB with LFB-LNE, addition of dietary oil for NE correction did not increase growth performance in pigs. Pigs fed LFB-LNE did not increased feed intake to compensate for reduced dietary value, indicating that young pigs can efficiently utilize barley grain to support growth.

Live BW at slaughter was 25.2, 26.9, 27.2, 25.0 and 25.6 kg for LFW, LFB, LFB-LNE, HFBB and HFAB diets, respectively, and did not differ among diets. Digesta pH and weights of kidneys, liver, spleen, carcass, non-carcass and EBW did not differ among pigs fed diets formulated to equal NE and SID lysine/NE ratio. However, EBW expressed as percentage of live BW was 2.1% lower ($P < 0.05$) for pigs fed HFBB than pigs fed LFB and LFW. Empty colon and rectum weight as percentage of EBW increased ($P < 0.05$) by 25% in pigs fed HFAB than pigs fed LFW and LFB, whereas empty cecum weight as percentage of EBW increased ($P < 0.05$) by 27% in pigs fed HFBB than pigs fed LFB and LFW. Although treatment did not affect live BW, carcass and non-carcass, pigs fed HFBB had a lower EBW as percentage of live BW than pigs fed LFW and LFB, indicating that high β-glucan may enlarge gut fill by decreasing passage rate and reducing nutrient digestion, thereby increasing the mass of undigested residue (Fouhse et al., 2017). The heavier large intestine as percentage of EBW in pigs fed high β-glucan and high amylose diets than pigs fed low fermentable wheat and hulled barley might be due to increased nutrient fermentation, resulting in adaptive changes in the size of the GIT (Fouhse et al., 2017). Digesta pH, organ weights, carcass, non-carcass and EBW did not differ between pigs fed LFB and LFB-LNE.

Conclusion

Dietary inclusion of barley grain to replace wheat did not affect ADFI, ADG, G:F, faeces consistency, and carcass weight in weaned pigs. Fermentable carbohydrates in barley reduced EBW and increased large intestine weight. Reducing dietary NE value while keeping SID AA content did not affect growth performance, EBW and organ weights in weaned pigs. Barley grain can be fed as energy source to starter pigs. Formulating low NE diets may reduce feed cost without affecting growth performance or carcass weight in weaned pigs.

References

Agyekum, A.K., and C.M. Nyachoti. 2017. Nutritional and metabolic consequences of feeding high-fiber diets to swine: A review. Engineering 3 716–725.

Agyekum, A.K., B.A. Slominski, and C.M. Nyachoti. 2012. Organ weight, intestinal morphology, and fasting whole-body oxygen consumption in growing pigs fed diets containing DDGS alone or in combination with a multienzyme supplement. J. Anim. Sci. 90 3032–3040.

Che, T.M., V.G. Perez, M. Song, and J.E. Pettigrew. 2012. Effect of rice and other cereal grains on growth performance, pig removal, and antibiotic treatment of weaned pigs under commercial conditions1. J. Anim. Sci. 90 4916–4924.

Diao, H., A.R. Jiao, B. Yu, X.B. Mao, and D.W. Chen. 2019. Gastric infusion of short-chain fatty acids can improve intestinal barrier function in weaned piglets. Genes Nutr. 14 4.

Fouhse, J.M., M.G. Gänzle, P.R. Regmi, T.A. Van Kempen, and R.T. Zijlstra. 2015. High amylose starch with low in vitro digestibility stimulates hindgut fermentation and has a bifidogenic effect in weaned pigs. J. Nutr. 145 2464–2470.

Fouhse, J.M., J. Gao, T. Vasanthan, M. Izydorczyk, A.D. Beattie, and R.T. Zijlstra. 2017. Whole-grain fiber composition influences site of nutrient digestion, standardized ileal digestibility of amino acids, and whole-body energy utilization in grower pigs. J. Nutr. 147 29–36.

Jha, R., J.M. Fouhse, U.P. Tiwari, L. Li, and B.P. Willing. 2019. Dietary fiber and intestinal health of monogastric animals. Front. Vet. Sci. 6 48.

Jørgensen, H., X.-Q. Zhao, and B.O. Eggum. 1996. The influence of dietary fibre and environmental temperature on the development of the gastrointestinal tract, digestibility, degree of fermentation in the hind-gut and energy metabolism in pigs. Br. J. Nutr. 75 365–378.

Le Goff, G., J. Van Milgen, abd J. Noblet. 2002. Influence of dietary fibre on digestive utilization and rate of passage in growing pigs, finishing pigs and adult sows. Anim. Sci. 74 503–515.

Molist, F., M. van Oostrum, J.F. Pérez, G.G. Mateos, C.M. Nyachoti, and P.J. van der Aar. 2014. Relevance of functional properties of dietary fibre in diets for weanling pigs. Anim. Feed Sci. Technol. 189 1–10.

Nasir, Z., L.F. Wang, M.G. Young, M.L. Swift, E. Beltranena, and R.T. Zijlstra. 2015. The effect of feeding barley on diet nutrient digestibility and growth performance of starter pigs. Anim. Feed Sci. Technol. 210 287–294.

Nyachoti, C.M., C.F.M. De Lange, B.W. McBride, S. Leeson, and H. Schulze. 2000. Dietary influence on organ size and in vitro oxygen consumption by visceral organs of growing pigs. Livest. Prod. Sci. 65 229–237.

Pluske, J.R., D.W. Pethick, D.E. Hopwood, and D.J. Hampson. 2002. Nutritional influences on some major enteric bacterial diseases of pig. Nutr. Res. Rev. 15 333–371.

Wang, L.F., H. Zhang, E. Beltranena, and R.T. Zijlstra. 2018. Diet nutrient and energy digestibility and growth performance of weaned pigs fed hulled or hull-less barley differing in fermentable starch and fibre to replace wheat grain. Anim. Feed Sci. Technol. 242 59–68.

Yen, J.T., J.A. Nienaber, D.A. Hill, and W.G. Pond. 1989. Oxygen consumption by portal vein-drained organs and by whole animal in conscious growing swine. Proc. Soc. Exp. Biol. Med. 190 393–398.

Zhou, X., E. Beltranena, and R.T. Zijlstra. 2016. Effect of feeding wheat- or barley-based diets with low or and high nutrient density on nutrient digestibility and growth performance in weaned pigs. Anim. Feed Sci. Technol. 218 93–99.

Feed Formulation for Sustainable Agriculture

La formulation des aliments dans un contexte d'agriculture durable

Greg Thoma
University of Arkansas, Fayetteville, AR, USA
gthoma@uark.edu

Sponsored by / Commandité par :

ROBINSON BIOPRODUCTS INC

Abstract

As the global population continues to grow and pressure on the food system to provide more with less increases coupled with the burgeoning global middle-class which is increasing demand for high-quality animal sourced food (ASF) leads to a situation which is extraordinarily challenging. We face unprecedented complexity in food production, distribution and consumption and in order to meet societies need for affordable and nutritious food we must understand the complex interacting factors which define sustainable food systems. The adage you manage what you measure has never been a more appropriate lens to view food system sustainability. The utility of the lifecycle assessment (LCA) framework as a means of providing quantifiable measures of sustainable performance is discussed.

In the context of ASF, LCA shows that an animal-centric perspective is useful for understanding sustainability. Specifically, feed conversion in the animal is a critical metric defining sustainable production systems. Two primary reasons for this are, first, improved FCR implies less feed must be produced in the upstream supply chain to produce certain quantity of ASF leading to reduced environmental effects, and second, improved FCR implies reduced production of manure and therefore reduced environmental effects associated with manure management.

Thus, feed formulation is a very important tool in farm management for sustainable production. We will consider some tools available for feed formulation which include an accounting of environmental sustainability and will consider some case studies of feed supplements in ration formulation.

Key words: animal sourced food, lifecycle assessment, feed formulation

Résumé

Jumelé à l'accroissement de la population de la planète et à la pression accrue exercée sur le système alimentaire pour produire davantage avec moins, l'essor de la classe moyenne mondiale, laquelle demande de plus en plus d'aliments d'origine animale de haute qualité, crée une situation extraordinairement complexe. Nous sommes confrontés à des difficultés sans précédent en matière de production, de distribution et de consommation alimentaires. Pour répondre aux besoins d'aliments nutritifs et abordables exprimés par les diverses communautés dans le monde, nous devons comprendre les facteurs d'interaction complexes qui définissent les systèmes alimentaires durables. L'adage selon lequel on ne peut gérer ce qu'on ne peut mesurer offre une perspective plus appropriée que jamais pour envisager la durabilité des systèmes alimentaires. Dans cet exposé, nous explorerons l'utilité du concept de l'analyse du cycle de vie (ACV) comme moyen de fournir des mesures quantifiables de la performance durable.

Dans le contexte des aliments d'origine animale, l'ACV montre qu'une perspective centrée sur l'animal est utile pour comprendre la durabilité. Plus précisément, l'efficacité alimentaire chez l'animal est une mesure essentielle pour définir les systèmes de production durables. Il y a deux raisons principales à cela : premièrement, l'amélioration de l'indice de conversion alimentaire (ICA) suppose que moins d'aliments devront être produits en amont pour obtenir une certaine quantité d'aliments d'origine animale, ce qui réduit les impacts environnementaux; deuxièmement, un meilleur ICA conduit à une réduction de la production de fumier et, donc, une réduction des effets environnementaux associés à la gestion du fumier.

Ainsi, la formulation des aliments est un outil très important à utiliser dans la gestion des exploitations pour parvenir à une production durable. Nous examinerons certains outils disponibles pour la formulation des aliments, dont une prise en compte de la durabilité environnementale, et nous examinerons quelques études de cas portant sur l'utilisation de compléments alimentaires dans la formulation des rations.

Introduction

The global population is anticipated to reach around 10 billion people by 2050. At the same time the global middle-class is growing significantly particularly in high populous countries such as China and India. The combination of these two factors places significant pressure on global food systems which must be capable of providing food and nutrition security in 2050. The next 30 years are arguably one of the most critical transitions in human history as we approach a plateau to global population while recognizing that agriculture, and thus the food system, are significant factors in the overall environmental pressures affecting geopolitical stability. The burgeoning middle class is expecting to have high quality animal sourced food readily available, thus there are exacerbated pressures on the global livestock sector to increase resource use efficiency and produce more high-quality foods using a smaller resource base. It is in this context that we recognize the key role of animal feed in the global context of food and nutrition security for humanity. Thus, a suite of management techniques focused on improving the capacity of the livestock sector increased production and reduce environmental impact simultaneously are critical for our future success in meeting the food and nutrition requirements of humanity.

Lifecycle Assessment

Life cycle assessment (LCA) is a systems framework for evaluating environmental sustainability metrics for goods and services (ISO 2006). It consists of four main stages: 1) goal and scope definition; 2) lifecycle inventory; 3) lifecycle impact assessment; 4) interpretation. The process of performing an LCA is generally iterative in nature and is useful for identifying "hotspots" in supply chains which are activities that offer significant leverage for improvement because of relatively large contribution to environmental impacts (Figure 1). Past work (Figure 2) as shown that feed is an important consideration in sustainability assessment of livestock systems (Thoma et al. 2013; Putman et al. 2017, 2019). It follows that approaches focused on improving feed utilization and livestock systems should have beneficial outcomes improving environmental sustainability of the systems. Thus, in broad terms efforts to improve feed conversion ratios in livestock via improved animal genetics as well as utilizing feed formulation and additives have high potential for beneficial effects.

Feed Formulation for Sustainability

The formulation of rations for livestock is a highly constrained activity where the nutritional requirements of the animal are paramount and in general a least cost formulation is desired in support of economic sustainability. Thus, any modifications of feed ration composition with additional targets of environmental improvement may lead to trade-offs with the least cost formulation since the nutritional requirements are not subject to modification. An early example of the investigation into feed formulation effects is reported by Mosnier et al. (2011) who identify substitution of soybean meal in swine and poultry rations with synthetic amino acids as an approach to reduce climate change impacts. They report that, depending upon the source of soy meal different benefits and costs can result. For soybean meal sourced from Brazil in a region where deforestation is occurring a 2% increase in costs can result in an approximately 5% reduction in greenhouse gas emissions while soybean meal sourced from a region without

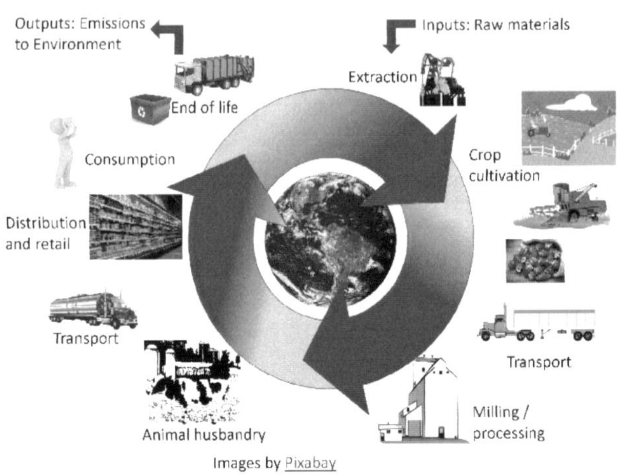

Figure 1. Example cradle-to-grave lifecycle

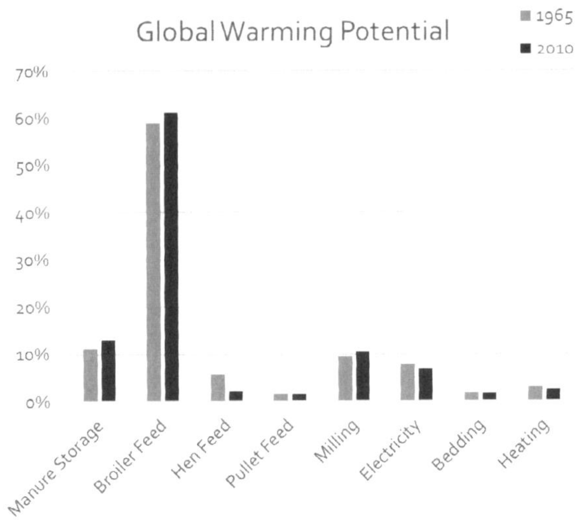

Figure 2. Feed is a dominant contributor to poultry production since modern systems became prevalent in the 1960's. Putman et al. 2017

deforestation saw a 6.4% decrease in greenhouse gas emissions associated with an increase of 8.7% in the price of the formulated ration.

Swine Production

Figure 4 presents a comparison of different ration formulation approaches focused on optimizing the ration itself for reduced environmental impact compared to these cost formulations. A standard corn and soy meal-based ration for different feeding phases (three nursery, two grow and three finish fractions). The least cost, least climate change, least water depletion, and at least land use contributing factors are shown in each of the vertical panels of the figure. The different rows of the panel represent to alternative approaches to ration formulation (note that for top 2 rows the ration for each phase is identical across each column-that is the baseline N1 ration is the same for each of the metrics evaluated). The middle row is constructed using synthetic amino acids, specifically lysine and methionine, to replace a fraction of the soy meal in the ration.

The alternate optimization row shows the effect of optimizing a ration for each of the metrics. Thus, in this case the ration formulated in each of the vertical panels is different. This analysis shows that we can achieve notable reductions of individual footprints, but that there are significant trade-offs when we optimize for land use versus climate change or cost. However, as discussed below for dairy production, this is an incomplete evaluation because the animals' performance on these rations has not been accounted (in theory, since nutrient levels are consistent, performance should be unaffected, but this needs to be tested).

Broiler Production

More recently, Blonk et al. (2021) reported on a series of dietary interventions based on different feed supplements which affect feed formulation for broiler production, summarized in **Table 1**.

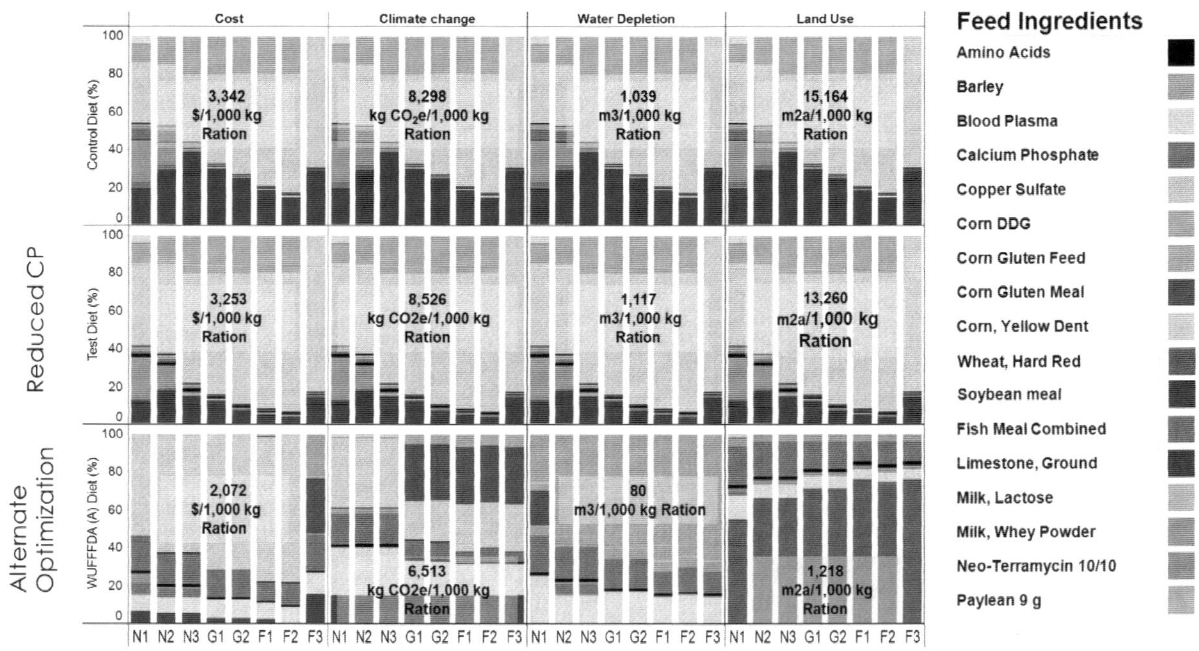

Figure 3. Multi-criteria optimization of swine rations

Table 1. Dietary interventions and expected effects for broiler performance. Blonk et al. (2021)

Additive	*Zootechnical effects*	*Changed flows and mechanisms of change*
Vitamin (25(OH)D3)	Muscle and bone development support via enhanced mineral homeostasis.	Reduction in mortality. Higher amount of breast meat leading to more valuable product sales at slaughterhouse.
Eubiotics (acid and phytogenic compounds)	Gut functionality support via acidification of the digesta, gut flora modulation and stimulation of the digestive enzymes.	Improvement in FCR. This can be modelled as an increase in liveweight output, reduction of feed input, shorter production cycle or a combination of the approaches.
Enzyme (Phytase)	Improved digestion of phytates.	Change in feed materials composition because of lower needs for phosphorus and nitrogen input.
Enzyme (Protease)	Improved digestion of proteins.	Change in feed composition (reduced nitrogen content) with higher overall feed digestibility.
Enzyme (Xylanase)	Increased hydrolysis of arabinoxylans.	Change in feed composition (higher level of wheat, lower levels of fat) with higher overall feed digestibility.

Various scenarios were considered with different combinations of dietary interventions. Because phytase is commonly used in broiler feed a counterfactual evaluation where phytase was removed from the formulation was included for comparison. The heat map shown in **Table 2** summarizes anticipated benefits associated with combinations of interventions. The authors caution that price variability can significantly affect the economic return associated with different formulation alternatives and specific studies accounting for current market conditions and feedstuff availability are representative of the production system under study are necessary.

Table 2. Environmental improvements associated with feed supplement-based ration reformulation.

	Scenario							
Impact Category	*B*	*A*	*C*	*D*	*E*	*F*	*G*	*H*
Climate change (w/o LUC)	2.3%	0.0%	-0.4%	-0.8%	-1.2%	-2.3%	-0.3%	-3.8%
Climate change	5.3%	0.0%	-2.2%	-3.8%	-5.6%	-2.9%	-0.3%	-8.6%
Ozone depletion	6.9%	0.0%	-0.1%	0.1%	0.2%	-2.4%	-0.1%	-2.5%
Photochemical ozone formation	4.2%	0.0%	-1.6%	-1.5%	-3.3%	-6.0%	-0.3%	-9.1%
Respiratory inorganics	3.2%	0.0%	-1.6%	0.1%	-1.5%	-3.6%	-0.4%	-5.5%
Non-cancer human health effects	1.1%	0.0%	-0.4%	-0.8%	-1.1%	-1.0%	-0.3%	-2.4%
Cancer human health effects	4.4%	0.0%	-0.4%	-0.6%	-0.8%	-0.8%	-0.3%	-2.0%

The scenarios in the headings are
A: Baseline, **B**: No phytase, **C**: A + protease, **D**: A + xylanase, **E**: A + all enzymes, **F**: E + eubiotics, **G**: E + 25(OH)D3,
H: All solutions. LUC = land-use change
Adapted from Blonk et al. (2021)

Beef Production

A recent study reported by Matlock et al.(2021) reports on the environmental benefits associated with use of Enogen® Feed corn. Enogen® Feed corn is a GMO product that has been designed to produce a heat stable and pH tolerant α-amylase, which improves the digestibility of the starch content of corn and therefore improves the feed conversion ratio. The results shown in **Table 3** show that the approximately 5% gain in feed conversion ratio associated with the improvement digestibility leads to approximately 5% reduction in environmental impact across the four categories reported in the study.

Table 3. Cradle-to-gate environmental benefits of improved corn digestibility in beef feedlot production systems

Impact Category	units	Conventional Corn	Enogen® Feed Corn	Enogen percent decrease in impact
Climate change	(kg CO_2e/1,000 kg LWG)	8,608[a]	8,109[b]	-5.80%
Land use	(m^2a/1,000 kg LWG)	15,405[a]	14,461[b]	-6.13%
Water use	(m^3/1,000 kg LWG)	1,384[a]	1,307[b]	-5.61%
Fossil Energy	(kg oil eq/1,000 kg LWG)	1,127[a]	1,060[b]	-5.99%

Values with different letters within a category (row) are significantly different ($p<0.01$). LWG= live weight gain.
Matlock et al.(2021)

Dairy Production

In a previous study, we evaluated the environmental effects of modifying dairy rations (Kim et al. 2019). **Table 4** presents the scenarios evaluated in simulated using the Integrated Farm System Model (Rotz et al. 2015). Figure 3 highlights the importance of full system evaluation when considering the impact of changes in livestock rations. Note that the contribution of the production of the feed itself (the cradle to farm gate emissions associated with delivering the feed to the animal) is generally larger for the alternative ration formulations for both the New York and Wisconsin dairies (red horizontal lines). However, the overall emissions associated with milk production, except for the increased corn silage scenario remain the same or are reduced. This is the result of the effect that the alternate rations have on animal performance, specifically enteric methane emissions and to a lesser extent manure-related emissions. It should be noted that the economics of these ration modifications has not been accounted in this evaluation and therefore a full sustainability assessment must always include a multi-criteria evaluation that fully accounts for implementation cost.

Table 4. Alternate feed management approaches for dairy production systems.

Scenario	Description
High corn silage	Increase corn silage to alfalfa/grass ratio from 1:1 to 3:1 in animal diets
Low forage	Forage rations reduced from 65% to 50% of (DMI) fed to lactating cows with cropland adjusted to provide increased corn grain and undegradable crude protein
High digestibility	NDF digestibility of feeds is increased by 2% from the baseline.
Increase FCR	Increase feed conversion efficiency from 1.5 to 1.65 kg milk/kg feed DMI
Low protein	Reduce protein content of lactating cows' ration from 17% to 14%
Increase fat content	Increase supplemental fat in the diet of lactating cows from 0.4 to 0.9 kg/day/cow

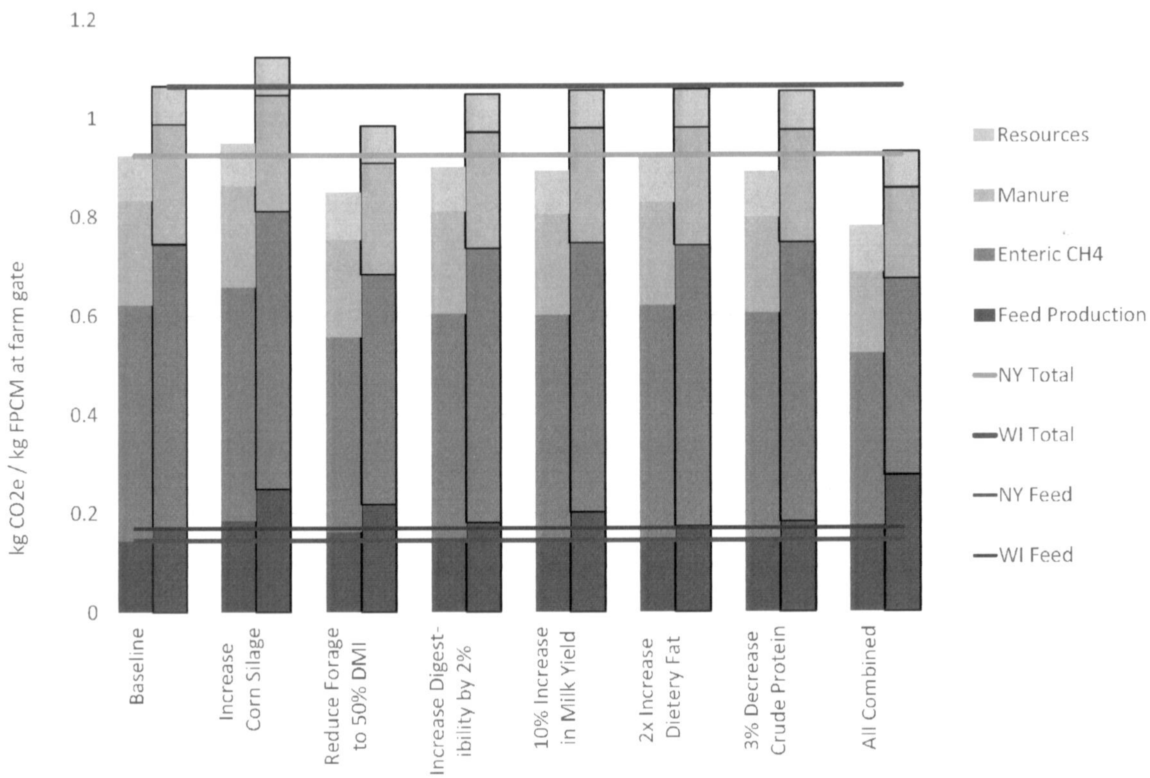

Figure 4. Greenhouse gas emissions reductions as a function of ration formulation for New York and Wisconsin dairy production systems

Conclusions

In summary, feed formulation and the use of additives/supplements is a promising avenue for improving the sustainability of livestock production systems. The primary caution is that the full system response should be accounted. That is, animal performance with modified diets must be included because alternate feed formulations may result in higher emissions for production of the feed, but lower overall emissions at the system level. Optimization of separate phases of the system will not always lead to an optimized system with improved sustainability characteristics.

Nutritional Strategies to Mitigate Enteric Methane Emissions from Dairy Cows: State of Knowledge and New Perspectives

Stratégies nutritionnelles pour atténuer les émissions entériques de méthane par les vaches laitières: État des connaissances et nouvelles perspectives

Chaouki Benchaar
Agriculture and Agri-Food Canada, Sherbrooke Research and Development Centre, Sherbrooke, QC
chaouki.benchaar@agr.gc.ca

Sponsored by / Commandité par :

CANOLAMAZ!NG
Discover the **amazing** value of canola meal

Abstract

Like any livestock production system, dairy production faces a major challenge, namely, to be environmentally sustainable while maintaining and/or enhancing animal productivity to ensure farms competitiveness and to provide the consumers with safe and high-quality products. The dairy sector contributes to greenhouse gas (GHG) emissions, mainly through the production of methane (CH_4) gas from enteric fermentation. The global warming potential of CH_4 is 28 times that of carbon dioxide. In addition, enteric CH_4 is also a loss of productive energy for lactating dairy cows (4 to 7% of gross energy intake). Thus, mitigation of enteric CH_4 is beneficial from both nutritional and environmental standpoints. Accordingly, several dietary strategies have been suggested to mitigate enteric CH_4 production. These strategies vary in terms of their effect (i.e., direct or indirect) on ruminal methanogenesis and the extent of CH_4 inhibition (i.e., low, moderate, high). Overall, individual dietary interventions have low to moderate (5 to 20%) mitigation effect with the exception of 3-nitrooxypropanol and red seaweed (e.g., *Asparagopsis taxiformis*) for which up to 40% decreases have been reported. Adding lipids (unsaturated) can also significantly reduce (up to 25%) enteric CH_4. However, at high inclusion level (> 4% of diet dry matter), animal productivity may be impaired, particularly when lipids are added in high-starch diets. It has been suggested that combining mitigation strategies with relatively small decrease potentials may allow to achieve larger reductions. However, this will be only achieved if the effects of the combined strategies are additive. Regardless of the type of the dietary intervention, it is important to ensure that the gain achieved via the reduction in enteric CH_4 is not offset by increased emissions elsewhere in the farming system (e.g., manure). The adoption of any mitigation strategy by dairy producers would only be possible if it is accompanied by an increase in milk production. Low-CH_4 diets are not usually low-cost and therefore financial incentives are needed to motivate producers to adopt mitigation. Consumers have a negative perception towards the use of feed antibiotics and chemical additives in dairy cow diets and therefore, alternatives to these substances (e.g., plant-extracts) are needed. The objective of this paper is not to discuss all dietary mitigation strategies available to date, but rather focusing on the potential of specific options not only on enteric CH_4 emissions, but also their possible impact on CH_4 emissions from manure and other GHG (e.g. N_2O).

Key words: enteric methane, mitigation, nutrition, dairy cow

Résumé

À l'instar de tout système de production animale, la production laitière est confrontée à un défi majeur, à savoir être écologiquement durable tout en maintenant et/ou en améliorant la productivité pour assurer la compétitivité des exploitations et fournir aux consommateurs des produits sûrs et de haute qualité. Le secteur laitier contribue aux émissions de gaz à effet de serre, principalement par la production de gaz méthane (CH4) issu de la fermentation entérique. Le potentiel réchauffement climatique du CH_4 est 28 fois supérieur à celui du dioxyde de carbone. De plus, le CH_4 entérique est également une perte d'énergie productive (4 à 7 % de l'apport énergétique brut) pour les vaches en lactation. Par conséquent, l'atténuation du CH_4 entérique serait bénéfique à la fois d'un point de vue nutritionnel et environnemental. Plusieurs stratégies alimentaires ont été suggérées pour atténuer la production de CH_4 entérique. Ces stratégies varient en fonction de leur effet (i.e., direct ou indirect) sur la méthanogenèse ruminale et le degré d'inhibition du CH_4 (i.e., faible, modéré, élevé). Dans l'ensemble, les interventions alimentaires individuelles ont un effet d'atténuation faible à modéré (5 à 20 %) à l'exception du 3-nitrooxypropanol et des algues rouges (e.g., *Asparagopsis taxiformis*) pour lesquels des diminutions allant jusqu'à 40 % ont été obsevées. L'ajout de lipides (insaturés) peut également réduire de façon importante (jusqu'à 25 %) les émissions de CH_4 entérique. Cependant, à un niveau d'inclusion élevé (> 4 % sur une base de matière sèche), la productivité animale peut être altérée, en particulier lorsque des lipides sont ajoutés dans des rations riches en amidon. Il a été suggéré que la combinaison de stratégies d'atténuation avec des potentiels de diminution relativement faibles pourrait permettre d'obtenir des réductions plus importantes. Cependant, cet objectif ne sera atteint que si les effets des stratégies combinées s'additionnent. Quel que soit le type d'intervention alimentaire, il est important de s'assurer que le gain obtenu via la réduction du CH_4 entérique n'est pas annulé par une augmentation des émissions ailleurs dans le système agricole (par exemple, le fumier). L'adoption d'une stratégie d'atténuation par les producteurs laitiers ne serait possible que si elle s'accompagnait d'une augmentation de la production laitière. Les régimes faibles émissions de CH_4 ne sont généralement pas bon marché et des incitations financières sont donc nécessaires pour encourager les producteurs à adopter des mesures d'atténuation. Les consommateurs ont une perception négative de l'utilisation d'antibiotiques alimentaires et d'additifs chimiques dans l'alimentation des vaches laitières et, par conséquent, des alternatives à ces substances (par exemple, des extraits de plantes) sont nécessaires. L'objectif de cet article n'est pas de discuter toutes les stratégies d'atténuation alimentaires disponibles à ce jour, mais plutôt de se concentrer sur le potentiel d'options spécifiques non seulement sur les émissions de CH_4 d'origine entérique, mais aussi sur leur impact possible sur les émissions de CH_4 provenant du fumier et d'autres GES (par exemple N_2O).

Mots clés: Méthane entérique, atténuation, nutrition, vache laitière

Introduction

Ruminants play a crucial role in food security. They supply 51% of all protein from the livestock sector; of which 67 and 33% are from milk and meat, respectively (Gerber et al., 2013). For many populations, livestock is a primary source of nutrition, and not simply a source of calories. Ruminants have the digestive particularity of being able to digest fibrous material (i.e., forages, agro-industrial by-products and crop residues) that is not edible to humans, and convert it into high-quality products (i.e., meat, milk). Thus, ruminants are able to valorize resources that would otherwise be wasted.

In 2018, The International Panel on Climate Change (IPCC) presented its Special Report on Global Warming of 1.5°C and concluded that "limiting warming to 1.5°C implies reaching net zero CO_2 emissions globally around 2050 with concurrent deep reductions in emissions, particularly those of CH_4". The report concluded that "24 to 27% reduction in methane (CH_4) is needed by 2050".

Ruminant production systems are an important source of anthropogenic CH_4. Reducing CH_4 emissions from ruminants is a hot topic because this greenhouse gas (GHG) contributes substantially to global warming. Consequently, there is an urgent need to reduce the concentration of CH_4 in the atmosphere in order to contribute to slowing down the global warming of the planet. All sectors, including the agricultural sector, are moving towards reducing their carbon footprints and words like "zero emission", "carbon neutral", or "low carbon economy" among others, are part of our daily reality.

The Agriculture GHG emissions accounted for 55 Mt, or 8.2% of total GHG emissions for Canada in 2020. Agriculture accounted for 30% of national CH_4 emissions and 75% of national N_2O emissions (ECCC, 2022). In 2020, emissions from enteric fermentation accounted for 43% of total agricultural emissions, and the application of inorganic nitrogen fertilizers accounted for 21% of total agricultural emissions (ECCC, 2022). According to ECCC (2022), 90% of CH_4 emissions are from enteric fermentation and the remaining 10% is from manure storage and management. Emissions from enteric fermentation originate almost entirely (96%) from cattle production. Beef cattle are the main contributor to these emissions (81%) followed by dairy cattle (15%), and other species (5%).

Methane has a global warming potential (GWP) 28 times higher than CO_2, when compared over a 100-yr period (Forster et al., 2007). However, CH_4 has a much shorter (12 years) lifetime than CO_2 (hundred years) in the atmosphere (Forster et al., 2007). This difference makes CH_4 an attractive target for short-term gains in global warming reduction.

"Cradle-to-farm gate" Life Cycle Assessments (LCA) of milk produced in confinement systems in Canada revealed that approximately 40 to 50% of the carbon footprint of milk is from enteric fermentation (Guyader et al., 2017; Little et al., 217; Holtshausen et al, 2021). Therefore, mitigation of enteric CH_4 from dairy cows could play an important role in stabilizing and reducing GHG emissions from the dairy sector.

In addition to be a potent GHG, CH_4 produced from enteric fermentation represents a loss of energy to ruminants (2 to 12% of gross energy intake; Johnson and Johnson, 1995). These losses vary from 4 to 7% in lactating dairy cows fed forage-based diets (Kebreab et al., 2008). Theoretically, a reduction in energy losses in the form of CH_4 should result in an improvement in animal productivity because the recovered energy would be used for production purposes (e.g., milk, meat).

Based on these considerations and given that enteric CH_4 is a not only a potent GHG but also of a loss of productive energy to the ruminant, mitigating enteric CH_4 emissions from dairy cows has both long-term environmental and short-term economic benefits. Accordingly, several dietary and nutritional strategies have been suggested to mitigate enteric CH_4 emissions from dairy cows.

Several comprehensive reviews on dietary options to mitigate enteric CH_4 production have been developed over the years (e.g., Hristov et al., 2013; Beauchemin et al., 2020). The objective of this paper is not to review all existing enteric CH_4 mitigation strategies, but rather focusing on the potential of specific mitigation options, including those recently investigated.

Enteric methane production

Methane arises primarily from enteric fermentation in the rumen and to a lesser extent in the hindgut. Murray et al. (1976) reported that approximately 87% of the CH_4 exhaled from the mouth and nose of the animal originates from the forestomach via eructation and absorption into the blood. Approximately 13% of CH_4 is produced in the hindgut, where 89% of that (11% of total CH_4 produced) is absorbed into the bloodstream and eliminated via expiration (Ricci et al., 2014). Contrarily to popular beliefs, CH_4 is not released through flatulence. In fact, 90-95% of the CH_4 produced is emitted via respiration and eructation and a very small amount (1-5%) is released from the rectum.

The key element in ruminal methanogenesis is H_2 (Figure 1). Rumen microbes ferment carbohydrates and to a lesser, proteins, to produce volatile fatty acids, mainly, acetate, propionate and butyrate. Production of acetate and butyrate liberates hydrogen, whereas propionate serves as a net hydrogen sink. Consequently, diets such as high-forage diets that increase acetate also increase CH_4 production because of the increasing availability of H_2 to methanogens. In contrast, diets that increase propionate production, such as high-grain diets are often associated with a reduction in ruminal CH_4 production, given that less H_2 is available to methanogens for reducing CO_2 to CH_4.

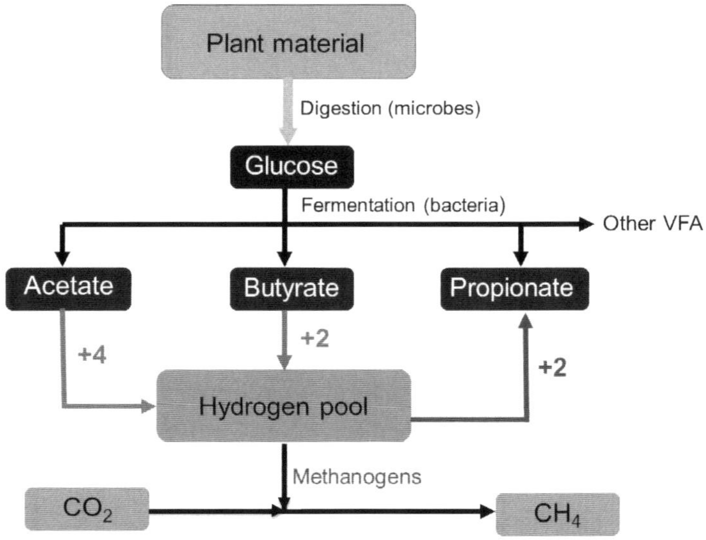

Figure 1. Production of methane in the rumen

Forage utilization

Forages are an integral part of dairy cow rations as they can represent 50-90% of the diet. Forages are important for the cow's digestive and animal health. They supply energy and protein for milk synthesis, thereby reducing costs of producing milk.

Cereal-forages (e.g., corn or barley silages) contain high starch concentrations, which favors the production of propionate over that of acetate and reduces CH_4 production in the rumen. Furthermore, intake of cereal-forages is often greater than that of legume/grass forages, which reduces ruminal residence time and hence, restricts ruminal fermentation and promotes post-ruminal digestion. Hassanat et al. (2013) reported lower CH_4 yield [g/kg dry matter intake (DMI); percentage of gross energy intake (GEI)] and CH_4 emission intensity for cows fed corn-silage based diets versus alfalfa silage-based diets. However, replacing alfalfa silage with corn silage increased emissions of CH_4 from stored manure due to increased volatile solids (i.e., organic matter: OM) in manure as a consequence of an inhibition of ruminal degradation of NDF (Massé et al. (2016). In an LCA study, Little et al. (2017) demonstrated that despite the 10% decrease in enteric CH_4 yield observed by Hassanat et al. (2013) when replacing alfalfa silage with corn silage, differences in CO_2 emission intensity between the two forage systems were minimal when soil carbon was taken into account. Thus, recommending the use of high-starch forages to mitigate enteric CH_4 production must consider possible effects (i.e., increases) on GHG emissions elsewhere in the production system, and this can be addressed using the LCA approach.

Not all cereal-forages are equal in terms of their ability to reduce enteric CH_4 production. For instance, Benchaar et al. (2014) showed that replacing barley silage (starch = 14 %; NDF = 52.3%) with corn silage (starch = 32%; NDF = 36.7%) in dairy cow diets decreased CH_4 yield (g/kg DMI) and CH_4 intensity (g/kg ECM) mainly because of increased DMI, ruminal propionate proportion, and milk production (Table 1). In the same study, urinary losses decreased as the proportion of corn silage increased in the diet at the expense of barley silage, suggesting a better N utilization and low potential emissions of ammonia and N_2O. With the increased availability of high yielding short season cultivars, corn silage may offer opportunity to reduce GHG emissions from dairy cows in Canada, but an LCA is necessary to determine the net emissions of GHG and the carbon footprint of milk.

Table 1. Barley silage (BS) versus corn silage (CS) in the diet of lactating dairy cows (Benchaar et al., 2014)

	Treatment			SEM	*P*	
	100% BS: 0%CS	50% BS: 50% CS	0%:BS 100% CS		Linear	Quadratic
DMI (kg/d)	21.9	24.9	26.2	0.66	<0.01	0.13
ECM (kg/d)	35.1	37.9	38.9	1.52	<0.01	0.29
Acetate:propionate	3.14	2.92	2.74	0.118	0.01	0.89
Methane						
g/d	487	540	523	17.2	0.11	0.07
g/kg of DMI	22.3	21.8	19.1	0.58	<0.01	0.07
% GE intake	6.60	6.49	5.67	0.174	<0.01	0.06
g/kg of ECM	14.0	14.4	13.5	0.51	0.33	0.15
Urinary N (% N intake)	38.8	34.3	33.4	0.83	<0.01	0.04

DMI = dry matter intake; ECM = Energy corrected milk; GE = gross energy

Choice of forage cultivars offers opportunities to improve forage quality and mitigate enteric CH_4 emissions. For example, compared with conventional corn silage (CCS) cultivar, brown midrib (BMR) corn silage is characterized by lower lignin concentration and higher rumen potentially digestible NDF (Ebling and Kung, 2004; Gehman et al., 2008). Hassanat et al. (2017) reported (Table 2) that replacing CCS with BMR in dairy cow diets increased DMI and milk production and lowered CH_4 yield (g/kg DMI; % GEI) and CH_4 emission intensity (g/kg ECM). Urinary N decreased with the replacement of CCS with BMCS, suggesting an enhanced efficiency of N utilization by the animal. Thus, the use of low-lignin cultivar may represent an option to mitigate enteric CH_4 emissions. However, field survivability of low-lignin forage cultivars and their ability to support higher milk production still need to be investigated.

Table 2. Conventional corn silage (CCS) versus brown midrib corn silage (BMCS) in the diet of lactating dairy cows (Hassanat et al., 2017)

	Treatment			
Item	CCS	BMCS	SEM	*P*
DMI (kg/d)	25.8	27.4	0.67	<0.01
ECM (kg/d)	37.5	40.4	1.60	<0.01
Methane				
g/d	483	470	21.4	0.43
g/kg DMI	18.4	17.2	0.62	0.05
% GE intake	5.52	5.16	0.186	0.05
g/kg ECM	13.1	11.8	0.62	0.02
Urinary N				
g/d	319	282	8.9	<0.01
% of N intake	48.1	41.3	0.95	<0.01

CCS: NDF = 34.3%; CP = 5.54%; lignin = 2.71%
BMCS: NDF = 34.4%; CP = 6.54%; lignin = 2.11%

At a similar physiological stage of maturity, legume forages contain less (NDF) and more non-structural carbohydrates than grasses. Thus, substituting grass forages with legume forages in dairy cows diets may represent an interesting means to mitigate enteric CH_4 production. Indeed, feeding legumes versus grasses increases DMI, which lowers CH_4 yield (g/kg DMI) because of faster passage rates from the rumen. In addition, very often feeding legumes versus grasses is associated with improved milk production, which is expected to lower enteric CH_4 emission intensity (g/kg milk). However, caution must be taken as plant maturity at the time of harvest can confound the impact of forage species (e.g., legumes versus grasses) on CH_4 production. Advancing maturity decreases the soluble carbohydrates content and increased lignification of plant cell walls, which promote the production of acetate in the rumen, thereby increasing the amount of CH_4 produced per unit of forage digested.

Other forage-based strategies such as using high-sugar grasses or tannin-containing forages (e.g., sainfoin, and birdsfoot trefoil) may have the potential to reduce CH_4 emissions (Guyader et al., 2016).

Feed additives

Several feed additives have been suggested to mitigate enteric CH_4 production. Among them, ionophores (e.g., monensin), plant bioactive compounds (e.g., condensed tannins, saponins, essential oils), yeast, direct-fed microbials, hydrogen sinks (e.g., nitrate), and inhibitors (e.g., 3-nitrooxypropanol: NOP).

3-nitrooxypropanol: NOP

In recent years, 3-NOP has attracted much attention as several studies have proven its effectiveness in reducing enteric CH_4 emissions from dairy cows. The 3-NOP is a structural analog to methyl-coenzyme M, a cofactor involved in the terminal step of ruminal methanogenesis. The 3-NOP is supposed to bind to the active site of the methyl-Co A reductase, causing an inhibition of CH_4 synthesis in the rumen. Hristov et al. (2015) observed a sustained (12 weeks) decrease (up to 30%) in emissions (g/d), yield (g/DMI), and intensity (g/kg ECM) of enteric CH_4 when 3-NOP was fed at 40, 60, and 80 mg/kg DM to lactating dairy cows. In that study, there were no effects on DMI, milk yield or milk composition. It has been reported (Thiel et al., 2019a, b) that 3-NOP is metabolized in the rumen to very low concentrations of nitrate, nitrite and 1,3-propanediol and presents low risks for human health (i.e., not detected residues in milk). To the best of our knowledge, 3-NOP is not currently commercially available in Canada because it is not approved by Health Canada, which considers inhibitors as a drug and not a feed additive.

Industrial production processes and transportation of feed additives may also contribute to GHG emissions and therefore, these emissions should be taken into account to determine the efficacy of a given feed additive to improve the carbon footprint (CO_2e/kg of milk) of milk production.

Seaweed (Macroalgae)

Recently, seaweeds (i.e., macroalgae) have been investigated for their potential to manipulate rumen microbial fermentation in a manner that inhibits ruminal methanogenesis. Of particular interest are red and brown algae. The main secondary metabolite produced by these algae species is bromoform ($CHBr_3$) and this compound has been shown to exhibit antimethanogenic properties (Machado et al., 2016). Bromoform interferes with methanogenesis by inhibiting the cobamide-dependent methyl transferase at the terminal step of the methanogenic pathway (Denman et al., 2007). Other seaweeds can contain polysaccharides, proteins, peptides, bacteriocins, lipids, phlorotannins, saponins, and alkaloids that have the potential to inhibit methanogenesis (Abbott et al., 2020). The concentration of $CHBr_3$ varies considerably depending on algae species (Table 3).

Table 3. Bromoform levels in brown, red and green seaweeds (from Abbott et al., 2020 based on Carpenter and Liss, 2000)

Seaweed species	$CHBr_3$ (ng/g/h fresh weight)
Brown	
Fucus vesiculosus	4.9
Fucus serratus	2.1
Ascophyllum nodosum	2.7
Laminaria digitata	49.7
Laminaria saccharina	32
Macrocystis pyrifera	125
Red	
Meristiella gelidium	25
Rhodymenia californica	47
Pterocladia capillacea	500
Cordllina officinalis	1.4–20
Gigartina stellata	4.1–26
Asparagopsis spp	3–1256
Chondrus crispus	0–1.3
Polysiphonia lanosa	2.1
Green	
Ulva intestinalis	87–192
Ulva linza	11
Ulva spp	150
Ulva spp. (formerly lactuca)	13.0–150
Cladophoria albida	0

Extensive research on the antimethanogenic effect of seaweed has been conducted in Australia with native red algae species such as *Asparagopsis taxiformis* and *Asparagopsis armata.* These seaweeds have been shown to decrease ruminal methanogenesis in vitro Kinley et al., 2016) and in vivo (Roque et al., 2019). Roque et al. (2019) reported that feeding *Asparagopsis armata* to dairy cows at 1% of the diet decreased enteric CH_4 production by 67%. However, in that study, feed intake and milk production dramatically decreased (11 and 38%, respectively) upon feeding the seaweed. Bromoform can potentially be harmful, particularly at high-concentrations and long-term oral exposure of animals to high concentrations of bromoform can cause liver and intestinal tumors (ATSDR, 2016). High levels of bromoform could pose risks to human health and Heath Canada (2020) considers this compound as a possible human carcinogen. In the study of Roque et al. (2019), milk bromoform concentrations of cows fed *Asparagopsis* were in the range of 0.11-0.15 µg/L, which much lower that the level of 100 µg/L set by Health Canada (2020) for drinking water. Because of high levels of inorganic minerals in *Asparagopsis*, accumulation of iodine and bromide in milk (Stefenoni et al., 2021). Thus, it would probably be necessary to process the algae in order to eliminate/reduce the concentration of inorganic minerals to achieve iodine concentrations in milk not exceeding the upper limits recommended by Health Canada.

Based on this information, seaweeds may represent an effective means to mitigate enteric CH_4 emissions from dairy cows. However, importing these algae into Canada requires farming, processing and shipping of the product, which will contribute to more GHG emissions and may, therefore, offset the gain achieved via the mitigation of enteric CH_4 emissions. Thus, seeking alternative Canadian seaweeds is likely to be beneficial. Also, more work is needed to determine the balance between the extent of reducing enteric CH_4, the cost, and safety of these supplements for animals and humans. Regardless of the source, the provenance (i.e., transporting), and the mode of production (i.e., harvesting, growing, processing, storing) of the algae, it is important to carry out an LCA to determine the net impact on the GHG intensity of milk production.

Biochar

Recently, biochar has been suggested as a means to reduce enteric CH_4 production. Biochar is a co-product obtained via pyrolysis by heating (350–600°C) plant biomass (agricultural or forestry) under oxygen-free or oxygen-limited conditions (Lehman and Joseph, 2015). The product obtained has a very porous structure, giving it a great capacity to absorb gases (e.g., CH_4) and liquids. Biochar may be beneficial for animal health due its detoxifying (Villalba et al., 2002), antidiarrheal (Watarai et al., 2008), and anthelmintic (Van et al., 2006) properties. Biochar has been reported to enhance biofilm formation and H_2 transfer among members within microbial communities (Chen et al., 2014). The effects in the rumen are highly dependent upon the biomass and condition of pyrolysis (e.g., temperature) used to produce the biochar, although in a recent in vitro study (Benchaar et al., unpublished), CH_4 production was not affected by biochar produced from different sources of biomass used to produce the biochar. The lack of an effect of biochar on enteric CH_4 production has also been reported in vivo (Terry et al., 2019). Most of the literature data available to date suggest that biochar is not a viable option to mitigate enteric CH_4 emissions from dairy cows (Honan et al., 2021).

Fat supplementation

There is a general consensus among the scientific community that diet supplementation with unsaturated fat is a potentially effective strategy for mitigating enteric CH_4 emissions (Beauchemin et al., 2020). A wide range of unsaturated fat supplements were evaluated for their potential to reduce enteric CH_4 production. Their effects are variable depending on many factors.

These include the level of fat supplementation, the form of the fat supplement (e.g., oils vs. seeds); seed processing (e.g., (extruded > whole seeds); the fatty acid (FA) composition of the supplement: medium (e.g.,12:0; 14:0) and long-chain FA (e.g., 18:3) very effective and the composition of the basal diet: high-grain diets (i.e., high-starch) are more responsive than high-forage diets.

Feeding high amounts of fat may impair diet digestibility and animal productivity. Beauchemin et al. (2020) suggested that a supplementation level of added fat lower than 4% of diet DM can reduce enteric CH_4 emissions by up to 20% without impairing animal productivity, although the authors warned that the effect may vary. Benchaar et al. (2015) showed that adding 4% linseed oil (LO) in a red clover silage-based diet reduced enteric CH_4 emissions and yield by 10%, with no negative effects on DMI, OM digestibility and milk production (Table 4). The same LO supplementation level in a corn silage-based diet markedly reduced (25%) enteric CH_4 emissions, but impaired DMI, OM digestibility and milk production. As a consequence of changes in the quantity and the composition of manure upon LO supplementation, CH_4 emissions from manure increased (Hassanat and Benchaar 2019).

Table 4. Including linseed oil (LO) at 4% of dietary DM in lactating dairy cows diets based on red clover or corn silage (Benchaar et al., 2015).

	Treatment				SEM	*P*		
	Red clover silage		Corn silage			LO	Silage	Inter.
	– LO	+LO	– LO	+LO				
DMI (kg/d)	35.8	37.7	35.1	34.2	1.59	0.56	0.02	0.13
ECM (kg/d)	37.5	38.1	37.9	32.7	1.57	<0.01	<0.01	<0.01
Milk fat (kg/d)	1.38	1.37	1.37	1.05	0.073	<0.01	<0.01	<0.01
Acetate:propionate	3.47	3.34	2.91	2.28	0.092	<0.01	<0.01	<0.01
NDF digestibility (%)	65.5	66.8	54.0	47.1	2.81	<0.01	<0.01	<0.01
Methane								
g/d	500	453	491	362	26.8	<0.01	<0.01	<0.01
g/kg DMI	20.6	18.9	20.1	16.1	0.83	<0.01	<0.01	0.03
% GE intake	6.07	5.43	6.13	4.70	0.236	<0.01	0.02	<0.01

Inter = LO × silage interaction; DMI = dry matter intake; ECM = energy-corrected milk.

One of the limitations of the use of fat to mitigate enteric CH_4 is the high cost. Alternative low-cost lipid sources are dry distillers' grain with solubles (DDGS), a by-product of the ethanol industry. Adding DDGS at 30% of dietary DM decreased enteric CH_4 yield (% GEI) by 14%, but adversely affected OM digestibility (Benchaar et al., 2013), which increased volatile solids in manure and increased (14%) fugitive CH_4 emissions from stored manure (Massé et al., 2014). Thus, the gain achieved in enteric CH_4 emission was completely offset by higher emissions from stored manure, suggesting that caution must be taken when using DDGS to mitigate enteric CH_4 emissions from dairy cows. A limitation to using DDGS to reduce CH_4 emissions is that it may contribute to increase nitrogen excretion, which could potentially increase ammonia and N_2O emissions (Benchaar et al., 2013).

Considering the potential impact of fat supplementation on CH_4 emission from manure, an LCA is necessary to determine the effectiveness of this mitigation option to improve the carbon footprint of milk.

Combination of mitigation strategies

It has been suggested that combining mitigation strategies may allow to achieve larger reductions (Beauchemin et al., 2020). However, this will be only achieved if the effects of the combined strategies are additive. The effectiveness of combining individual strategies may be further increased if the strategies have different mode of action (e.g., direct and indirect effects). Additive effects between fat and starch supplementation were observed (Benchaar et al., 2015; Hassanat and Benchaar, 2021). Additive effects of diet supplementation with fat on CH_4 mitigation were reported when canola oil was combined with 3-NOP (Zhang et al., 2021) and linseed oil with nitrate (Guyader et al., 2015). There is a lack of information about the effect of combining more than two dietary strategies to mitigate enteric CH_4 emissions from dairy cows. It is important to ensure that the combination of individual enteric CH_4 strategies does not have a negative effect on animal productivity and does not lead to increased GHG emissions (e.g., N_2O).

Perspectives/Conclusions

A number of dietary strategies have been suggested to mitigate enteric CH_4 production from dairy cows. Fat supplementation, forages (i.e., legumes vs. grasses; cereal forages versus legumes/grasses; corn silage vs. barley silage; more digestible forages: BMR vs. regular corn), seaweeds (i.e., red); and 3-NOP have been shown to be effective mitigation option. However, focusing only on suppressing enteric CH_4 production would not ensure that the carbon footprint of milk would be improved because DMI, nutrient digestibility and animal productivity may be impaired when high levels of fat, starch, or seaweed are included in the diet. Manure CH_4 emissions and other potential emissions (e.g., N_2O, ammonia) may also increase if the implementation of these strategies increases volatile solids and/or change the chemical composition of manure.

Production of forages is greatly affected by several factors, including the climate conditions (e.g., rainfall, water availability, temperature), the agronomic conditions (e.g., soil quality, use of fertilizers), management practices (e.g. harvest, preservation), soil carbon sequestration. Such factors affect not only the yield/quality of forages, animal productivity, and manure nutrient excretion but also GHG emissions. The use of forages to mitigate enteric CH_4 emissions needs to account for net GHG emissions, including soil carbon sequestration in the farming system.

The use of seaweeds, particularly the red ones (e.g., *Asparagopsis*) may represent an opportunity. However, these algae species are non-native to Canada and importing them from countries as far away as Australia, would not only increase the cost of their use, but would contribute to increased GHG emissions associated with processing, packing and transportation. The production at large scale of this type of seaweed in Canada requires their cultivation, growth, harvest, processing, storage, etc., which may increase sale prices. The identification of native seaweeds that contain required concentrations of bromoform to inhibit ruminal methanogenesis may represent a viable option if it is cost-effective. Because bromoform can be harmful (carcinogenic) at high concentrations, more work is warranted to provide more information on the extent of its transfer into milk and meat.

The feed additive 3-NOP is apparently very effective in mitigating enteric CH_4, but its use could be limited because of its high cost and its lack of improving animal productivity (i.e., milk production). Therefore, financial incentives (e.g., governmental) may be necessary to encourage its adoption by the dairy farmers. The environmental impact associated with manufacturing feed additives and their use along the supply chain for dairy cows must be also considered when evaluating the potential of a given feed additive to reduce enteric CH_4 emissions. This can be addressed using LCA based on guidelines developed for this purpose by the FAO (2020).

Combination of individual strategies with relatively small potential mitigation may help to achieve larger decreases in enteric CH4 emissions. However, as for any diet intervention, a LCA is necessary to account all emissions from the dairy production system.

The adoption of any mitigation strategy by dairy producers would only be possible if it is accompanied by an increase in milk production. Low-CH_4 diets are not usually low-cost and therefore financial incentives are needed to motivate producers to adopt mitigation. Consumers have a negative perception towards the use of feed antibiotics and chemical additives in dairy cow diets and therefore, alternatives to these substances (e.g., plant-extracts) are needed.

It is difficult to indicate the extent of decrease in enteric CH_4 emissions for a given dietary mitigation because there is no "one-size-fits-all approach" to emissions reduction. Several mitigation approaches exist, being developed and others will be developed. This will allow the agriculture sector to play an important role in curbing GHG emissions, thereby helping Canada to achieve Canada to 30% reduction by 2030 and a pledge to become net-zero by 2050.

References

Abbott, D.W., I.M. Aasen, K.A. Beauchemin, F. Grondahl, R. Gruninger, M. Hayes, S. Huws, D. A. Kenny, S. J. Krizsan, S. Kirwan, V. Lind, U. Meyer, M. Ramin, K. Theodoridou, D. von Soosten, P. Walsh, S. Waters and X. Xing. 2020. Seaweed and seaweed bioactives for mitigation of enteric methane: Challenges and opportunities. Animals 10 2432.

Agency for Toxic Substances and Disease Registry (ATSDR). 2016. https://www.epa.gov/sites/default/files/2016-09/documents/bromoform.pdf.

Beauchemin, K.A., E.M. Ungerfeld, R.J. Eckard and M. Wang. 2020. Fifty years of research on rumen methanogenesis: lessons learned and future challenges for mitigation. Animal 14(S1) s2–s16.

Benchaar, C., F. Hassanat, R. Martineau and R. Gervais. 2015. Linseed oil supplementation to dairy cows fed diets based on red clover silage or corn silage: Effects on methane production, rumen fermentation, nutrient digestibility, N balance, and milk production. J. Dairy Sci. 98 7993-8008.

Benchaar, C., F. Hassanat, R. Gervais, , P.Y. Chouinard, H.V. Petit and D.I. Massé. 2014. Methane production, digestion, ruminal fermentation, nitrogen balance, and milk production of cows fed corn silage or barley silage based diets. J. Dairy Sci. 97 961–974.

Benchaar, C., F. Hassanat, R. Martineau and R. Gervai2015. Linseed oil supplementation to dairy cows fed diets based on red clover silage or corn silage: Effects on methane production, rumen fermentation, nutrient digestibility, nitrogen balance, and milk production. J. Dairy Sci. 98 7993-8008.

Carpenter, L.J. and P.S Liss. 2000. On temperate sources of bromoform and other reactive organic bromine gases. J. Geophys. Res. 105 20539–20547**.**

Chen, S., A-E. Rotaru, P.M. Shrestha, N.S. Malvankar, F. Liu, W. Fan, K. P. Nevin and D.R. Lovley. 2014. Promoting interspecies electron transfer with biochar. Sci. Rep. 4 5019.

Denman, S.E., N.W. Tomkins and C.S. Mcsweeney. 2007. Quantitation and diversity analysis of ruminal methanogenic populations in response to the antimethanogenic compound bromochloromethane. FEMS Microbiol. Ecol. 62 313-322.

Ebling, T. L. and L. Kung, Jr. 2004. A comparison of processed conventional corn silage to unprocessed and processed brown midrib corn silage on intake, digestion, and milk production by dairy cows. J. Dairy Sci. 87 2519-2526.

Environment and Climate Change Canada. 2022. National Inventory Report 1990-2020. Greenhouse Gas Sources and Sinks in Canada. Canada's submissions to the United Nation Framework Convention on Climate Change. Available at: https://www.canada.ca/fr/environnement-changement-climatique/services/changements-climatiques/emissions-gaz-effet-serre/inventaire.html.

FAO (Food and Agriculture Organization of the United Nations). 2020. Environmental performance of feed additives in livestock supply chains – Guidelines for assessment – Version 1. Livestock Environmental Assessment and Performance Partnership (FAO LEAP), Rome, Italy.

Forster, P., V. Ramaswamy, P. Artaxo, T. Berntsen, R. Betts, D.W. Fahey, J. Haywood, J. Lean, D.C. Lowe, G. Myhre, J. Nganga, R. Prinn, G. Raga, M. Schulz and R. Van Dorland. 2007. Changes in Atmospheric Constituents and in Radiative Forcing. In: Climate Change 2007: The Physical Science Basis. Contribution of Working Group I to the Fourth Assessment Report of the Intergovernmental Panel on Climate Change [Solomon, S., D. Qin, M. Manning, Z. Chen, M. Marquis, K.B. Averyt, M. Tignor and H.L. Miller (eds.)]. Cambridge University Press, Cambridge, United Kingdom and New York, NY, USA.

Gehman, A.M., P.J. Kononoff, C.R. Mullins and B.N. Janicek. 2008. Evaluation of nitrogen utilization and the effects of monensin in dairy cows fed brown midrib corn silage. J. Dairy Sci. 91 288-300.

Gerber, P.J., H. Steinfeld, B. Henderson, A. Mottet, C. Opio, J. Dijkman, A. Falcucci and G. Tempio. 2013. Tackling Climate Change through Livestock: A Global Assessment of Emissions and Mitigation Opportunities. Food and Agriculture Organization of the United Nations, Rome, Italy.

Guyader, J., M. Eugène, B.M. Doreau, D.P. Morgavi, M. Silberberg, Y. Rochette, C. Gerard, C. Loncke and C. Martin. 2015. Additive methane-mitigating effect between linseed oil and nitrate fed to cattle. J. Anim. Sci. 93 3564–3577.

Guyader, J., H.H. Janzen, R. Kroebel, and K.A. Beauchemin. 2016. Forage use to improve environmental sustainability of ruminant production. J. Anim. Sci. 94 3147–3158,

Guyader, J., S. Little, R. Kröbel, C. Benchaar and K.A. Beauchemin. 2017. Comparison of greenhouse gas emissions from Canadian dairy production systems using corn or barley silage. Agric. Syst. 152 38-46.

Hassanat, F. and C. Benchaar. 2019. Methane emissions of manure from dairy cows fed red clover-or corn silage-based diets supplemented with linseed oil. J. Dairy Sci. 102 11766–11776.

Hassanat, F., and C. Benchaar. 2021. Corn silage-based diet supplemented with increasing amounts of linseed oil: Effects on methane production, rumen fermentation, nutrient digestibility, N utilization, and milk production of dairy cows. J. Dairy. Sci. 104 5375-5390.

Hassanat, F., R. Gervais and C. Benchaar. 2017. Methane production, ruminal fermentation characteristics, nutrient digestibility, nitrogen excretion, and milk production of dairy cows fed conventional or brown midrib corn silage. J. Dairy Sci. 100 2625-2636.

Hassanat, F., R. Gervais, C. Julien, P.Y. Chouinard, D.I. Massé, A. Lettat, H.V. Petit and C. Benchaar. 2013. Replacing alfalfa silage with corn silage in dairy cow diets: Effects on enteric methane production, ruminal fermentation, digestion, N balance, and milk production. J. Dairy Sci. 96 4553-4567.

Health Canada. 2020. Guidelines for Canadian Drinking Water Quality—Summary Table. Water and Air Quality Bureau, Healthy Environments and Consumer Safety Branch, Health Canada, Ottawa, Ontario. Available at: https://www.canada.ca/content/dam/hc-sc/migration/hc-sc/ewh-semt/alt_formats/pdf/pubs/water-eau/sum_guide-res_recom/summary-table-EN-2020-02-11.pdf.

Holtshausen, L., C. Benchaar, R. Kröbel and K.A. Beauchemin. 2021. Canola meal versus soybean meal as protein supplements in the diets of lactating dairy cows affects the greenhouse gas intensity of milk. Animals 2021, 11 1636.

Honan M., X., Feng, J.M. Tricarico and E. Kebreab. 2021. Feed additives as a strategic approach to reduce enteric methane production in cattle: modes of action, effectiveness and safety. Anim. Prod. Sci. Available at: https://doi.org/10.1071/AN20295.

Hristov, A. N., J. Oh, J.L. Firkins, J. Dijkstra, E. Kebreab, G. Waghorn, H.P. Makkar, A.T. Adesogan, W. Yang, C. Lee, P.J. Gerber, B. Henderson and J.M. Tricarico. 2013. Mitigation of methane and nitrous oxide emissions from animal operations: I. A review of enteric methane mitigation options. J. Anim. Sci. 91 5045–5069.

Hristov, A.N., J. Oh, F. Giallongo, T.W. Frederick, M.T. Harper, H.L. Weeks, A.F. Branco, P.J. Moate, M.H. Deighton, S.R.O. Williams, M. Kindermann and S. Duval. 2015. An inhibitor persistently decreased enteric methane emission from dairy cows with no negative effect on milk production. Proc. Natl. Acad. Sci. 112 10663–10668.
Identification of bioactives from the red seaweed *Asparagopsis taxiformis* that promote antimethanogenic activity in vitro. J. Appl. Phycol. 28 3117-3126.

IPCC (International Panel on Climate Change). 2018. Global Warming of 1.5°C. An IPCC Special Report on the impacts of global warming of 1.5°C above pre-industrial levels and related global greenhouse gas emission pathways, in the context of strengthening the global response to the threat of climate change,sustainable development, and efforts to eradicate poverty [Masson-Delmotte, V., P. Zhai, H.-O. Pörtner, D. Roberts, J. Skea, P.R. Shukla, A. Pirani, W. Moufouma-Okia, C. Péan, R. Pidcock, S. Connors, J.B.R. Matthews, Y. Chen, X. Zhou, M.I. Gomis, E. Lonnoy, T. Maycock, M. Tignor, and T. Waterfield (eds.)]. Available at: https://www.ipcc.ch/sr15/download/.

Johnson, K.A. and D. E. Johnson. 1995. Methane emissions from cattle. J. Anim. Sci. 73 2483-2492.

Kebreab, E., K.A. Johnson, S.L. Archibeque, D. Pape and T. Wirth. 2008. Model for estimating enteric methane emissions from United States dairy and feedlot cattle. J. Anim. Sci. 86 2738–2748

Kinley, R.D., R. de Nys, M.J. Vucko, L. Machado and N.W. Tomkins. 2016. The red macroalgae *Asparagopsis taxiformis* is a potent natural antimethanogenic that reduces methane production during in vitro fermentation with rumen fluid. Anim. Prod. Sci. 56 282-289.

Lehman, J. and S. Joseph. 2015. Biochar for Environmental Management: Science, Technology and Implementation. Routledge, New York, USA.

Little S.M., C. Benchaar, H.H. Janzen, R. Kröbel, E.J. McGeough and K.A. Beauchemin. 2017. Demonstrating the effect of forage source on the carbon footprint of a Canadian dairy farm using whole-systems analysis and the Holos model: alfalfa silage vs. corn silage. Climate 5 87.

Machado, L., M. Magnusson, N.A. Paul, R.D. Kinley, R. de Nys and N. Tomkins. 2016.

Massé, D. I., G. Jarret, F. Hassanat, C. Benchaar and N.M. Cata Saady. 2016. Effect of increasing levels of corn silage in an alfalfa-based dairy cow diet and of manure management practices on manure fugitive methane emissions. Agric. Ecosyst. Environ. 221 109–114.

Massé, D.I., G. Jarret, C. Benchaar and N. Cata Saady. 2014. Effect of Corn Dried Distiller Grains with Soluble (DDGS) in Dairy Cow Diets on Manure Bioenergy Production Potential. Animals 4(1) 82-92.

Murray, P.J., A. Moss, D.R. Lockyer and S.C Jarvis. 1999. A Comparison of Systems for Measuring Methane Emissions from Sheep. J. Agric. Sci. 133 439-444.

Ricci, P., M.G.G. Chagunda, J. Rooke, J.G Houdijk, C.-A. Duthie, J. Hyslop, R. Roehe and A. Waterhouse. 2014. Evaluation of the laser methane detector to estimate methane emissions from ewes and steers. J. Anim. Sci. 92 5239–5250.

Roque, B. M., J. K. Salwen, R. Kinley and E. Kebreab. 2019. Inclusion of *Asparagopsis armata* in lactating dairy cows' diet reduces enteric methane emission by over 50 percent. J. Clean. Prod. 234 132–138.

Stefenoni, H.A., S.E. Räisänen, S. F. Cueva, D. E. Wasson, C. F. A. Lage, A. Melgar, M. E. Fetter, P. Smith, M. Hennessy, B. Vecchiarelli, J. Bender, D. Pitta, C. L. Cantrell, C. Yarish, and A. N. Hristov. 2021. Effects of the macroalga *Asparagopsis taxiformis* and oregano leaves on methane emission, rumen fermentation, and lactational performance of dairy cows J. Dairy Sci. 104 4157–4173.

Terry, S.A., G.O. Ribeiro, R.J. Gruninger, A.V. Chaves, K.A. Beauchemin, E. Okine and T.A. McAllister. 2019. A pine enhanced biochar does not decrease enteric CH_4 emissions, but alters the rumen microbiota. Front. Vet. Sci. 6 1-12.

Thiel, A., R. Rümbeli, P. Mair, H. Yeman and P. Beilstein. 2019a. 3-NOP: ADME studies in rats and ruminating animals. Feed Chem. Tox. 125 528-539.

Thiel, A., A.C.M. Schoenmakers, I.A.J. Verbaan, E. Chenal, S. Etheve and P. Beilstein. 2019b. 3-NOP: mutagenicity and genotoxicity assessment. Feed Chem. Tox. 123 566–573.

Van, D.T.T., N.T. Mui and I. Ledin. 2006. Effect of method of processing foliage of A*cacia mangium* and inclusion of bamboo charcoal in the diet on performance of growing goats. Anim. Feed Sci. and Technol. 130 242-256.

Villalba, J.J., F.D. Provenza and R.E. Banner. 2002. Influence of macronutrients and activated charcoal on intake of sagebrush by sheep and goats. J. Anim. Sci. 2002. 80 2099-2109.

Watarai, S., Tana and M. Koiwa. 2008. Feeding activated charcoal from bark containing wood vinegar liquid (nekka-rich) is effective as treatment for cryptosporidiosis in calves. J. Dairy. Sci. 91 1458-63.

Zhang, X.M., M.L. Smith, R.J. Gruninger, L. Kung Jr., D. Vyas, S.M. McGinn, M. Kindermann, M. Wang, Z.L. Tan and K.A. Beauchemin. 2021. Combined effects of 3-nitrooxypropanol and canola oil supplementation on methane emissions, rumen fermentation and biohydrogenation, and total tract digestibility in beef cattle. J. Anim. Sci. 99 1–10.

Understanding Feed Efficiency in the Feedlot

Comprendre l'efficacité alimentaire au parc d'engraissement

Katie M. Wood[1]
[1]*Associate Professor, Department of Animal Biosciences, University of Guelph, Ontario, Canada. kwood@uoguelph.ca*

Sponsored by / Commandité par :

MORE THAN JUST FEED

Abstract

With near record high commodity prices, the importance of feed efficiency in feedlot production is more economically relevant than ever before. However, feed efficiency is a complex trait which can be influenced by a variety of factors. Cattle type (British vs Continental vs Dairy) and genetic merit, growth promotants, and ionophores also have been shown to improve feedlot efficiency measurements. However, a number of dietary factors also can improve feedlot efficiency, including increasing energy density of the diet, advancements in grain processing (steam-flaking vs rolled grains), and better understanding of nutritional components of the diet. In a recent meta-analysis, we found that using uNDF was more accurately able to predict performance and carcass traits than when using NDF. In addition, one of the challenges in efficient feedlot management is determining the optimum time to market cattle. With the increasing trend towards marketing cattle at increasingly heavier carcass weights, there is a trade-off with efficiency of feed conversion. Although, this reduction in feed efficiency is partially attributed to the differences in the composition of gain, there may also be metabolic differences contributing to reduction in feed efficiency in finishing cattle. In a serial slaughter experiment, we found linear decreases with metabolic markers associated with impaired glucose utilization efficiency over the finishing period. In addition, these results were not impacted when starch in the ration was partially replaced with fat, suggesting these differences are metabolically driven. In this paper, these and other recent advances in improving feed efficiency will be discussed.

Key words: feedlot, feed efficiency, RFI, feed conversion

Résumé

Considérant que les prix des matières premières n'ont jamais été aussi élevés, l'importance de l'efficacité alimentaire dans la production en parc d'engraissement est plus pertinente que jamais sur le plan économique. Cependant, l'efficacité alimentaire est un paramètre complexe qui peut être influencé par divers facteurs. Il a été démontré que le type de bovin (britannique, continental ou laitier) et leur valeur génétique, les stimulateurs de croissance et les ionophores améliorent également les mesures de l'efficacité alimentaire. Cependant, un certain nombre de facteurs alimentaires peuvent aussi améliorer l'efficacité des parcs d'engraissement, notamment l'augmentation de la densité énergétique du régime alimentaire, les progrès réalisés dans le traitement des grains (floconnage à la vapeur ou grains roulés) et une meilleure compréhension des composants nutritionnels du régime alimentaire. Dans une récente méta-analyse, nous avons constaté que l'utilisation des uNDF permettait de mieux prévoir les performances et les caractéristiques de carcasse que les NDF. En outre, une des difficultés que présente la gestion efficace des parcs d'engraissement est de déterminer le moment optimal pour commercialiser les animaux. Considérant la tendance actuelle à commercialiser les bovins à des poids de carcasse de plus en plus élevés, on doit accepter un certain recul d'efficacité alimentaire. Bien que cette réduction de l'efficacité alimentaire soit partiellement attribuable aux différences dans la composition du gain, des écarts métaboliques peuvent aussi contribuer au phénomène chez les bovins en finition. Dans une expérience d'abattage en série, nous avons observé des diminutions linéaires avec des marqueurs métaboliques associés à la dégradation de l'efficacité de l'utilisation du glucose au cours de la période de finition. De plus, ces résultats n'ont pas changé lorsque l'amidon de la ration a été partiellement remplacé par du gras, ce qui suppose que ces écarts sont d'origine métabolique. Dans cet article, ces résultats et d'autres percées récentes portant sur l'amélioration de l'efficacité alimentaire seront abordés.

Introduction

Near record high feed, fuel, and fertilizer prices, and high inflation rates all have put increasing economic burden on beef producers. The increased cost of feeding cattle means that increasing feed efficiency is critical to feedlot production. Historically, feed costs represent about 70% of the total cost of production for feedlot operators and therefore, feed efficiency has 4 x greater economic return when compared to the same improvement in feedlot growth (Gibb and McAllister, 1999). In addition to economic benefits, improved efficiency of production also has some positive environmental benefits, including lowered greenhouse gas emissions, and reduced water demand and manure output (Terry et al., 2020). This paper will investigate concepts in feed efficiency as it pertains to feedlot productions systems, as well as discuss some practical approaches which can be used to improve feed efficiency on commercial operations.

What is feed efficiency?

One of the challenges with feed efficiency it that it can refer to a broad scope of definitions, where efficiency can be used to describe impacts on everything from whole-farm systems to individual animal differences. More generally, feed efficiency relates to the ratio between production system inputs (usually feed) to system outputs (usually growth or gain), but refinements can be made to this definition for systems approaches, individual animal efficiencies, or to unit of saleable product. For feedlot operations often feed conversion ratios (either feed to gain (F:G) or the inverse, gain to feed (G:F)) or residual feed intake (RFI) are routinely used to describe feed efficiency for these production systems. Feed conversion ratios are simple to calculate, dependency on body weight and free choice intake favours animals with high mature bodyweights, feed intake, animal growth rate, and mature size, as these factors are all highly correlated (Terry et al., 2020). Therefore, feed conversion ratios may not actually identify animals that are truly metabolically more efficient because of these co-dependencies. As a result, residual feed intake (RFI) was developed to help normalize for growth rates and body weight and therefore may better reflect true metabolic efficiency in growing feedlot animals. In general, RFI is calculated by determining the "residuals" between an individual animal's observed feed intake, subtracted from their predicted feed intake (Koch et al., 1963). Predicted intake is most often determined by linear regression using data from a contemporary group of animals, and normalizing for growth (ADG), midpoint BW (or midpoint$BW^{0.75}$) and may also contain other relevant factors like estimations of body composition (ultrasound backfat measurements) in feedlot animals. Therefore, animals with negative RFI score represent the efficient phenotype, and consumes less feed than is predicted when BW and rate of gain are held constant. However, since RFI depends on measurement of free-choice intake, it can be difficult to measure in commercial operations largely due to requirement of specialized feeding equipment needed to obtain these individual intake measurements. However, RFI has been commonly used in genetics and research applications for feedlot production.

Genetic improvement in feed efficiency is an important strategy in meeting beef industry feed efficiency improvement goals. In particular, RFI in growing feedlot cattle is considered a moderately heritable trait, with h^2 reported between 0.14 and 0.43 % (Berry and Crowley, 2013). Genetics approaches and efficiency traits are nicely summarized in a review by Terry et al., (2020). While different beef breed types (British, Continental, dairy) may reflect differences in growth, lean gain, and mature body size –traits which are often correlated with efficiency traits, research into benefits of heterosis in complex traits like efficiency, longevity, and resiliency are continuing in hopes of better understanding how crossbreeding impacts these traits. More advanced genomic techniques like identification of single nucleotide polymorphisms (SNPs) and genome-wide association studies (GWAS) are also helping to identify new genetic traits associated with efficiency measures. With identification of new genetic markers, molecular breeding values for feedlot efficiency traits can more easily and accurately be developed.

Beyond the genetic composition of the animal itself, the composition of the rumen microbiome can also account for as much as 20% of the variation in performance and efficiency between animals (Paz et al., 2018). In particular, a more diverse rumen microbiome has been associated with improved efficiency traits in cattle (Guan et al., 2008, Lam et al., 2018). This may lead to new areas of research where strategies to manipulate the rumen environment may help improve efficiency in the feedlot and further understanding on how the host interacts with the microbial community.

Although RFI has been a useful tool for genetic improvement of cattle, this measure of efficiency is also more likely to reflect metabolic differences in efficiency and energy partitioning between individual animals. Richardson and Herd (2004) identified that traits like body composition, feeding patterns, digestibility, heat increment of feeding, metabolic factors and stress, which all contribute to animal-to-animal variation in feed efficiency (Figure 1). However, many of these metabolic factors remain poorly defined for feedlot animals. A closer look at these biological factors influencing inter-animal variation in feed efficiency and are well reviewed in Cantalapiedra-Hijar et al., (2018) and Kenny et al., (2018). However these reviews demonstrate how complex and multi-factorial traits influencing feed efficiency can be from a metabolic point of view.

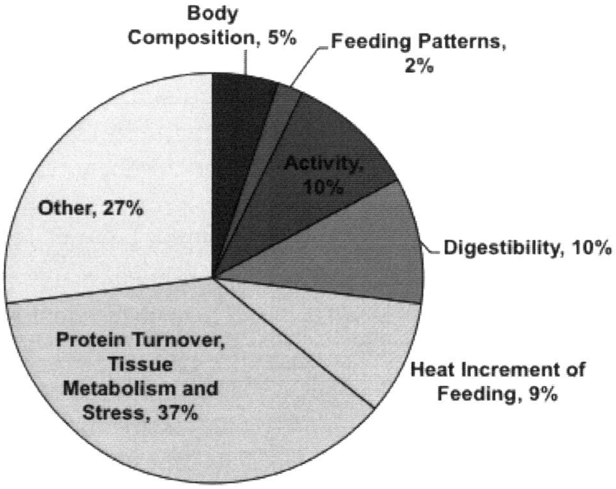

Figure 1: Factors contributing to individual animal differences in residual feed intake. (adapted from Richardson and Herd, (2004)

Energy partitioning is another component of understanding efficiency in feedlot cattle. With growing cattle and increasing time on feed, cattle begin to reach their mature size and begin to shift their composition of weight gain. As the growth curve begins to plateau as the animal matures, protein accretion begins to decrease and fat accretion increases. As the energy density of adipose tissue is greater than that of lean tissue, greater energy intake is needed to maintain the same level of gain as an animal within the lean gain phase of their growth curve. Steer growth and composition of gain is reviewed in detail by Owens et al., (1995) and Pethick et al., (2004). Although the decrease in feed efficiency with increasing time on feed is largely attributed to these changes in composition of gain, other metabolic factors may also be contributing to feed inefficiencies. In a serial slaughter experiment by our group (Kim et al., 2017), steers were slaughtered in 42 day increments until d 162, and performance traits along with tissues were collected to look at potential metabolic changes influencing efficiency. Results suggest that insulin sensitivity was impaired with increasing time on feed, as linear decreases in circulating insulin, and glucose to insulin ratio were observed. In addition, increased hepatic protein expression of insulin receptor, in line with blood results. Hepatic citrate synthase activity also decreased linearly with increasing time on feed. As citrate synthase is a common biomarker for mitochondrial efficiency (Trounce et al., 1996), this data suggests that mitochondria efficiency may also be reduced with increasing time on feed. This suggests that metabolic changes may also be responsible for reductions in feed efficiency with increasing time on feed.

Although research into increasing the understanding of mechanisms behind feed efficiency in feedlot cattle continues, there are numerous more practical strategies which are easily adaptable to feedlot operations which can help to improve overall feed efficiency.

Feed management

Although research often focuses on more technical aspects of metabolism and genetics in understanding feed efficiency, other simple feed management techniques also can greatly help reduce feed waste and spoilage. These include proper feed storage and harvest techniques to reduce spoilage and feed shrink, proper calibration of TMR mixers to optimize feed mixing, and a consistent bunk management strategy once feed is delivered to animals (Van Schaik and Wood, 2020; Figure 2). In addition, routine feed analysis can help to ensure rations are in line with diet formulations so that feed formulations are as accurate as possible.

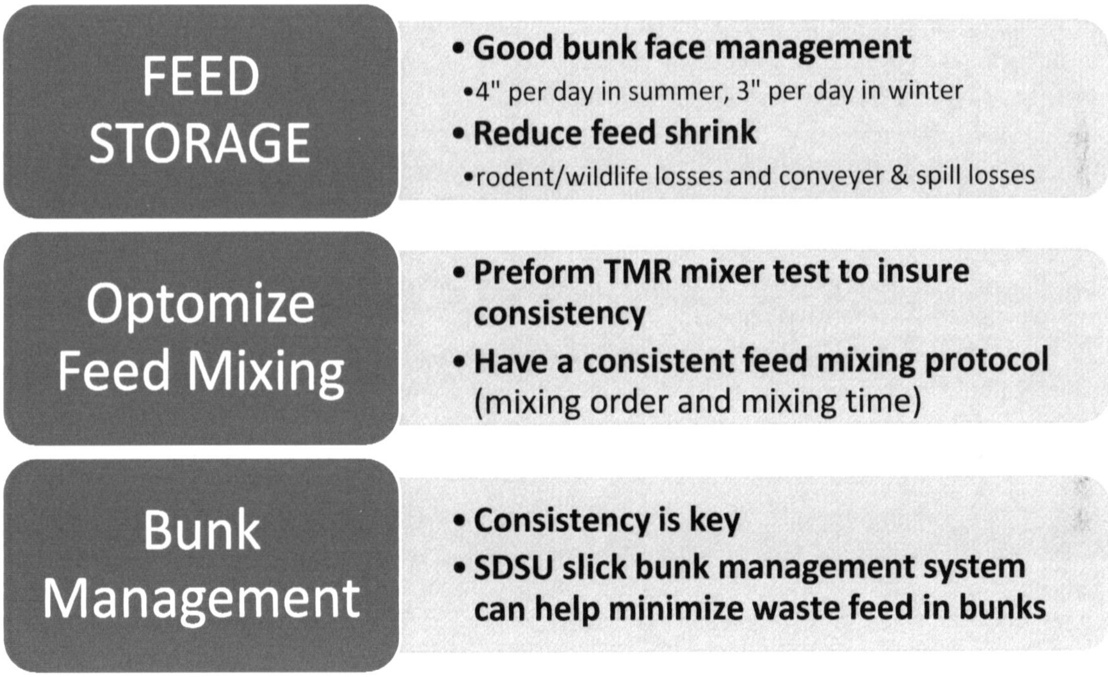

Figure 2: Summary of on farm feed management strategies which can improve overall on-farm feed efficiency. Adapted from Van Schaik and Wood, 2020

Improvements to Feeds and Feeding

Independent of animal factors influencing feed efficiency, improvements in feed, feed formulation, feed digestibility, and better-defined nutritional requirements can also help improve efficiencies in the feedlot (Terry et al., 2020).

Firstly, increased understanding of the intricacies of animal nutritional requirements and the growth prediction models from nutrients ingested will enable nutritionists and producers to more accurately align animal requirements and feed supplied. For example, unpublished work from our group (Williams et al., 2022) has investigated how better defining fibre in feedlot rations can help to increase feedlot performance prediction. Using a meta-analysis approach, a dataset of 58 high-grain (up to 10% forage) feedlot treatment means from 22 experiments were used to fit growth prediction models using diet parameters, ionophore, tylosin inclusion etc. In particular, the model looked at replacing dietary NDF in the model with undigestible NDF estimates (from feed library averages) and found that uNDF estimates improved steer growth model fit (lower RMSPE and increased CCC) and were also able to effectively predict HCW, feed efficiency, dressing percentage. This suggests that uNDF is a parameter which may be more effective at predicting feedlot performance, and nutritionist should consider including uNDF in feed analysis to generate more accurate performance models.

For high-grain diets, improvements in feed digestibility, either through plant breeding or feed processing technologies also can help improve feed efficiency. In feedlots, where high-grain diets are fed, feed processing index and particle size often has strong associations with rumen fermentability and pH, and gut health. Therefore, feed processing should aim for an optimization rather than maximization approach. Feed should be processed enough to increase digestibility, but not so processed that gut health becomes a risk factor for decreasing animal performance. Feed processing may include everything from simple grinding and/or tempering, to more complex feed processing methods like steam-flaking or other treatments which can increase digestibility and improve feed utilization by the animal (Terry et al., 2020). With high-grain diets, monitoring fecal starch may help producers optimize feed processing to ensure maximal starch digestion and efficiency. Although traditionally this analysis is conducted using wet chemistry approaches, new applications of near infrared (NIR) spectroscopy have been shown to accurately estimate fecal starch (Jancewicz et al., 2016). As NIR technology decreases in cost, it may be possible for feedlots to use this technology on farm, to make near real-time adjustments to grain processing, and regularly monitor fecal starch to ideally keep fecal starch below 5%.

Conclusion

In summary, although the economic and environmental need for improved feed efficiency is great within the cattle industry, the complex and multifaceted aspects of feed efficiency are challenging. There are still many unknown mechanisms controlling feed efficiency in cattle production, however researchers are beginning to have a better understanding of some of these biological mechanisms and genetic and feed management technologies continue to be developed. Despite these unknowns, there are many opportunities for improved efficiency on commercial farms through relatively simple management changes.

References

Berry, D.P., and Crowley, J.J. 2013. Cell biology symposium: genetics of feed efficiency in dairy and beef cattle. J. Anim. Sci. 91(4): 1594–1613. doi:10.2527/jas.2012-5862. PMID:23345557.

Cantalapiedra-Hijar, G., Abo-Ismail, M., Carstens, G. E., Guan, L. L., Hegarty, R., Kenny, D. A., ... & Ortigues-Marty, I. 2018. Biological determinants of between-animal variation in feed efficiency of growing beef cattle. *animal*, *12*(s2), s321-s335.

Gibb, D.J., and McAllister, T.A. 1999. The impact of feed intake and feeding behaviour of cattle on feedlot and feedbunk management. Agriculture and Agri-Food Canada.

Guan, L.L., Nkrumah, J.D., Basarab, J.A., and Moore, S.S. 2008. Linkage of microbial ecology to phenotype: correlation of rumen microbial ecology to cattle's feed efficiency. FEMS Microbiol. Lett. 288(1): 85–91. doi:10.1111/j.1574-6968.2008.01343.x. PMID:18785930.

Jancewicz, L. J., Swift, M. L., Penner, G. B., Beauchemin, K. A., Koenig, K. M., Chibisa, G. E., ... & McAllister, T. A. 2016. Development of near-infrared spectroscopy calibrations to estimate fecal composition and nutrient digestibility in beef cattle. *Canadian Journal of Animal Science*, *97*(1), 51-64.

Kenny, D. A., Fitzsimons, C., Waters, S. M., & McGee, M. 2018. Invited review: Improving feed efficiency of beef cattle–the current state of the art and future challenges. *animal*, *12*(9), 1815-1826.

Kim, J. J. M., Penner, G. B., Cant, J. P., & Wood, K. M. 2017. The effect of partial replacement of corn with a high-lipid, high-fiber byproduct pellet on hepatic indicators of metabolic efficiency and insulin sensitivity in beef steers throughout the finishing period. *Journal of Animal Science*, *95*, 377.

Koch, R.M., Swiger, L.A., Chambers, D., and Gregory, K.E. 1963. Efficiency of feed use in beef cattle. J. Anim. Sci. 22(2): 486–494. doi:10.2527/jas1963.222486x.

Lam, S., Munro, J. C., Zhou, M., Guan, L. L., Schenkel, F. S., Steele, M. A., ... & Montanholi, Y. R. 2018. Associations of rumen parameters with feed efficiency and sampling routine in beef cattle. *Animal*, *12*(7), 1442-1450. doi: 10.1017/S1751731117002750. PMID: 29122053.

Nkrumah JD, Okine EK, Mathison GW, Schmid K, Li C, Basarab JA, Price MA, Wang Z, Moore SS. 2006. Relationships of feedlot feed efficiency, performance, and feeding behavior with metabolic rate, methane production, and energy partitioning in beef cattle. J Anim Sci. 2006 Jan;84(1):145-53. doi: 10.2527/2006.841145x. PMID: 16361501.

Owens, F. N., Gill, D. R., Secrist, D. S., & Coleman, S. W. 1995. Review of some aspects of growth and development of feedlot cattle. *Journal of animal science*, *73*(10), 3152-3172.

Paz, H.A., Hales, K.E., Wells, J.E., Kuehn, L.A., Freetly, H.C., Berry, E.D., et al. 2018. Rumen bacterial community structure impacts feed efficiency in beef cattle. J. Anim. Sci. 96(3): 1045–1058. doi:10.1093/jas/skx081. PMID:29617864.

Pethick, D. W., Harper, G. S., & Oddy, V. H. 2004. Growth, development and nutritional manipulation of marbling in cattle: a review. *Australian Journal of Experimental Agriculture*, *44*(7), 705-715.

Richardson, E. C. A., & Herd, R. M. 2004. Biological basis for variation in residual feed intake in beef cattle. 2. Synthesis of results following divergent selection. *Australian Journal of Experimental Agriculture*, *44*(5), 431-440.

Terry, S. A., Basarab, J. A., Guan, L. L., & McAllister, T. A. 2020. Strategies to improve the efficiency of beef cattle production. *Canadian Journal of Animal Science*, *101*(1), 1-19.

Trounce, I. A., Kim, Y. L., Jun, A. S., & Wallace, D. C. 1996. [42] Assessment of mitochondrial oxidative phosphorylation in patient muscle biopsies, lymphoblasts, and transmitochondrial cell lines. In *Methods in enzymology* (Vol. 264, pp. 484-509). Academic press.

Van Schaik, M. and Wood, K.M. 2020. "Improving efficiencies in a feedlot feeding program" Virtual Beef, Aug 18, 2020. Online: http://www.omafra.gov.on.ca/english/livestock/beef/news/vbn0820a4.htmaccessed. Accessed: May 9, 2022.

Precision Mineral Nutrition for Dairy Cows

Nutrition minérale de précision chez les vaches laitières

Bill Weiss
Department of Animal Sciences, Ohio Agricultural Research and Development Center, The Ohio State University, Wooster, OH, USA
weiss.6@osu.edu

Sponsored by / Commandité par :

PROTEKTA®

Abstract

The recently released 8th Revised Edition of the Nutrient Requirements for Dairy Cattle gives up-to-date requirements for most essential minerals and those recommendations should be the starting point when formulating diets. The new recommendations for most minerals are similar to the previous standards except that magnesium and manganese requirements were about doubled, requirements for copper for high producing cows was reduced about 50% but increased about 40% for dry cows. The dietary requirements assume normal absorption and users must adjust the diet when antagonists are present (e.g., water with high concentration of sulfur will increase the amount of copper needed). Furthermore, equations calculate average requirements which means for a specific type of cow (e.g., 600 kg cow producing 35 kg of milk), the calculated requirement will be too high for half the cows and too low for half the cows. Modest deficiencies of minerals can increase health problems and reduce milk production and perhaps reproduction. These problems are usually worse than the problems associated with modest overfeeding. Therefore, nutritionists need to formulate diets for greater than the average requirements. However, overfeeding minerals increases feed costs, may cause environmental harm, and can, depending on the mineral and degree of overfeeding, harm the animal. Precision mineral nutrition means that the nutritionist considers each situation and applies the appropriate safety factor. Nutritionists must consider the presence of potential antagonists in diet and water, variability in feed composition and among cows within a group, source of mineral, risk of toxicity, and supplementation costs.

Key words: minerals, supplementation, toxicity, health

Résumé

La 8e édition révisée du guide Nutrient Requirements for Dairy Cattle, qui vient de paraître, présente les besoins actualisés pour la plupart des minéraux essentiels, et ces recommandations devraient constituer le point de départ de la formulation des régimes alimentaires. Les nouvelles recommandations pour la plupart des minéraux sont similaires aux normes précédentes, sauf que les besoins en magnésium et en manganèse ont été approximativement doublés et que les besoins en cuivre pour les vaches fortes productrices ont été réduits d'environ 50 %, mais augmentés d'environ 40 % pour les vaches taries. Les besoins alimentaires supposent une absorption normale, et les utilisateurs doivent adapter le régime alimentaire en cas de présence d'antagonistes (par exemple, une eau à forte teneur en soufre augmentera la quantité de cuivre nécessaire). De plus, les équations tiennent compte des besoins moyens, ce qui signifie que pour un type de vache spécifique (par exemple, une vache de 600 kg produisant 35 kg de lait), les besoin calculés seront trop élevés pour la moitié des vaches et trop faibles pour l'autre moitié. De faibles carences en minéraux peuvent accroître les problèmes de santé et nuire à la production de lait, voire à la reproduction. Ces problèmes sont généralement pires que ceux associés à une légère suralimentation. Par conséquent, les nutritionnistes des animaux doivent formuler les régimes pour répondre à des besoins supérieurs à la moyenne. Cependant, la suralimentation en minéraux augmente le coût des aliments, risque de nuire à l'environnement et peut, selon le minéral et le degré de suralimentation, causer du tort à l'animal. La nutrition minérale de précision signifie que le nutritionniste considère chaque situation et applique le facteur de sécurité approprié. Les nutritionnistes des bovins laitiers doivent tenir compte de la présence d'antagonistes potentiels dans la ration et dans l'eau, de la variabilité de la composition des aliments et des vaches d'un même groupe, de la source du minéral, du risque de toxicité et des coûts de la supplémentation.

Introduction

Precision nutrition means diets are formulated as close as possible to requirements while avoiding any deficiency issues. Benefits of precision mineral nutrition can include reduced diet costs, reduced environmental impact, and reduced risk of toxicity. The major problem with precision mineral nutrition is increased risk of deficiency. Deficiency can be manifested as a clinical deficiency (e.g., grass tetany caused by inadequate magnesium), or as numerous subclinical responses including impaired health, reproduction, and production. Diets should never be formulated below the average mineral requirements for the pen. The main question is, how much greater than pen average mineral requirement should the formulation goals be? The answer to that question is farm and mineral specific. The nutritionist job is to consider all factors affecting requirements and absorption of minerals for each specific situation and formulate accordingly.

Mineral Requirements

The NRC (now NASEM) published updated requirements for dairy cattle in late 2021 (NASEM, 2021). All known essential minerals were considered by the committee and then based on data, either requirements or adequate intakes (AI) were established for each mineral. A requirement is the amount of mineral needed to meet the needs of an average animal within a defined population (e.g., the average for 650 kg cows producing 35 kg of milk). To establish a requirement adequate data had to be available to have high confidence in the values. An AI is similar to a requirement except that it means the committee did not have the same degree of confidence because of limited data. Often this was because of limited number of studies or studies that did not include multiple levels of the mineral. An AI is not necessarily the optimal amount that should be fed but in the opinion of the committee, meeting the AI will prevent problems and data are available showing that positive effects occurred compared with feeding lower amounts. For diet formulation a requirement and an AI can usually be considered the same thing. The committee evaluated mineral needs for calves, heifers, and dry and lactating cows but this paper will concentrate on lactating cows. An overview of changes in requirements (or AI) compared with NRC (2001) is shown in Figure 1.

Mineral	**Heifer**	**Dry cow**	**Lact. cow**
Ca	+	+	+
P	**0**	**0**	**0**
Mg	+++	+++	+++
K	**0**	**0**	**0**
Na	**0**	**0**	**0**
Cl	**0**	**0**	**0**
S	**0**	**0**	**0**
Co	+	+	+
Cu	+	++	- -
Fe	**0**	**0**	**0**
I	**0**	**0**	**0**
Mn	+++	+++	+++
Se	**0**	**0**	**0**
Zn	+	+	+

Figure 1. Comparison of requirements of minerals calculated using NASEM (2021) to those calculated using NRC (2001). 0 = essentially no change, - = new requirements are less than previous and + = requirements are greater than previous. The number of symbols indicate the degree of change.

Discussing all the changes made to mineral requirements in the NASEM (2021) is beyond the scope of this paper and they will only be discussed briefly. The NASEM set requirements for calcium (Ca), phosphorus (P), magnesium (Mg), sodium (Na), chloride (Cl), sulfur (S), copper (Cu), and zinc (Zn), and AI were established for cobalt (Co), iodine (I), iron (Fe), manganese (Mn) and selenium (Se). For most minerals (exceptions are S, Co, I, and Se) requirements or AI were estimated using the factorial approach. The amounts of absorbed mineral needed for maintenance plus the amount of minerals secreted in milk, retained in new body tissue, and retained in conceptus are summed and then divided by an absorption coefficient (AC) to get dietary requirements or dietary AI. Almost all the equations used to estimate requirements or AI were updated from NRC (2001) but in several cases the final dietary values changed very little for 'average cows' fed 'typical diets'. However, the committee thinks the new equations are more biologically correct and therefore should be better when cows are not average and typical diets are not fed. Compared with NRC (2001) the greatest changes in requirements or AI were for Mg, Cu, and Mn (discussed below).

For most of the other mineral requirements (or AI) did not change greatly from NRC (2001). Because of substantial reductions in the AC for Ca supplements, diets with low basal Ca (e.g., high corn silage diets) will probably require more Ca supplementation than NRC (2001) estimated. Requirement for P and the method used to estimate AC were changed slightly usually resulting in slightly lower dietary requirements. The AC calculation should improve accuracy over a wider range of diets. Zinc requirements were modestly increased by about 15%. The AI for Co is about twice the requirement estimate in NRC (2001); however, many diets likely contain adequate basal Co so this probably will not affect diet formulation greatly.

Significant Changes to Mineral Requirements

Based on newer data, the maintenance requirement for absorbed Mg more than doubled in NASEM (2021) largely because it is now a function of DMI rather than body weight as was done in NRC (2001). Endogenous fecal losses of mineral should be a function of DMI so the new equations are biologically correct and should be more accurate. The AC for basal and supplemental Mg also changed markedly. The greatest change occurred for Mg supplements. In NRC (2001) the AC for Mg supplements were likely calculated incorrectly resulting in very high values (e.g., 70% for Mg oxide and 90% for Mg sulfate. The new values fit available data very well and are much lower (approximately 23% for Mg oxide and 27% for Mg sulfate. In addition, the new software (NASEM, 2021) uses an equation to reduce AC for Mg as dietary K increases. Overall, because of changes in absorbed requirements and AC, dietary requirements for Mg are about twice as high using NASEM (2021) compared with NRC (2001).

Because of increased concerns about Cu toxicity, the committee conducted a very extensive review of Cu requirements. Adequate data were found to estimate metabolic fecal Cu using DMI rather than body weight which resulted in an approximate doubling of the maintenance requirement. This will result in an approximate 40% increase in dietary Cu requirement for dry cows compared with NRC (2001). However, newer data on concentration of Cu in milk (used for lactation requirement) was 0.04 compared with 0.14 mg/L which greatly reduces the lactation requirement. Overall the dietary Cu requirement for cows producing about 35 kg of milk did not change very much but for cows producing 45 kg of milk, NASEM (2021) dietary requirement for Cu is about 45% lower than NRC (2001).

Manganese was also evaluated carefully because pregnant beef heifers fed to meet the NRC (2001) Mn requirement produced calves that were clinically deficient in Mn (Hansen et al., 2006). Adequate data were not available to establish a requirement but rather Mn has an AI. Based on a single experiment, maintenance AI was increased 30% compared with NRC (2001), and some new information on AC of Mn supplements resulted in a substantial decrease in the AC for Mn. Combining both changes, the dietary AI for Mn about doubled compared with NRC (2001).

Formulation Goals

The equations used to estimate requirements or AI are based on more data and the new data better reflects current cows and diets; therefore, the estimates should be more accurate than values from NRC (2001). However, even if the equations were perfect (which they are not because all statistical equations have prediction error) the estimates would still be for the average which means half the cows with similar characteristics would be overfed and half would be underfed. The potential problems caused by underfeeding are almost always more expensive than the costs associated with overfeeding. Several factors must be considered when determining how much overfeeding should be done including cost of supplemental mineral, source of supplemental mineral, maximum tolerable level of the mineral, accuracy of the estimates, variation in requirements and dietary supply and potential problems if inadequate mineral is fed. Assuming typical supplement costs, based mostly on variation in diet composition but also some on variation in requirements (or AI) most diets should be formulated to provided about 20% more than NASEM (2021) requirements. That degree of overfeeding has essentially no toxicity risk, will not add too much extra costs to diets and will meet the needs of most of the cows in a group. For example, if the average Ca requirement for a pen of cow was 110 g/d, the diet should be formulated to provide 110 x 1.2 = 132 g/d at pen average dry matter intake (DMI). This approach will work for most, but not all, minerals and in most, but not all, situations.

Adjustments to standard safety factors

For most minerals on most farms, including a safety factor of 20% above average NASEM requirements (or AI) is appropriate (Table 1). However, there are exceptions; S, P, Se, Cu, and Mg. For S and P, no safety factor is needed or should be applied. Moderately excess sulfur reduces the absorption of Cu, Mn, and Se so in the long term (months) excess S can cause deficiencies of those minerals. In addition, excess S can reduce dietary cation anion difference (DCAD) which is detrimental to lactating cows. Because of the risk of excess S, it should be fed at about the NASEM requirement (i.e., 2 g/kg of diet DM). Diets with excess S are often fed during the prefresh period to lower DCAD and reduce hypocalcemia. These diets are only fed for a few weeks which is not long enough to cause problems with trace minerals. The requirements for P are very well defined and because of recycling within the cow, P deficiencies almost never will occur when animals are fed at NASEM (2021) requirements. A safety factor often cannot be applied to Se because supplementation may be limited by regulation. In most areas of the world, nutritionists should formulate dairy diets to the maximum legally allowable Se concentration.

Formulation goals for Cu and Mg need to be considered carefully. Modest overfeeding of Cu increases liver Cu concentrations which increases the risk of Cu toxicity which is most cases is fatal. This risk should encourage nutritionists to feed very close to the requirement. On the other hand, diets and water often contain antagonists that reduce Cu absorption which should encourage nutritionists to overfeed Cu to a greater extent. The most common antagonists to Cu absorption are excess S (includes both diet and water), excess molybdenum (Mo), and grazing on high clay soils. Dietary S (including S from water) at 3 and 4 g/kg of diet DM may reduce expected Cu absorption by 15 and 30% when dietary Mo is around 1 mg/kg diet DM. If diets contain 5 mg Mo/kg diet DM, Cu absorption may decrease by 18 and 37% for diets with 3 and 4 g/kg S). This means that if diets have 4 g/kg of S and 5 mg/kg Mo, an appropriate safety factor may be close to 1.5 times NASEM requirement. Diets with 4 g/kg S cause other problems so the first goal should be to try to reduce S concentrations in diet and water. If that is not possible you may need to formulate diets to provide 1.5 times as much as the average Cu requirement as calculated by NASEM (2021). That recommendation is for Holsteins. Jersey cattle accumulate more Cu in their liver than do Holsteins fed the same diet. Therefore, Cu should be fed closer to requirement for Jerseys. Although data are limited, based on hepatic accumulation differences safety factors probably should be about half as much for Jerseys as for Holsteins.

Magnesium has a very wide safety margin, so overfeeding is usually not a concern, and several factors can inhibit Mg absorption encouraging overfeeding. In addition, the AC for Mg from Mg oxide (most common Mg supplement) can vary greatly. Magnesium absorption decreases as dietary K concentration increases. This antagonism starts occurring even before the K requirement is met. The NASEM (2021) model includes an equation to adjust the AC for Mg based on diet K so if using that model, the safety factor (i.e., 1.2X) does not need to be increased for higher K diets. Monensin inhibits absorption of Mg from Mg sulfate by about 25% but it increases Mg absorption when Mg oxide is fed (Tebbe et al., 2018). The AC of Mg from Mg oxide averages about 25% but can be close to 0 for some low-quality sources (large particle size, improper calcination procedures). Increased concentrations of dietary fat can inhibit Mg absorption. Considering all the potential antagonists, the wide safety margin, and the problems associated with inadequate Mg, diets should probably include a safety factor of about 1.4X (assuming the AC is already adjusted for high K).

"Non-requirement" effects

Feeding some minerals above requirements can elicit responses such as reduced risk for some diseases or increased milk production or milk components. These effects are considered responses, rather than requirements (i.e., feeding these levels are not necessary to obtain high milk yield or healthy cows). NASEM (2021) did not include a response model for minerals (it does include a response model for amino acids). No requirement or AI was established for chromium (Cr) because data are not available to determine it. However, several studies have shown a positive milk response when transition and early lactation are supplemented with Cr at a rate of about 0.5 mg/kg of diet DM. Feeding Mg at about twice the requirement to dry and prefresh cows reduces the risk of hypocalcemia. To meet the requirements for Na, K, S and Cl of a lactating dairy cow, DCAD (calculated as concentration of Na + K minus concentrations of (Cl +S) where concentrations are in mEq/kg) needs to be about 175 mEq/kg of diet DM. However increasing DCAD above that value is expected to increase milk yield, milk fat yield and DMI ((Iwaniuk and Erdman, 2015). Replacing sulfate forms of trace minerals with hydroxy or certain types of organic trace minerals can increase fiber digestibility (Faulkner and Weiss, 2017) or alter gut microbiome in ways that may improve hoof health (Faulkner et al., 2017). In a meta-analysis, organic trace mineral from a single manufacturer increased milk yield (Rabiee et al., 2010). Dependent on economics and some other factors, feeding extra mineral to elicit a production response can be warranted.

Table 1. Dietary concentrations (DM basis) of minerals that approximately meet the average requirements (NASEM, 2021) for a Holstein cow producing 35 kg of milk, and recommended formulation goals based on expected variation in mineral supply and requirements[1].

Mineral	Concentrations to meet NASEM (2021)	Recommended concentrations
Ca, g/kg	5.7	6.8 – 7.5
P, g/kg	3.2	3.2 – 3.5
Mg (1.2% K without supplemental fat), g/kg	1.6	2.2
Mg (1.2% K with supplemental fat), g/kg	1.6[2]	2.6
Mg (2% K with supplemental fat), g/kg	2.0	3.0
K, g/kg	10.2	12.0[3]
Na, g/kg	2.0	2.4[3]
Cl, g/kg	2.8	3.4[3]
S, g/kg	2.0	2.0
Co, mg/kg	0.20	0.24
Cu (2 g/kg S and 1 mg/kg Mo), mg/kg	10	12 – 15
Cu (4 g/kg S and 5 mg/kg Mo), mg/kg	10[3]	15 – 18
Fe, mg/kg	16	20
I, mg/kg	0.4	0.5
Mn, mg/kg	27	33 – 40
Se[4], mg/kg	0.3	0.3
Zn, mg/kg	55	66 – 70

[1] Concentrations based on DMI estimated using NASEM equation. Recommended concentrations are usually 1.2 times NASEM requirement (with some exceptions discussed in the text) and reflect the author's opinion, not NASEM.

[2] The NASEM model does not adjust Mg requirement when supplemental fat is fed.

[3] Because of potential positive effects of increased DCAD, dietary concentrations of K and Na may be substantially greater than these (without a concomitant increase in Cl or S).

[4] The U.S. Food and Drug Administration (FDA) limits supplemental Se concentrations to 0.3 mg/kg. In general, supplement Se to the highest concentration allowed by regulations.

References

Faulkner, M. J. and W. P. Weiss. 2017. Effect of source of trace minerals in either forage- or byproduct–based diets fed to dairy cows: 1. Production and macronutrient digestibility. Journal of Dairy Science 100:5358-5367.

Faulkner, M. J., B. A. Wenner, L. M. Solden, and W. P. Weiss. 2017. Source of supplemental dietary copper, zinc, and manganese affects fecal microbial relative abundance in lactating dairy cows. Journal of Dairy Science 100:1037-1044.

Hansen, S. L., J. W. Spears, K. E. Lloyd, and C. S. Whisnant. 2006. Feeding a Low Manganese Diet to Heifers During Gestation Impairs Fetal Growth and Development. Journal of dairy science 89,:4305-4311.

Iwaniuk, M. E. and R. A. Erdman. 2015. Intake, milk production, ruminal, and feed efficiency responses to dietary cation-anion difference by lactating dairy cows. Journal of Dairy Science 98:8973-8985.

National Academies of Science, Engineering, and Medicine. 2021. Nutrient Requirements of Dairy Cattle, 8th rev. ed. National Acad Press, Washington DC.

National Research Council. 2001. Nutrient Requirements of Dairy Cattle. 7th rev. ed. ed. Natl. Acad. Press, Washington DC.

Rabiee, A. R., I. J. Lean, M. A. Stevenson, and M. T. Socha. 2010. Effects of feeding organic trace minerals on milk production and reproductive performance in lactating dairy cows: A meta-analysis. J. Dairy Sci. 93:4239-4251.

Tebbe, A. W., D. J. Wyatt, and W. P. Weiss. 2018. Effects of magnesium source and monensin on nutrient digestibility and mineral balance in lactating dairy cows. Journal of Dairy Science 101(2):1152-1163.

High mycotoxin levels in wheat grain and their effects on beef cattle ruminal fermentation, performance, and carcass traits

Teneurs élevées en mycotoxines dans le grain de blé et effets sur la fermentation ruminale, les performances et les caractéristiques de carcasse chez les bovins de boucherie

Bierworth, R.M.[1,2], Ribeiro, G.O.[1], Terry, S.A[2], Penner, G.B.[1], McKinnon, J.J.[1], Hucl, P.[1], Randhawa, H.[2], Beauchemin, K.A. [2], Schwartzkopf-Genswein, K.[2], Yang, W.Z.[2], Gruninger, R.[2], Guan, L.L.[3], Gibb, D.[4], and McAllister, T.A.[1]

[1]Department of Animal and Poultry Science, University of Saskatchewan, Saskatoon, SK

[2]Agriculture and Agri-Food Canada, Lethbridge Research and Development Centre, Lethbridge, AB

[3]Agricultural Food and Nutritional Science, University of Alberta, Edmonton, AB

[4]Gowan's Feed Consulting, Raymond, AB

rmf658@usask.ca

Sponsored by / Commandité par :

Cbs Bio Platforms

Abstract

With the rising impact of climate change on grain crops in North America, more wheat grain has failed to meet milling grade standards and has been diverted to animal feed. A common reason for rejection is infection with *F. graminearum* and *C. purpurea,* which produce deoxynivalenol (DON) and ergot alkaloids (EA), respectively. Current allowable limits in cattle for DON are 5 ppm in Canada, while EA limits are 2 to 3 ppm. The objectives of this study were to investigate the effects of high mycotoxin levels on ruminal fermentation, growth performance, and carcass characteristics of finishing steers. The primary mycotoxin assessed was DON (5 ppm and 10 ppm), but EA were also present in the same wheat at 2.1-4.3 ppm. Forty crossbred steers (8 cannulated) were housed in individual pens, blocked by weight, and randomly assigned to 1 of 4 treatments; control-low (0 ppm), control-high (0 ppm), MYC-low (5 ppm-DON; 2.1 ppm-EA), and MYC-high (10 ppm-DON; 4.2 ppm-EA). Wheat screenings were added to control diets so as to generate diets with a chemical composition that was similar to MYC diets. Steers were fed a finishing diet consisting of 88% dry rolled wheat based-concentrate and 12% barley silage on a dry matter basis for 112 days. Ammonia and volatile fatty acid concentrations in rumen fluid did not differ among treatments. Final body weight, dry matter intake (DMI), average daily gain (ADG) and gain to feed ratio (G:F) of steers fed MYC were reduced when compared to controls ($P<0.01$), with DMI for controls averaging 9.67-9.85 kg/d, MYC-low at 8.38 kg/d, and MYC-high at 6.29 kg/d. ADG for controls ranged between 1.90-1.92 kg/d, MYC-low at 1.36 kg/d and MYC-high at 0.84 kg/d. Hot carcass weight and fat cover were also reduced ($P<0.01$), but lean meat yield was increased in MYC steers ($P<0.01$). No differences were noted in carcass grades. In conclusion, a combination of DON and EA, negatively impacted growth performance and carcass traits of steers. These findings illustrate the importance of knowing what toxins are present in feed wheat for finishing cattle. Adjusting approved levels of mycotoxins when two or more are present may help avoid adverse impacts on growth performance of finishing feedlot cattle.

Keywords: beef cattle, wheat, mycotoxins, performance, DON, ergot

Résumé

Considérant l'impact croissant du changement climatique sur les cultures céréalières en Amérique du Nord, davantage de grains de blé ne répondant pas aux normes de qualité meunière sont détournés vers l'alimentation animale. Une des principales raisons de rejet du grain est l'infection par *F. graminearum* et *C. purpurea*, des champignons qui produisent respectivement du déoxynivalénol (DON) et des alcaloïdes de l'ergot (EA). La limite actuellement permise chez les bovins pour le DON a été fixée à 5 ppm au Canada, tandis que pour les EA elle est de 2 à 3 ppm. Cette étude visait à examiner les effets de concentrations élevées de mycotoxines sur la fermentation ruminale, la performance de croissance et les caractéristiques de carcasse des bouvillons de finition. La principale mycotoxine évaluée a été le DON (5 ppm et 10 ppm), quoique ce même blé contenait aussi de 2,1 à 4,3 ppm d'EA. Quarante bouvillons croisés (8 portant une fistule) ont été logés dans des enclos individuels, regroupés en fonction de leur poids et assignés de manière aléatoire à un de quatre traitements : témoin-faible (0 ppm), témoin-élevée (0 ppm), MYC-faible (5 ppm-DON; 2,1 ppm-EA) et MYC-élevée (10 ppm-DON; 4,2 ppm-EA). Des criblures de blé ont été ajoutées aux régimes témoins afin d'obtenir des rations de composition chimique similaire à celle des régimes MYC. Les bouvillons ont reçu un régime de finition composé de 88 % de concentré sec à base de blé roulé et de 12 % d'ensilage d'orge (sur une base de matière sèche) pendant 112 jours. Les concentrations d'ammoniaque et d'acides gras volatils dans le liquide ruminal n'ont présenté aucune différence selon les traitements. Une réduction du poids corporel final, de la consommation volontaire de matière sèche (DMI), du gain moyen quotidien (ADG) et de l'indice de conversion alimentaire (G:F) a été observée chez les bouvillons qui ont reçu les rations MYC, comparativement aux animaux ayant consommé les rations témoins ($P<0,01$), soit une DMI moyenne de 9,67-9,85 kg/j pour les bouvillons témoins, de 8,38 kg/j pour les bouvillons MYC-faible et de 6,29 kg/j pour les sujets MYC-élevée. Les sujets témoins ont enregistré un ADG compris entre 1,90 et 1,92 kg/j, les animaux MYC-faible ont obtenu 1,36 kg/j et les bouvillons MYC-élevée ont atteint 0,84 kg/j. Une réduction a également été notée pour le poids de carcasse chaude et le gras dorsal ($P<0,01$), mais le rendement en maigre était plus élevé chez les bouvillons MYC (*P*<0,01). Aucune différence n'a été enregistrée dans le classement des carcasses. En conclusion, une combinaison de DON et d'EA a eu un impact négatif sur la performance de croissance et les caractéristiques de carcasse des bouvillons. Ces résultats montrent l'importance d'identifier les toxines présentes dans le blé fourrager pour les bovins de finition. Modifier les concentrations de mycotoxines approuvées lorsque deux mycotoxines ou plus sont présentes pourrait aider à éviter les effets négatifs sur la performance de croissance des bovins en parc de finition.

Introduction

Changing climatic conditions in Western Canada has led to increased amounts of wheat grain failing milling grade standards and as a result more is available as feed (He et al., 2015). One reason for wheat grain failing grading standards is infection with *Fusarium graminearum* and/or *Claviceps purpurea,* which produce deoxynivalenol (DON; Bianchini et al., 2015) and ergot alkaloids (EA; Tittlemier et al., 2015), respectively. Due to the ability of the rumen microbial population to metabolize DON to de-epoxy DON (DOM-1; Guerre, 2020), ruminants are considered less sensitive to the toxin than monogastrics (Dänicke et al., 2005; Roberts et al., 2021). Ruminants are more sensitive to EA (Rahimabadi et al., 2022) and can exhibit variable symptoms among individuals. This variation in response is attributable to complex interactions among the plant, pathogen, animal and the environment (Klotz, 2015). Generally, there are 3 types of manifestations to ergot toxicity; [1] convulsive (Rahimabadi et al., 2022) [2] gangrenous (Klotz, 2015), and [3] other (Coufal-Majewski et al., 2016). Research on possible interaction between DON and EA is scarce. The objectives of this study were to assess the effects of a mixture of DON and EA in feed wheat on ruminal fermentation, growth performance, and carcass traits of feedlot cattle.

Methods and Materials

Forty crossbred steers (8 cannulated) were housed in individual pens, blocked by weight, and randomly assigned to 1 of 4 treatments; control-low (0 ppm), control-high (0 ppm), MYC-low (5 ppm-DON; 2.1 ppm-EA), and MYC-high (10 ppm-DON; 4.2 ppm-EA). Steers were fed a finishing diet consisting of 88% dry rolled wheat-based concentrate, 12% barley silage on a dry matter basis for 112 days. Wheat screenings were added to control diets so as to generate diets with a chemical composition that was similar to MYC diets, consequently producing two different control diets (high and low). This was to account for poor quality of the mycotoxin infected wheat in the treatment diets in a non-MYC control. Weigh-backs were collected weekly from troughs to estimate dry matter intake (DMI). Steers were weighed on 2 consecutive days at the beginning and end of the trial, and at 28-d intervals. Ruminal pH loggers were placed in the rumen for 7 days and upon removal rumen digesta was collected (~300 ml) and strained through two layers of cheese cloth and the filtrate was placed in either sulfuric acid for NH_3, or metaphosphoric acid for VFA and stored at -20 ^oC. Steers were marketed after 112 d on feed to Cargill Foods (High River, AB, Canada). Hot carcass weight, dressing percentage, back fat thickness, rib-eye area, lean meat yield, and quality grade were determined. Livers were scored and checked for cirrhosis, hyperplasia and tumours. PROC MIXED and PROC GLIMMIX were used to analyze data (SAS Int. Inc., Cary, NC). Models included DON and EA as fixed effects, and individual animal as a random effect.

Results and Discussion

The DMI of steers fed MYC-L and MYC-H treatments was reduced ($P < 0.001$) by 14.1% and 35.6%, respectively, when compared to control fed steers. Consequently, ADG ($P < 0.001$) and G:F ($P = 0.001$) for MYC (L and H) fed steers were linearly reduced, showing a 28.8% (L) and 56.0% (H) decrease for ADG, and 16.5% (L) and 31.8% (H) decrease for G:F. These findings are in agreement with Coufal-Majewski et al. (2017), where ergot alkaloids fed at a concentration of 0.4 ppm reduced ADG and G:F of lambs. However, when comparing similar studies in dairy cattle using DON, performance losses were not reported (Charmley et al., 1993; Ingalls, 1996), even when it was present in feed at 12 ppm. Ammonia and VFA concentrations did not vary among treatments, although total VFA concentrations approached significance ($P = 0.06$). This tendency could be explained by reduced DMI (Schären et al., 2016) observed in MYC fed steers. Mean ruminal pH was higher ($P = 0.018$) for MYC steers, which could be explained by lower DMI, due to reduced digestion and production of VFA's. Higher pH could also be a result of feed sorting (Miller-Cushon and DeVries, 2017) and refusals as these steers favoured the consumption of silage. Refusal of the wheat-based concentrate resulted in a reduced intake of fermentable carbohydrates and consequently higher ruminal pH (Owens and Pioneer, 2015). Hot carcass weights were linearly reduced ($P < 0.001$) in MYC steers as mycotoxin concentrations increased. Steers fed MYC diets also exhibited decreased ($P = 0.001$) average fat cover as compared to control steers. These observations reflected the reduced ADG and G:F of these steers. Meat yield ($P = 0.041$) and lean meat yield ($P = 0.011$) were increased in MYC fed steers as compared to control fed steers, which is likely a result of the reduced DMI, promoting lower ADG and leaner carcasses due to less energy available for fat deposition. There were no significant differences in carcass grades or liver scores.

Conclusions

Feed wheat contaminated with DON and EA decreased growth performance and carcass quality in feedlot steers but increased mean ruminal pH. When comparing similar levels of DON or EA in previous studies, DON was reported to have little to no negative impacts on cattle, even at concentrations of 12 ppm. EA was reported to negatively impact ruminants at 0.4 ppm and this impact increased linearly as concentrations increased. The results seen in previous EA studies are in agreement with the results of our study. However, more research is needed to determine if these results reflect the impact of a combination of DON and EA, or just EA. These findings illustrate the importance of knowing what the types and concentrations of mycotoxins are in feed wheat fed to finishing cattle, so as to avoid substantial monetary losses for producers.

References

Bianchini, A., R. Horsley, M. M. Jack, B. Kobielush, D. Ryu, S. Tittlemier, W. W. Wilson, H. K. Abbas, S. Abel, and G. Harrison. 2015. DON occurrence in grains: a North American perspective. Cereal Foods World. 60:32–56.

Charmley, E., H. L. Trenholm, B. K. Thompson, D. Vudathala, J. W. G. Nicholson, D. B. Prelusky, and L. L. Charmley. 1993. Influence of level of deoxynivalenol in the diet of dairy cows on feed intake, milk production, and its composition. J. Dairy Sci. 76:3580–3587.

Coufal-Majewski, S. 2017. Characterising the impact of ergot alkaloids on digestibility and growth performance of lambs.

Coufal-Majewski, S., K. Stanford, T. McAllister, B. Blakley, J. McKinnon, A. V. Chaves, and Y. Wang. 2016. Impacts of cereal ergot in food animal production. Front. Vet. Sci. 3:15.

Dänicke, S., K. Matthäus, P. Lebzien, H. Valenta, K. Stemme, K. H. Ueberschär, E. Razzazi-Fazeli, J. Böhm, and G. Flachowsky. 2005. Effects of Fusarium toxin-contaminated wheat grain on nutrient turnover, microbial protein synthesis and metabolism of deoxynivalenol and zearalenone in the rumen of dairy cows. J. Anim. Physiol. Anim. Nutr. (Berl). 89:303–315. doi:10.1111/j.1439-0396.2005.00513.x.

Guerre, P. 2020. Mycotoxin and gut microbiota interactions. Toxins (Basel). 12:769.

He, M. L., J. Long, Y. Wang, G. Penner, and T. A. Mcallister. 2015. Effect of replacing barley with wheat grain in finishing feedlot diets on nutrient digestibility , rumen fermentation , bacterial communities and plasma metabolites in beef steers. 176:104–110. doi:10.1016/j.livsci.2015.03.024.

Ingalls, J. R. 1996. Influence of deoxynivalenol on feed consumption by dairy cows. Anim. Feed Sci. Technol. 60:297–300.

Klotz, J. L. 2015. Activities and effects of ergot alkaloids on livestock physiology and production. Toxins (Basel). 7:2801–2821.

Miller-Cushon, E. K., and T. J. DeVries. 2017. Feed sorting in dairy cattle: Causes, consequences, and management. J. Dairy Sci. 100:4172–4183.

Owens, F. N., and D. Pioneer. 2015. Acidosis in Cattle : A Review. doi:10.2527/1998.761275x.

Rahimabadi, P. D., S. Yourdkhani, M. Rajabi, R. Sedaghat, D. Golchin, and H. Asgari Rad. 2022. Ergotism in feedlot cattle: clinical, hematological, and pathological findings. Comp. Clin. Path. 1–11.

Roberts, H. L., M. Bionaz, D. Jiang, B. Doupovec, J. Faas, C. T. Estill, D. Schatzmayr, and J. M. Duringer. 2021. Effects of Deoxynivalenol and Fumonisins Fed in Combination to Beef Cattle: Immunotoxicity and Gene Expression. Toxins (Basel). 13:714.

Schären, M., G. M. Seyfang, H. Steingass, K. Dieho, J. Dijkstra, L. Hüther, J. Frahm, A. Beineke, D. von Soosten, and U. Meyer. 2016. The effects of a ration change from a total mixed ration to pasture on rumen fermentation, volatile fatty acid absorption characteristics, and morphology of dairy cows. J. Dairy Sci. 99:3549–3565.

Tittlemier, S. A., D. Drul, M. Roscoe, and T. McKendry. 2015. Occurrence of ergot and ergot alkaloids in western Canadian wheat and other cereals. J. Agric. Food Chem. 63:6644–6650.

pHix-up®
POWERFUL INNOVATIVE
& NATURAL
Powerful, fast & sustainable action on rumen pH, naturally supporting performance in ruminants.
INNOVATION
TIMAB
magnesium
CONTACT CANADA:
Soraya Seghir (+1) 514 820 2063
soraya.seghir@timabusa.com

PROTEKTA®
Pioneering Solutions to Improve Animals' Everyday Health.
We believe that prevention is the most effective and profitable model for raising animals. To learn about the solutions offered through Protekta for your poultry, swine and dairy productions contact us!
457 Campbell Street
Lucknow, ON
N0G 2H0
w: www.protekta.com
p: 1.888.822.3135
e: info@protekta.com

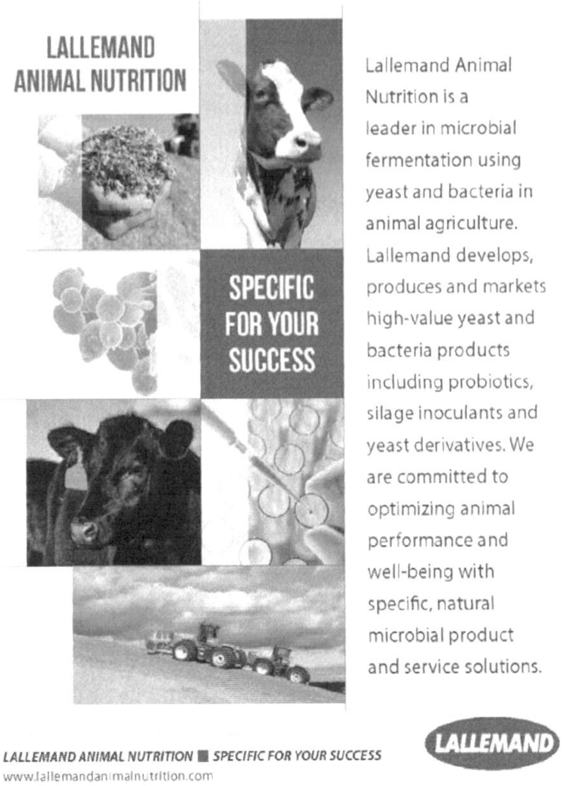
LALLEMAND ANIMAL NUTRITION
SPECIFIC FOR YOUR SUCCESS
Lallemand Animal Nutrition is a leader in microbial fermentation using yeast and bacteria in animal agriculture. Lallemand develops, produces and markets high-value yeast and bacteria products including probiotics, silage inoculants and yeast derivatives. We are committed to optimizing animal performance and well-being with specific, natural microbial product and service solutions.
LALLEMAND ANIMAL NUTRITION ■ SPECIFIC FOR YOUR SUCCESS
www.lallemandanimalnutrition.com
LALLEMAND

Feed Formulation in The Future: Quantum Change or Incremental Steps Forward

La formulation des aliments du futur : changement radical ou avancées progressives

John F. Patience
Department of Animal Science, Iowa State University, Ames, IA, USA
jfp@iastate.edu

Abstract

Feed formulation is not done in isolation but rather occurs in the context of feeding programs. Feeding programs, in turn, integrate the dietary needs of the animal into a proper quantitative and qualitative sequence from birth until market, or throughout their adult life in the instance of the breeding herd. Feeding programs must satisfy an array of needs: to achieve optimal biological performance, to meet carcass and meat quality expectations, to support environmental sustainability, and to fulfil animal health and welfare objectives, all at a price that the marketplace can afford. In this context, diet formulation will continue to evolve, probably not in terms of quantum change but most certainly at an ever-increasing pace. The industry will progress technologically, responding to higher productivity expectations arising from advances in genetics, housing, husbandry, and physiology. The impact of the marketplace, which includes not only individual consumer expectations but also food industry requirements, trade regulations and legislative obligations, will also find its way onto the desk of the nutritionist. Perhaps the biggest change which has already started, is the need to formulate diets on the basis of energy and nutrient requirements concurrent with the harnessing of the functional properties of ingredients, to enhance animal health, improve environmental impacts and benefit animal welfare. All of these big picture issues will be the backdrop for the common day-to-day challenges of the nutritionist associated with supply chain disruption, ingredient variability, individual customer demands, and processing and delivery troubles, just to name a few.

Résumé

La formulation des aliments ne se fait pas de manière isolée, mais plutôt dans le cadre de programmes d'alimentation. Les programmes d'alimentation, quant à eux, intègrent les besoins alimentaires de l'animal dans une séquence quantitative et qualitative appropriée, de la naissance à la commercialisation, ou tout au long de sa vie adulte dans le cas d'un sujet reproducteur. Les programmes d'alimentation doivent répondre à toute une série de besoins : atteindre une performance biologique optimale, répondre aux attentes en matière de qualité de carcasse et de viande, soutenir la durabilité environnementale et atteindre les objectifs de santé et de bien-être des animaux, le tout à un prix abordable pour le marché. Dans ce contexte, la formulation des régimes alimentaires continuera à évoluer, probablement pas de façon radicale, mais très certainement à un rythme toujours plus soutenu. L'industrie progressera sur le plan technologique, répondant aux attentes de productivité plus élevées que permettent les progrès réalisés dans les domaines de la génétique, du logement, des techniques d'élevage et de la physiologie. Les spécialistes de la nutrition animale devront également tenir compte de l'impact du marché, c'est-à-dire des attentes de chaque consommateur, mais aussi des exigences de l'industrie de l'alimentation, des réglementations applicables au commerce et des obligations législatives. Le changement le plus important, déjà en cours, est peut-être la nécessité de formuler les régimes alimentaires sur la base des besoins en énergie et en nutriments, tout en exploitant les propriétés fonctionnelles des ingrédients, afin d'améliorer la santé des animaux, de réduire l'empreinte environnementale et de favoriser le bien-être animal. Toutes ces questions d'ordre général serviront de toile de fond aux nutritionnistes des animaux qui devront relever les défis quotidiens liés aux perturbations de la chaîne d'approvisionnement, à la variabilité des ingrédients, aux demandes individuelles des clients et aux problèmes de traitement et de livraison, pour n'en citer que quelques-uns.

Basic Role of Feed Formulation

Feed formulation is a process by which our knowledge of the nutrition of the animal can be translated into feeding programs which achieve our production objectives. These objectives typically include optimizing growth performance and producing a final carcass that can be converted into safe, healthy and appealing consumer meat products. This must be done in a profitable manner that concurrently contributes to environmental sustainability and animal welfare. Ideally, feeding programs developed in this manner will also result in predictable performance outcomes (Patience, 2017).

Historically, feed formulation has been a process in which the nutrients supplied by available ingredients are matched in required proportions to meet energy and nutrient levels required by the pig for maintenance or maintenance plus some productive purpose such as growth, lactation or pregnancy. In all instances, there is an implied level of performance which is assumed in establishing energy and nutrient requirements. For example, NRC (2012) provides specific growth rates and daily ME intakes levels attached to their requirement tables. For pigs over 25 kg, assumed protein deposition rates are also provided. Feed efficiency can be calculated from the available information. Notably, NRC (2012) does not provide energy requirements, but rather provides the energy levels (DE, ME and NE) that correspond to a typical corn soybean meal; these energy values can be used to determine lysine:energy ratios, phosphorus:energy ratios, etc. to allow adjustment when local conditions dictate that a higher or lower energy level will be more economical. The important point here is that the nutrient requirements are not defined for a pig of a certain body weight range, but rather for a pig of a certain body weight range performing at a specific level. If pigs are performing at different levels, such as in terms of protein deposition rate or daily energy intake, the requirements must be adjusted accordingly.

To illustrate this point, consider a 50 to 75 kg pig growing 975 g/d and with a protein deposition rate of 165 g/d (17% of gain as body protein) might have an SID lysine requirement of 0.90% while a similar pig growing only 800 g/d (PD = 120 g/d; 15% of gain as body protein) might have a lysine requirement of only 0.75%. The importance of economic analysis becomes immediately apparent. Feeding a diet containing 0.90% SID lysine when the pig is only going to grow 800 g/d is not only a waste of nutrients, but the unnecessarily high cost of the diet compounds the economic penalty suffered by the producer whose pigs are unable to grow at the higher level.

To be fully successful in feed formulation, the nutritionist requires at least a basic understanding of many of the biological sciences, including physiology, chemistry, biochemistry and increasingly microbiology and immunology. For example, when a nutritionist is presented with a novel fat source, a number of issues come into play. They must understand its chemistry, in order to determine its quality and composition. They must understand its biochemistry to be able to estimate a reasonable energy value as well as evaluate if the fat source will possibly impact the metabolic oxidative status of the animal. And by understanding both chemistry and physiology, the nutritionist will understand the impact of this particular fat source on de novo lipid synthesis (Shurson et al., 2015; Kellner et al., 2017; Kellner and Patience, 2017).

Looking at the topic more broadly, nutritionists rarely formulate an individual diet without context, and that context is widely known as the feeding program or feed budget – the sequence of diets that sees the growing pig from weaning through to market or the glit/sow from gilt development through mating and weaning. In the case of the newly weaned pig, the nutrient and ingredient composition of the phase 2 diet will be strongly influenced by the composition and feed budget for the phase 1 diet. And the nutrient and ingredient composition of the phase 3 diet will similarly be impacted by that of the phase 2 diet, and so on. Consequently, it is the sequence of diets and not individual diets, which defines the true value of the nutritionist to the feed company or to the pork producer. All of this has to be accomplished within the context of restrictions imposed by a myriad of external forces such as feed mill capability and capacity, restrictions on delivery volumes and supply chain irregularities which have become so common in the past 18 months, to name a few.

Current Challenges in Feed Formulation

Diet formulation is fraught with many challenges, some of which are obvious and others which are quite subtle. For example, when developing a feeding program, should the focus be on the needs of the average pig, the superior pig, somewhere between the two, or even below the average pig? How is this decision affected by the fact that greater feed intake is at least in part supporting the greater performance of the superior pig compared to the average pig. For example, we have previously shown that up to half of these greater requirements are satisfied by above average feed intake (Patience, unpublished data). If formulating to the requirements of the superior pig, should this be the same in all circumstances, when feed prices are low and market prices are high as well as when feed prices are high and market prices are low? In other words, which is more important: maintaining maximal growth rate because barn throughput is so critical in most production systems or optimizing the return over feed cost, expressed on a per pig basis or barn turn basis? Is the decision different in a continuous flow barn compared with a barn operating on an all-in-all-out basis?

Another challenge is acquiring data on the pigs being fed. There is a clear lack of data on the nutrient requirements of animals with different genetic backgrounds. There is also a lack of understanding of how nutrient requirements may have changed due to genetic improvement over the past two or three decades. Do the more modern genotypes actually use nutrients with greater or less efficiency than their predecessors? Do modern genotypes have different maintenance requirements than their predecessors? The real question is whether differences in productivity (sows) and growth rate and feed efficiency (market hogs) explain higher nutrient requirements today compared with the past, or are they due to something more fundamental, such as efficiency of digestion, pace of basal metabolism or the efficiency of post-absorptive utilization. For such an important question, there is scant research upon which to base an opinion. Yet, nutritionists are dealing with this issue pretty much every day of the week. Another crucial bit of information which is frequently unavailable is the feed intake curve for specific batches of pigs being fed; if this information became available, nutritionists could be much more specific in their recommended feeding programs.

The source of ingredient data are anything but homogeneous. Some data have been generated by direct wet chemistry while others have been determined using NIR. Some of the data have been generated on only one or two samples of a particular ingredient, expecting it to be representative of all sources available in the marketplace. To illustrate this point, and using soybean hulls as an example, NRC (2012) has a value for its gross energy based on 1 sample. Dr. Hans Stein maintains an extensive database of nutrient profiles of some 200+ ingredients and has the results of testing only one sample for the DE, ME and NE on soy hulls. Moving closer to home in western Canada, NRC (2012) shows no values for the ME and NE of field peas and Stein shows an NE value based on 1 sample and an ME value based on 5 samples. For a more common ingredient, such as barley, NRC (2012) presents a DE value based on 8 samples and no values for ME or NE. Stein shows DE values based on 17 samples, ME based on 7 samples and NE based on 6 samples. Part of the reason for the low numbers in the NRC (2012) was their policy of using only values presented in refereed sources, which excluded many values presented on websites and in trade magazines; they did this, obviously, to maintain a certain level of quality control. However, the NRC and Stein databases provide the number of samples upon which their numbers are based, which is relatively unique in the work of ingredient evaluation.

In any event, such small numbers for such commonly used ingredients like barley is disappointing, especially when it is well known that its DE, for example, varies by about 15% (Fairbairn et al., 1999). Lopez et al. (2020) reported that, based on four or five samples from each nation, the ME value of soybean meal varies by 10% within Brazil and by 7% within the U.S. Soybean meal from Brazil, on average, had 2.6% greater ME than that sourced from the U.S. Li et al. (2014) evaluated 100 corn samples from the main corn growing regions of China and reported that the ME content varied by 8%, making it one of the most uniform basal ingredients available to the pig industry.

There is also the issue of combining data from different methodologies. For example, some NE values are based on indirect calorimetry and others are based on the use of prediction equations. Usually, the prediction equations are based on data derived from indirect calorimetry (Noblet et al., 1994). Some NE values have been derived from direct calorimetry, although this is rarely used today. Still other NE values have been determined using graded amounts of the test ingredient in growth studies based on feed efficiency (Boyd et al., 2010). Consequently, nutritionists must carefully vet the sources of their ingredient information to ensure they are not mixing methodologies. Evenso, within a methodology, there are differences in how the study was conducted, the nature of the animals used, the preparation of the diets and the interpretation of the results (Noblet et al., 2022).

Another key challenge in diet formulation is the assumption that nutrient utilization changes in a linear fashion, when there are now data showing that this is not completely true. For example, as fiber increases in the diet, the digestibility of other nutrients decline, but not in a linear fashion. Consequently, nutrient loading values employed in diet formulation may be correct at one level of fiber content, but not at another.

It is apparent that what outwardly appears to be a fairly straight forward mathematical process is, in fact, very complex due to the volume of information required to achieve success (Black, 2000). Despite these challenges, feed formulation has helped to move the feeding of animals forward in a very impressive way. For example, it has been estimated that between 1974 and 2020, average market weight increased 22% to 128 kg, average daily gain increased by 33% to 865 g/d and feed conversion has improved 18% to 2.80. In fact, market weight has increased by 23 kg but time to market has actually decreased by 5 days. While genetics, housing and other factors have certainly contributed to this success, nutrition and diet formulation have obviously played a critical role as well.

While feed formulation technology has served the industry well in the past, many challenges remain. Changes will occur in feed formulation as these, and other problems, get corrected. It will also evolve because the industry, and the world it operates in, is changing. The rest of this presentation will identify what I suggest, with all due humility, are some of the changes that we might expect in feed formulation in the coming decades. Please remember this is only one person's opinion, although I discussed the topic with a number of colleagues that I hold in very high regard. I will not attempt to offer timelines for any of the proposed changes; some are already underway and will accelerate while others will transpire over many decades, but most will become part of our lives in the next decade or two.

Energy Systems vs Energy Modelling

Dietary energy represents perhaps the greatest challenge to nutritionists. It is by far the most expensive component of the diet, and it drives performance, including rate and efficiency of gain and carcass composition. The energy content of most ingredients is highly variable, as previously described. Compounding the challenge of energy is the fact that it is utilized with different efficiencies depending on the source of energy (non-starch polysaccharides vs starch vs protein vs fat); independent of source, its efficiency of utilization is also dependent on its fate in the body – maintenance vs protein accretion vs lipid accretion in the carcass (Black, 1995; Birkett and de Lange, 2001).

While measuring energy may be the most repeatable and accurate in the modern nutrition laboratory, such as by isoperibol bomb calorimetry, the most common method in use today, this type of gross measurement fails to provide the detail required to predict performance with a high degree of accuracy. For example, considering a diet containing 79% corn, 16% soybean meal and 2% choice white grease, approximately 71% of its net energy will be derived from starch, 14% from amino acids, 14% from fat, etc. If this diet is reformulated to include higher fiber ingredients such as 20% corn DDGS, 10% wheat middlings and 10% corn (corn declines to 41%, soybean meal drops to 11% and choice white grease increases to 5% to maintain constant net energy), only 54% of the net energy now comes from starch, 26% comes from fat and 18% from amino acids; it is difficult to estimate the energy derived from fibre, but it is quite low. Given our knowledge of how different energy sources are used with differing efficiencies, no classical energy system, not even one as refined as the net energy system, has the capability to provide the information needed to fully understand energy supply and energy utilization.

One possible solution could be an energy model which would be more dynamic than an energy system and thus provide the insight necessary to predict animal performance over a wide array of energy sources and levels, and pig genotypes. The value of such a system is apparent, because it will address the utilization of the most expensive component of the pig's diet. As such, it has the greatest capacity to manage the cost of energy to deliver the most economical feeding program to the pig. Some of this is already accomplished to a greater or lesser extent by growth simulation models. This may be the most advantageous path to follow, because such models bring amino acids and phosphorus into the calculation along with energy (Birkett and de Lange, 2001; Van Milgen et al., 2008). Regrettably there are less than a handful of researchers operating in the public domain worldwide whose focus is on growth simulation modeling; fortunately, there is considerable activity in the private sector at the present time. This should be a matter of priority, because feed formulation needs this elevated level of sophistication to fulfil its commitment to optimum performance, maximum net income, minimal impact on the environment and accurate prediction of animal outcomes.

Expectation for the Future

The net energy system, or its relatives, will continue to expand in market share in North America, although DE and ME are unlikely to completely disappear. In the longer term, energy models which quantify how energy is supplied to the pig and used by the pig will become more sophisticated and more commonly utilized by the industry to most effectively achieve predictable outcomes. Progress and adoption will be impaired by the small number of people working on this topic and the need for detailed training to achieve effective implementation. Energy models could be incorporated into growth models as a means of adoption as well.

Modelling Outcomes Against Input Costs

There are many, many variables affecting pig performance. As one example, surveys have shown that feed intake varies by at least 30% among farms. It is also clear that the utilization of energy and nutrients within feedstuffs is a dynamic phenomenon, as is the energy requirement of pigs for production and maintenance (van Milgen et al., 2008). Traditional diet formulation assumes the opposite – that the available energy and nutrient content of ingredients can be listed as unique and discrete values in tables of feed composition and that amino acid requirements, for example, can be list as a single value for a given weight of pig (PIC, 2016). This approach has achieved near universal application within the pork industry for a variety of reasons, chief among them: 1) insufficient information is available on how to adjustment nutrient utilization and nutrient requirements, 2) difficulty in acquiring information from farms upon which to make adjustments, 3) an industry structure that favours high volume, low cost throughput, 4) absence of tools to handle data needed to make necessary adjustments, 5) lack of training on how to develop and implement feeding programs that are tailored to specific genetics, environment, management, and financial circumstances, and 6) limited quantity of data, especially that collected under commercial conditions, which clearly show the advantages of tailored feeding programs. The situation is perhaps illustrated by the implementation of the Nutrient Requirements of Swine (NRC, 2012) which provides both discrete requirement values based on body weight and values based on three levels of performance (and a model to fully tailor requirements); in practice, the discrete values tend to be the ones most frequently used.

As mentioned above, pressure will continue on nutritionists to find ways to further reduce the cost of production, optimize net income and minimize environmental impacts of pork production. Growth simulation models represent one means by which to achieve these objective (de Lange et al., 2001). The biggest hurdle will be demonstration of the effectiveness of models under practical conditions.

Expectation for the Future

Growth simulation and economic models will increase in use, but progress will be slow, due to the many barriers to their immediate application. As they evolve, they will include not only growth performance and financial outcomes, but also project environmental impact.

Beyond Nutrients - Functional Properties of Feed Ingredients

The choice of dietary ingredients increasingly goes well beyond the need to provide energy and nutrients to the pig; there is growing recognition that the functional properties of ingredients is also important (Shurson et al., 2021). For example, ingredient selection can impact resistance to disease (or exacerbate illness), viscosity of the digesta, oxidative status, rate of passage, the microbial profile, the immune system and bulkiness of the digesta and feces, to name a few. The list is rather long and will become even more critical in antibiotic free production systems. One example of an ingredient possessing unique functional properties is limestone, which has the ability to buffer the pH of the gastrointestinal tract, notably in the stomach. Another example would be the report by Wilberts et al. (2014) that corn DDGS may increase a growing pig's susceptibility to swine dysentery, while replacement of this insoluble fiber source to one which is more readily fermented appears to provide protection (Helm et al., 2021).

Fiber is a well-known functional ingredient, or a critical component of functional ingredients. While fibre has traditionally been viewed as a nuisance in pig diets, it is now viewed as having functional properties that can provide benefit to the pig, especially one experiencing gastrointestinal pathologies. As one example, Li et al. (2019) found that sugar beet pulp, a highly fermentable fibre, when fed along with a carbohydrase enzyme, improves performance of pigs challenged with *Escherichia coli*; the authors also reported improvements in markers of gut barrier integrity and lowered markers of inflammation.

Expectation for the Future

Nutritionists have already been selecting specialty ingredients for the non-nutritive benefit they provided to pigs; this occurred most frequently in phase 1 and 2 starter diets (eg. whey powder, plasma proteins, etc). The practice will become more widespread – and beyond the nursery phase - and will provide more predictable benefits in the future. This will be particularly true in antibiotic free pork production.

Enhanced Understanding of Dietary Fibre and Fibre Assays

From a <u>physiological</u> perspective, fibre is the carbohydrate and lignin fractions of feeds and ingredients that are indigestible by endogenous enzymes; however, the fibre may be fermented in the lower small intestine and large intestine of mammalian species (Fahey et al., 2019). Fiber can also be defined <u>chemically</u> as the sum of all non-starch polysaccharides, oligosaccharides and lignin. This will include celluloses, hemicelluloses, lignin, gums, modified cellulloses, mucilages, oligosaccharides, pectins, β-glucans, waxes, cutin and suberin (DeVries et al., 1999).

Chemical assays to define the quantity of fibre in the diet, or specific fractions of the total fibre, have evolved tremendously over the past 50 years, and include crude fibre (CF), neutral detergent fibre (NDF), amylase-treated NDF (aNDF), acid detergent fibre (ADF), total dietary fibre (TDF), insoluble dietary fibre (IDF), soluble dietary fibre (SDF) and non-starch polysaccharides (NSP; Fahey et al., 2019).

Based on the above assays, arithmetic can be used to provide greater detail on fiber composition. For example, NDF - ADF = hemicellulose and ADF – ADL = cellulose (where ADL is acid detergent lignin). Total NSP can be calculated, rather than doing the wet chemistry, from TDF – ADL and non-cellulosic NSP can be calculated from TDF – (ADF - ADL).

One of the great challenges of fibre analyses is the difficulty of relating in chemical terms, what is happening physiologically. Part of the problem is that cellulose is not a single entity, but rather a structure that varies in composition and thus in its role in the gut. In the same way, hemicellulose is also not a single entity but also varies in its detailed chemical structure based on its origin within the plant kingdom. Given this reality, it is not surprising to learn that the impact of insoluble dietary fibre in the gastrointestinal tract varies because the cellulose and hemicellulose, which are its main constituents, differ. In the same way, soluble dietary fibre is quite diverse in its composition, resulting in quite different impacts in the gastrointestinal tract. As one example, some soluble dietary fibres are quite fermentable while others are very poorly fermented; β-glucan, guar gum and pectins are soluble and fermentable, while psyllium is soluble but not fermentable. Even within soluble dietary fibres, there is quite a range in the rate of fermentability.

It is inevitable that fibre will play an increasing role in swine diets of the future; cereal grains will be used in greater quantities for industrial purposes, leaving co-products to be used where wheat, barley and corn used to be fed. What will be the best methods for quantifying fibre to most effectively predict its function in the gut?

Crude Fibre Method

Crude fibre was developed in the 1860s at the Weende Agricultural Experiment Station in Germany (Henneberg and Stohmann, 1864). It includes most cellulose, but only captures a portion of lignin and hemicellulose and therefore is not a candidate for future quantification of the fibre content of feeds and ingredients. This inability to quantify meaningful portions of dietary fibre means that it should no longer be used in animal nutrition. Unfortunately, it is currently a required method for fibre guarantees according to many regulatory agencies.

Detergent Method

The detergent system of fibre analysis (ADF, NDF, ADL) was developed more than 50 years ago by Dr. Pete Van Soest at Cornell University (Van Soest, 1963). This system has achieved considerable acceptance by swine nutritionists even though it was originally developed for use with forages and for ruminant species.

Perhaps the greatest limitation of the detergent system, as it relates to swine nutrition, is the fact that it does not include soluble dietary fibre. However, the soluble dietary fibre concentration of corn and corn co-products is less than about 1.5% (Navarro et al., 2018; Abelilla and Stein, 2019b). Therefore, the detergent system is a suitable method to quantify the fiber concentration of corn and its co-products. Another limitation of the detergent system is the fact that it is not corrected for the ash and protein remining in the assay residues. This may explain why NDF is frequently found to be higher in corn and its co-products, as compared to TDF.

Nonetheless, the detergent method has proven itself to be effective in explaining the digestibility of energy in numerous ingredients, even those containing significant amounts of soluble dietary fibre. As one example, Fairbairn et al. (1999) found that ADF explained the variation in the apparent total tract digestibility of gross energy in 20 diverse barley samples with R^2 of 0.85. As a second example, Zijlstra et al. (1999) found that a combination of NDF and crude protein explained the variation in the apparent total tract digestibility of gross energy in wheat with R^2 of 0.75.

Going forward, the detergent system has served swine nutritionists well in the past. It is a relatively inexpensive assay; however, there appear to be large differences in assay results among labs, placing suspicion on the veracity of assay results. Knowing that NDF ignores the soluble portion of dietary fibre, it is clear that NDF works reasonably well for diets containing corn and corn co-products, because soluble dietary fibre is present in low amounts. However, when using ingredients containing significant amounts of soluble dietary fibre, NDF has limitations which cannot be ignored.

Total Dietary Fibre Method

Unlike the previous methods, the total dietary fibre method came to animal nutrition from human nutrition, following work that started in the late 1970s and early 1980s. Two methods evolved from this research, one being based on an enzymatic-gravimetric approach. This early approach, referred to as AOAC methods 985.29 and 991.43, could be used to determine total dietary fibre and to separate the result into insoluble and soluble components. However, these methods miss or underestimate a number of important fibre components from the perspective of pig nutrition, namely resistant starch, inulin, fructooligosaccharides and galactooligosaccharides, the latter of which includes raffinose and stachyose which represent between 3 and 5% of soybean meal. The problem was resolved with the development of AOAC 2009.01 and 2011.25, both of which are enzymatic-gravimetric-liquid chromatographic methods. The TDF methods have the obvious advantage over other fibre assay methods, and that is their inclusion of both soluble and insoluble fibre components. However, simply defining the soluble and insoluble content of a fibre source does not necessarily provide an accurate estimation of fermentability or other function in the gut. The biggest impediment to wider adoption is cost, which is much, much greater than that of ADF and NDF, as well as the limited number of laboratories offering this assay. Nonetheless, in order to make progress in the use of dietary fibre, both as a source of nutrients and as a functional ingredient, there is little option than to transition from the detergent methods to the total dietary fibre method.

Non-Starch Polysaccharide Method

One option to the total dietary fibre method is the measurement of non-starch polysaccharides, which provide more detailed information and can still separate assay outcomes into soluble and insoluble components. Knowing the level of individual sugars – and their solubility and insolubility - in the polysaccharide chain will help to distinguish certain functional aspects of fibre beyond that provided by total soluble and total insoluble fibre content. However, the NSP assay is complex and requires sophisticated equipment that many labs lack. It also does not yet have the endorsement by AOAC International, because a standard procedure which consistently demonstrates full recovery of all NSPs has not yet been approved.

Physiocochemical Properties of Fibre

Another approach to predicting the physiological impact of fibre in the diet is to quantify its physicochemical properties. Water holding capacity provides information on stool bulk as well as intestinal transit time. Viscosity will slow gastric emptying and possibly reduce nutrient digestion. Monosaccharide composition and chain conformation can help to estimate the rate and degree of fermentation. There is a lengthy list of physicochemical properties which can be measured; the question remains which ones are relevant swine nutrition. An important but as of yet outstanding issue is the absence of any approved methods for water holding capacity, water binding capacity and viscosity. Without some form of standardized methods, including physiocochemical properties in the evaluation of fibre products for use in pig diets will be extremely problematic – although the concepts have the potential to be valuable.

Other Considerations

One of the most common complaints about all fibre assays is inconsistency of results within and among labs. For this reason, and to achieve cost savings, many academic labs undertake their own fiber analyses. Because even minor deviations from standard protocols can compromise fiber assay outcomes, it is imperative that all procedures are followed with absolute care and dedication.

Expectation for the Future

Expect, or maybe hope, that crude fibre disappears completely. The detergent system will be gradually replaced by the TDF system, with diets formulated on a total dietary fibre and soluble:insoluble ratio basis, with the ratio depending on specific circumstances. The latter will evolve to fermentable:unfermentable ratio. Consideration of physicochemical properties of fibre sources will come into play as well. NSP methods could also become more common as standardized procedures achieve approval.

Enhanced tables of nutrient composition

It has long been known that growing pigs and adult pigs digest energy and nutrients with differing efficiency (Noblet and Shi, 1993). Most comparisons reported in the literature use ad libitum-fed growing pigs and limit-fed gestating sows, due to the challenges of undertaking such studies with lactating sows; this leaves the reader wondering if the difference is due to age, or due to the differences in feed intake. Stein et al. (2001) compared ad libitum-fed growing pigs and lactating sows with limit-fed gestating sows and reported that the biggest differences in amino acid digestibility existed between the growing pigs and gestating sows, with smaller differences with lactating sows. To achieve maximal precision in diet formulation, digestible nutrients should be defined specifically for growing pigs and for sows. The differences are not large and the impact on performance outcomes remains unclear. However, with greater differences in fibrous ingredients, developing separate matrices makes sense.

Expectation for the Future

Over time, separate ingredient matrices will be adopted for growing pigs and for sows; some nutritionists may go so far as to include young pigs as distinct from growing pigs. While some nutritionists already make this distinction, the pace of adoption across the industry will be small, impeded by limited data on the topic, and uncertain financial and performance advantage.

Limited water resources

Water is the "forgotten nutrient" in many ways, one of which is in the assumption that drinking water supplies are unlimited. In reality, increasing pressure is being placed on available potable water sources by urban dwellers and industrial users, placing agriculture on a collision course with other consumers in parts of the U.S. and around the world (Patience, 2012). In the future, water will become increasingly expensive and regulated, something which is already observed in the west coast of the US. Nutritionists have limited control over water utilization by the pig, although some aspects of diet composition influences ad libitum intake (Schiavon and Emmans, 2000; Shaw et al., 2006).

Expectation for the Future

As a limited natural resource, water will become a topic of increasing concern, and nutritionists will be expected to play whatever role they can to minimum water utilization by the pig industry.

Other Anticipated Changes

The following lists other changes which may occur in feed formulation but were not discussed in detail due to space limitations:

1. Big data is coming and whoever said that knowledge is power were right on the money. For example, imagine a situation where data from the genetic nucleus through multiplication, production, harvest, processing and consumer sales could be integrated into an all-encompassing data management/analysis system. The information would transform production by focusing attention – and investment – on processes that really matter to profitability, to environmental sustainability and to consumer preference for the final pork products.

2. We have been trained to develop feeding and management systems that maximize growth rate. Yet, situations will most certainly occur when producers must slow or stop growth, due to interruptions in harvest or animal movement. Research has shown that growth rate can be dramatically slowed or virtually stopped by lowering amino acid intake or feeding an

acidogenic diet, respectively (Helm et al., 2021a, b). Slowing or stopping growth is far superior to mass euthanasia which has been shown to be widely criticized by animal rights organizations and consumers.

3. Attention to the whole subject of climate change and environmental sustainability is growing. Nutritionists would be well advised to work as a national entity to quantify the positive impact they have on the environmental sustainability of pork production. Topics could include conservation of non-renewable resources such as phosphorus and reduction of greenhouse gases.

4. Antibiotic resistance is a critical issue in human medicine and is also involved in veterinary medicine as well. The high profile of this topic has led to consumers requesting pork from animals that have not received antibiotics – in the feed or otherwise. We know that it is impossible to raise all animals without antibiotics; it is not practical and certainly not friendly to the welfare of pigs. Through changes in diet formulation and the careful selection of non-antibiotic feed additives, nutritionists can play a central role in contributing to systems that are compatible with at least the majority of pigs reaching market without antibiotics.

References

Abelilla, J.J., and H.H. Stein. 2019. Degradation of dietary fiber in the stomach, small intestine, and large intestine of growing pigs fed corn-or wheat-based diets without or with microbial xylanase. J. Anim. Sci. 97:338-352. doi:0.1093/jas/sky403

Babcock, B.A. and J.F. Fabiosa. 2011. The impact of ethanol and ethanol subsidies on corn prices: Revisiting history. CARD Policy Brief 11-PB5, Iowa State University, Ames, IA.

Birkett, S. and C.F.M. de Lange. **2001.** A computational framework for a nutrient flow representation of energy utilization by growing monogastric animals. Brit. J. Nutr. 86:661–674. doi:10.1079/BJN2001442.

Black, J.L. 1995. Modelling energy metabolism in the pig – critical evaluation of a simple reference model. In: P.J. Moughan, M.W.A. Verstegen and M.I. Visser-Reyneveld, editors, Modelling growth in the pig, EAAP Publication no. 78. Wageningen Press, Wageningen, The Netherlands. pp. 87–102.

Black, J.L. 2000. Principles behind feed formulation. In: P. J. Moughan, M.W.A. Verstegen, and M. I. Visser-Reyneveld, editors, Feed evaluation – principles and practice. Wageningen Academic Publishers, Wageningen, The Netherlands. p. 209-220.

Boyd, R.D., C.E. Zier-Rush and C.E. Fralick. 2010. Practical method for estimating productive energy (NE) of wheat midds for growing pigs. J. Anim. Sci. 88(E Suppl 3): 89.

De Lange, C.F.M., B.J. Marty, S. Birkett, P. Morel and B. Skotnicki. 2001. Application of pig growth models in commercial pig production. Can. J. Anim. Sci. 81:1-8. doi:org/10.4141/A00-006.

De-Oliveira, L.D., F. S. Takakura, E. Kienzle, M.A. Brunetto, E. Teshima, G.T. Pereira, R.S. Vasconcellos and A.C. Carciofi. 2012. Fibre analysis and fibre digestibility in pet foods - a comparison of total dietary fibre, neutral and acid detergent fibre and crude fibre. J. Anim. Physiol. Anim. Nutr. 96:895-906. doi:org/10.1111/j.1439-0396.2011.01203.x.

DeVries, J.W., L. Prosky, B. Li and S. Cho. 1999. A historical perspective on defining dietary fiber. Cereal Foods Worldwide 5:367-369.

Fahey, Jr., G.C., L. Novotny, B. Layton and D.R. Mertens. 2019. Critical factors in determining fiber content of feeds and foods and their ingredients. J. A.O.A.C Int'l. 102:52-62. doi:org/10.5740/jaoacint.18-0067.

Fairbairn, S.L., J.F. Patience, H.L. Classen and R.T. Zijlstra. 1999. The energy content of barley fed to growing pigs: characterizing the nature of its variability and developing prediction equations for its estimation. J. Anim. Sci. 77:1502-1512. doi:org/10.2527/1999.7761502x.

Gutierrez, N.A., N.V.L. Serao and J.F. Patience. 2016. Effects of distillers' dried grains with solubles and soybean oil on dietary lipid, fiber, and amino acid digestibility in corn-based diets fed to growing pigs. J. Anim. Sci 94:1508-1519. doi:10.2527/jas.2015-9529.

Helm, E.T., N.K. Gabler and E.R. Burrough 2021. Highly fermentable fiber alters fecal microbiota and mitigates swine dysentery induced by brachyspira hyodysenteriae. Animals 11:396. doi:org/10.3390/ani11020396.

Helm, E.T., J.F. Patience, M.T. Romoser, C.D. Johnson, J.W. Ross and N.K. Gabler. 2021a. Evaluation of increased fiber, decreased amino acids, or decreased electrolyte balance as dietary approaches to slow finishing pig growth rates. J. Anim. Sci. 99:skab164. Doi:org/10.1093/jas/skab164.

Helm, E.T., J.W. Ross, J.F. Patience, S.M. Lonergan, E. Huff-Lonergan, L.L. Greiner, L.M. Reaver, C.W. Hastad, K.J. Prusa, C.W. Hastad, E.K. Arkfeld and N.K. Gabler. 2021b. Nutritional approaches to slow late finishing pig growth: implications on carcass composition and pork quality. J. Anim. Sci. 99:skaa368. doi:10.1093/jas/skaa368.

Hen0neberg, W. and F. Stohmann. 1864. Beiträge zur Begründung einer rationellen Fütterung der Wiederkäuer, Vol. 2, Schwetschtke, Braunschweig, Germany.

Kellner, T.A., N.K. Gabler and J.F. Patience. 2017. The composition of dietary fat alters the transcriptional profile of pathways associated with lipid metabolism in the liver and adipose tissue in the pig. J. Anim. Sci. 95:3609-3619. doi:10.2527/jas2017.1658.

Kellner, T.A. and J.F. Patience. 2017. The digestible energy, metabolizable energy, and net energy content of dietary fat sources in thirteen and fifty-kilogram pigs. J. Anim. Sci 95:3984-3995. doi:10.2527/jas2017.1824.

Li, Q., J. Zang, D. Liu, X, Piao, C. Lai and D. Li. 2014. Predicting corn digestible and metabolizable energy content from its chemical composition in growing pigs. J. Anim. Sci. Biotechnol. 5:1. doi:org/10.1186/2049-1891-5-11.

Li, Q.Y., E.R. Burrough, N.K. Gabler, C.L. Loving, O. Sahin, S.A. Gould and J.F. Patience. 2019. A soluble and highly fermentable dietary fiber with carbohydrases improved gut barrier integrity markers and growth performance in ETEC challenged pigs. J. Anim. Sci. 97:2139-2153. doi:10.1093/jas/skz093.

Lopez, D.A., L.V. Lagos and H.H. Stein. 2020. Digestible and metabo;lizable energy of soybean meal sources from different countries and fed to pigs. Anim. Feed Sci. Technol. 268:114600. doi:org/10.1016/j.anifeedsci.2020.114600.

Moughan, P. 1989. Simulation of the daily partitioning of lysine in the 50 kg liveweight pig-A factorial approach to estimating amino acid requirements for growth and maintenance. Res. Dev. Agric. 6:7–14.

Navarro, D.M., E.M. Bruininx, L. de Jong and H.H. Stein. 2018. Effects of physicochemical characteristics of feed ingredients on the apparent total tract digestibility of energy, DM, and nutrients by growing pigs. J. Anim. Sci. 96:2265-2277. doi:10.1093/jas/sky149.

Noblet, J., H. Fortune, X. Shi and S. Dubois. 1994. Prediction of net energy value of feeds for growing pigs. J. Anim. Sci. 72:344-354. doi:org/10.2527/1994.722344x.

Noblet, J. and X.S. Shi. 1993. Comparative digestibility of energy and nutrients in growing pigs fed ad libitum and adults sows fed at maintenance. Livestock Prod. Sci. 34:137-152. doi:org/10.1016/0301-6226(93)90042-G.

Noblet, J., S-B. Wu and M. Choct. 2022. Methodologies for energy evaluation of pig and poultry feeds: A review. Anim. Nutr. 8:185-203. doi:org/10.1016/j.aninu.2021.06.015.

NRC. 2012. Nutrient requirements of swine. 11th rev. ed. Natl. Acad. Press, Washington, DC.

Patience, J.F. 2012. The importance of water in pork production. Anim. Frontiers 2(2):28-35. doi:org/10.2527/af.2012-0037.

Patience, J.F. 2017. The theory and practice of feed formulation. In P. Moughan, K. de Lange and W. Hendriks, editors, Feed evaluation science. Wageningen Academic Press, Wageningen, The Netherlands. p. 457 – 490.

Patience, J.F. and A.L. Petry. 2019. Susceptibility of fibre to exogenous carbohydrases and impact on performance in swine. In: G. Gonzalez-Ortiz, M.R. Bedford, K.E. Bach Knudsen, C. Courtin and H.L. Classen, editors, The value of fibre. Engaging the second brain for animal nutrition. Wageningen Academic Press, Wageningen, The Netherlands. p. 99-115. doi:10.3920/978-90-8686-3_5.

PIC. 2016. Nutrient specifications manual. Pig Improvement Company, Hendersonville, TN.

Remus, A. L. Hauschild, M.-P. Létourneau-Montminy and C. Pomar. 2021. Estimating amino acid requirements in real-time for precision-fed pigs: the challenge of variability among individuals. Animals 11:3354. doi:org/10.3390/ani11123354.

Remus, A., J. van Milgen and C. Pomar. 2020. Precision feeding optimises efficiency of protein utilisation in pigs. In Proceedings of the 71st Annual Meeting of the European Federation of Animal Science (EAAP), Porto, Portugal.

Schiavon, S. and G.C. Emmans. 2000. A model to predict water intake of a pig growing in a known environment on a known diet. Brit. J. Nutr. 84:873-883. doi:org/10.1017/S000711450000249X.

Schinckel, A.P. and C.F.M. de Lange. 1996. Characterization of growth parameters needed as inputs for pig growth models. J. Anim. Sci. 74:2021-2036. doi:10.2527/1996.7482021x.

Shaw, M.I., A.D. Beaulieu and J.F. Patience. 2006. Effect of diet composition on water utilization in growing pigs. J. Anim. Sci. 84:3123-3132. doi:org/10.2527/jas.2005-690.

Shurson, G.C., B.J. Kerr and A.R. Hanson. 2015. Evaluating the quality of feed fats and oils and their effects on pig growth performance. J. Anim. Sci. Biotechnol. 6:10. doi:org/10.1186/s40104-015-0005-4.

Shurson, G.C., Y-T. Hung, J.C. Jang and P.E. Urriola. 2021. Measures matter—determining the true nutri-physiological value of feed ingredients for swine. Animals 11:1259. doi:org/10.3390/ani11051259.

Stein, H.H., S.W. Kim, T.T. Nielsen and R.A. Easter. 2001. Standardized ileal protein and amino acid digestibility by growing pigs and sows. J. Anim. Sci. 79:2113-2122. doi:org/10.2527/2001.7982113x.

van Milgen, J., A. Valancogne, S. Dubois, J-Y. Dourmad, B. Sève and J. Noblet. 2008. InraPorc: A model and decision support tool for the nutrition of growing pigs. Anim. Feed Sci. Technol. 143:387-405. doi;org/10.1016/j.anifeedsci.2007.05.020.

Van Soest, P. J. 1963. Use of detergents in the analysis of fibrous feeds. I. Preparation of fiber residues of low nitrogen content. J. Assoc. Off. Anal. Chem. 46:825–829. doi:org/10.1093/jaoac/46.5.825. doi:org/10.1093/jaoac/46.5.825.

United Nations. 2011. The global food crisis. In The global social crisis. Dept. of Economic and Social Affairs, New York, NY. p. 61-74.

Wilberts, B.L., P.H. Arruda, J.M. Kinyon, T.S. Frana, C. Wang, D.R. Magstadt, D.R. Madson, J.F. Patience and E.R. Burrough. 2014. Investigation of the impact of increased dietary insoluble fiber through the feeding of distillers dried grains with solubles (DDGS) on the incidence and severity of Brachyspira-associated colitis in pigs. PLoS ONE. 9:e114741. doi:org/10.1371/journal.pone.0114741.

Zijlstra, R.T., C.F.M. de Lange and J.F. Patience. 1999. Nutritional value of wheat samples for growing pigs: Chemical composition and digestible energy content. Can. J. Anim. Sci. 79:187-194. doi:org/10.4141/A98-103.

Graduate Student Posters
Affiches des étudiants

ANCC
CNAC

Application of Near-Infrared Spectroscopy for Rapid Analysis of Glucosinolate Content in Canola Meal

Andy Q. Che[1], Anna Rogiewicz[1], Veronique J. Barthet[2], Nicole Pogorzelec[2] and Bogdan A. Slominski[1]
[1]Department of Animal Science, University of Manitoba, Winnipeg, MB
[2]Canadian Grain Commission, Winnipeg, MB
anna.rogiewicz@umanitoba.ca
andy.che@umanitoba.ca

Abstract

The objective of this project is to assess glucosinolates (GLS) qualitatively and quantitatively in canola meal (CM) using near-infrared spectroscopy (NIRS) calibration models.

The variation in chemical composition and nutritive value of CM among the processing plants across Canada has been demonstrated, and it was concluded that they are directly related to the processing conditions, mainly due to heat and moisture treatment. The highest variability has been observed in the lysine digestibility and in content of GLS, which are categorized as anti-nutritive factors (ANF) for poultry and swine. Therefore, a reliable technology is needed for rapid determination of CM quality and to facilitate an effective use of CM in poultry and swine diets. Near-infrared spectroscopy has been used by the feed industry, however, CM analysis by NIRS is often restricted to very few measurements. Our research focuses on the development of advanced calibration curves for the direct and reliable assessment of CM quality. Use of NIRS to analyze GLS is a relatively new concept, and thus, requires development of advanced calibration curves for precision and reliability. Current research project involves determination of GLS content using gas chromatography (GC) that allows for NIR calibration for total GLS. A satisfactory correlation between determined GLS values to NIR spectra (r=0.768) using over 200 CM samples collected from all Canadian canola crushing plants was achieved. Samples of CM collected in Asia, Europe and Australia with higher GLS content were added to this calibration to improve the precision. Future research will focus on validation and further improvement of calibration. Already a well-balanced source of amino acids, successful analysis of GLS with NIRS could greatly improve potential of CM for higher inclusion in feed without adverse effects to livestock animals.

Key words: glucosinonates, near-infrared spectroscopy, canola meal

Apparent Metabolizable Energy and Plasma Biochemical Profile in Broiler Chickens Fed Diets Containing Grape (*Vitis vinifera*), Cranberry (*Vaccinium macrocarpon*), Wild Blueberry (*Vaccinium angustifolium*) and Apple (*Malus pumila/domestica*) Pomaces Without or With Multi-Enzymes Supplement

Munene Kithama[§], Kelly Ross[‖] and Moussa S. Diarra[*], and Elijah G. Kiarie[§/]*
[*]Guelph Research and Development Centre, AAFC, Guelph, ON
[‖]Summerland Research and Development Centre, AAFC, Summerland, BC
[§]Animal Biosciences, University of Guelph, Guelph, ON
mkithama@uoguelph.ca

Abstract

Fruit pomaces are rich in polyphenols and carbohydrates including fibers which could be a good source of energy in broiler diets. The aim of this study was to evaluate the effect of multi-enzyme supplement (**MES**) on the digestibility of apple (**APL**), low bush wild blueberry (**LBP**), organic cranberry (**CRP**) and grape (**GRP**) fruit pomaces and plasma metabolites in broiler. A total of 360-d old Ross 708 male chicks were placed in 72 cages (n=8), fed a commercial starter diet from d 0 to 13, then from d14 to 21 to a TiO_2-marker containing basal diet supplemented or not with 30% of studied pomaces in the presence or absence of MES. Excreta samples were collected on d 17 to 20 and 1 bird/cage bled on d 21 for plasma. APL pomace had significant effect on AME (P=0.008) but no enzyme effect (P=0.109) on AME. Interaction was significant for plasma alkaline phosphatase (P=0.04). Highest levels of cholesterol and bile acids (P<0.01) were observed in birds fed APL-diet. In conclusion, this study demonstrated that the use of the four pomaces could be used as energy sources in poultry feed and did not have detrimental effects on the physiology of the birds compared to the baseline profile of plasma. Additionally, MES did not improve the AME of pomaces, therefore, MES would not be taken into account in APL, LBP, CRP and GRP pomaces when formulating broiler diets.

Key words: broiler chicken, enzymes, fruit pomaces, metabolizable energy, plasma

Effects of *In Ovo* Seaweed Polyphenols and Fucoidans on Performance and Gut Health of Broiler Chickens Challenged with Diet-induced Oxidative Stress

Jing Lu[1], Stephanie Collins[1]
[1] Department Animal Science and Aquaculture, Dalhousie University, Truro, NS
s.collins@dal.ca , jinglu@dal.ca

Abstract

Eliminating in-feed antibiotics while maintaining the gastrointestinal and overall health of chickens is a strong priority of the poultry industry. Demonstrated immunomodulatory, antioxidant and antimicrobial activities of seaweed extracts could be used to enhance chicken gut health. The present study evaluated an early nutritional intervention to enhance the intestinal health of chickens. Seaweed polyphenols or fucoidans (*Fucus spp.)* were administered *in ovo* to chicken embryos via the air sac route on incubation day 17.5. In a randomized complete block design, 504 incubating eggs (9 treatments * 4 blocks * 14 eggs per experimental unit) were used to test five experimental and four control treatments. The experimental treatments administered (0.5ml injection dose for all) were: 0 mg/ml bioactive (saline), 30 mg/ml polyphenols, 60 mg/ml polyphenols, 30 mg/ml fucoidans or 60 mg/ml fucoidans via *in ovo* injection. The chicks were raised to day 14 on a standard grower diet, then oxidative stress was induced from days 15-28 with an oxidized dietary fat (5%) challenge (oxidized diet). Two challenged controls were fed a standard diet until day 14 then switched to oxidized diets (one medicated, one non-medicated) from days 15-28. Two non-injected, unchallenged controls were fed standard diets (one medicated, one non-medicated) throughout the entirety of the trial. The hatchability and mean hatch weight of all treatments were similar ($p>0.05$), except the 60 mg/ml fucoidans treatment, which had a significantly lower hatch rate, but not mean hatch weight, than the other treatments. During the grow-out stage, the mean body weight, feed intake and feed conversion ratio were similar among all treatments, despite the diet-induced dietary stress some treatments experienced. This was expected, as the challenge was intended to induce intestinal stress, not impair growth. Ileal tissue oxidative status and intestinal microbiome data, as assessments of the gut health status of the chickens administered the experimental treatments, are pending and will be presented at the conference.

The Prebiotic Effect of Enzymatically Released Bioactive Components of Canola Meal Fibre on Gut Health and Growth Performance of Broiler Chickens

Yanxing Niu[1], Anna Rogiewicz[1], Joao Ferreira[1], Rob Patterson[2], and Bogdan A. Slominski[1]
[1]Department of Animal Science, University of Manitoba, Winnipeg, MB
[2]CBS Bio-Platforms., Calgary, AB
anna.rogiewicz@umanitoba.ca

Abstract

Canola meal (CM) contains significant amount of non-starch polysaccharides (NSP), which might be beneficial for monogastric animals when modified with enzymes. We hypothesize that bioactive components enzymatically released from CM NSP have prebiotic effect on monogastric animals. To obtain the bioactive components, CM was pretreated with enzymes, which were selected after advanced *in vitro* and *in vivo* evaluations. The enzymatically modified CM (ECM) and enzymatically modified CM solubles (ECMS) were produced and their prebiotic effect was evaluated. Through a series of *in vitro* screening studies, the most effective enzyme combination was selected. It showed ability to hydrolyze 47.9% of CM NSP. The results of study with broiler chickens showed improvement in growth performance and NSP digestibility when diets were supplemented with selected multicarbohydrase. The same enzyme blend was used to produce ECM and ECMS.

Improvement in the total tract digestibility of NSP was observed when ECM replaced CM in the broiler diet. *Lactobacillus spp.* were present in much larger counts ($P<0.01$) in ileal and cecal digesta of birds fed the ECM diets when compared to the control. The AMEn value of ECM was 30.9% higher than CM. The standardized ileal amino acid digestibility of CM and ECM was similar ($P < 0.05$). Results of another study showed that supplementation with ECMS improved growth performance of broilers, modulated the GIT microbiome and increased the total SCFA in ileum ($P<0.05$). The abundance of *Lactobacillus spp.* and *Enterococcus spp.* were higher in the ileal and cecal digesta of birds fed ECMS diets than in the control.

Modification of CM with carbohydrases significantly changed the composition and structure of CM fiber. Both ECM and ECMS products expressed their prebiotic properties and improved the gut health of broilers. The enzymatic modification of CM can create the value added feed additives, which could enhance the health of the flock, and support the antibiotic-free feeding programs in broiler chickens.

Key words: carbohydrase, canola meal, enzymatically modified canola meal (ECM), broiler chicken

Research on Evaluation of Expeller/Cold-pressed Canola (EPC) as a Valuable Feed Ingredient for Poultry

Tetiana Sessingnong[1], Oluwakemi Omotosho[1], Anna Rogiewicz[1], Yanxing Niu[1], Joao Ferreira[1], Rob Patterson[2], and Bogdan A. Slominski[1]
[1]Department of Animal Science, University of Manitoba, Winnipeg, MB
[2]CBS Bio-Platforms, Calgary, AB
anna.rogiewicz@umanitoba.ca

Abstract

Expeller/cold-pressed canola (EPC) is a co-product of oil production obtained by the mechanical pressing/expelling of canola seed. The differences in techniques of cold-press extraction highly affect the variability among EPC. The objective of this research is to develop a thorough nutrient profile of the EPC to identify its best possible applications and to optimize nutrient utilization in formulating poultry and swine diets. Twenty-two samples of EPC were collected across Western Canada and analyzed using advanced wet chemistry procedures. On average and DM basis, EPC contains 36.6% of crude protein, 16.8% of fat, 31.7% of dietary fiber, and 6.15% of ash. Variation in the chemical composition of EPC samples was observed. Variability in fat and glucosinolates contents (range from 8.5 to 24.1%, DM, and 4.7 to 14.7 µmol/g DM, respectively) reflects the difference in processing methods. Differences in the content of other chemical components of EPC samples would not be qualitative but quantitative due to the variable fat content. It also influenced the variation in the AMEn of EPC which was determined in the broiler chicken study. The AMEn averaged 2322 kcal/kg DM but ranged between 1877.8 and 2642.6 kcal/kg DM. Ongoing research is carried out to determine the amino acid digestibility, to develop the equations for prediction on the nutritive value of EPC, and to develop a nutritional strategy for efficient utilization of EPC in monogastric animal nutrition.

Key words: expeller/cold-pressed canola (EPC), broiler chicken, chemical composition, glucosinolates, AMEn

Effects of Substituting Barley Grain and Silage with High-moisture Corn Products in Western Canadian Beef Finishing Diets on Ruminal Fermentation Patterns and Site and Extent of Starch Digestibility

Rachel E. Carey[1], Mikaela G. Evans[1], Gabriel O. Ribeiro[1], Diego Moya[2], Herbert (Bart) A. Lardner[1], Zachary D. Paddock[3], Timothy A. McAllister[4], and Gregory B. Penner[1]
[1]Department of Animal and Poultry Science, University of Saskatchewan, Saskatoon, SK
[2]Department of Large Animal Clinical Sciences, University of Saskatchewan, Saskatoon, SK
[3]Feedlot Health Management Services, Ltd., Okotoks, AB
[4]Lethbridge Research and Development Centre, Agriculture and Agri-Food Canada, Lethbridge, AB
rachel.carey@usask.ca

Abstract

With the introduction of short-season hybrids, corn production in western Canada is increasing with the majority being used for corn silage or standing corn grazing. High-moisture corn products (high-moisture corn and snaplage) could be harvested to replace traditional barley-based components in feedlot diets, but there are no studies evaluating the use of these products from short-season corn hybrids. The objective of this study was to investigate the impact of partially replacing barley grain and barley silage with high-moisture corn products on ruminal fermentation and the site and extent of starch digestion for heifers fed finishing cattle diets. Six heifers (420 ± 21 kg body weight) equipped with ruminal and duodenal cannulas were used in a replicated 3×3 Latin square with 25-d periods. The control diet (**BG**; DM basis) contained dry-rolled barley grain (85%) and barley silage (9.7%). For the high-moisture corn treatment (**HC**), high-moisture corn replaced 50% of the barley grain, and for the snaplage treatment (**SN**), the silage and 12.5% of the barley grain were replaced by snaplage relative to BG. Ruminal digesta samples were collected over 24 h and analyzed for short-chain fatty acids (SCFA) and ammonia. Indwelling ruminal pH meters recorded pH every 5 minutes. Duodenal and fecal samples were collected to determine the site and extent of digestibility. Dry matter intake (6.9 ± 0.4 kg) and starch intake (3.9 ± 0.2 kg) did not differ among treatments, but NDF intake was greater for SN than HC ($P = 0.02$), although neither differed from BG. Ruminal starch digestibility was greater ($P = 0.05$) for SN than BG, while these treatments did not differ from HC. Intestinal and total tract starch digestibility were not affected. Mean ruminal pH averaged 6.19 ± 0.13 and did not differ among treatments. Likewise, concentrations of total SCFA and the molar proportion of individual SCFA did not differ. This research indicates that high-moisture corn can be used as a partial replacement for barley grain and that although snaplage increases the digestion of starch in the rumen, it can be used to partially replace barley grain and silage in finishing cattle diets without impacting total tract starch digestibility.

Key words: starch, snaplage, high-moisture corn, feedlot

Feeding Supplemental Energy to Pregnant Beef Cows During Late Gestation to Support Cow-Calf Performance

E. Croft, K. Hare, K. M. Wood, M. Steele,
Department of Animal Biosciences, University of Guelph, Guelph, ON
ecroft@uoguelph.ca

Abstract

Nutrient requirements increase rapidly during late gestation due to exponential fetal growth; yet, changes in dry matter intake and nutrient digestion may limit the cow's capability to meet her late gestation requirements when fed a forage-based diet. Thus, the objective of this study was to evaluate how supplementing metabolizable energy (ME) during late gestation impacts cow performance and calf growth. At 53 d prior to calving, primiparous (PP; n = 45) and multiparous (MP; n = 107) Angus-Simmental cattle were blocked by their expected calving date and randomly assigned to diets formulated to provide 80 (LME; n = 52), 100 (CME; n = 51), or 120% (HME; n = 49) of their predicted ME requirements. Postpartum, all cows were fed the same ration. Cattle were weighed on d -53, -39, -25, -10, and -3 and on d 7, 13, 27, and 55 relative to calving. Rib and rump fat depths were measured by ultrasound on d -54, -40, -26 and -and on d 13, 27, and 55 relative to calving. Calves were weighed at birth before nursing and on d 7, 15, 28, 55, and 207 (weaning). Preweaning average daily gain (ADG) was calculated using weight gain from birth to d 55. Weaning weight was adjusted to 205 d of age. From d -53 to -3, HME and CME cows had greater ($P < 0.01$) body weight gain than LME cows. Treatment did not affect change in body weight from calving to d 55 postpartum ($P = 0.62$). From d -53 to d -3, HME cows lost less rump fat (treatment×time; $P < 0.01$), but rib fat depth was not affected by treatment ($P = 0.62$). Neither rump nor rib fat depth were affected ($P \geq 0.13$) by treatment postpartum. Multiparous cows had greater ($P < 0.01$) rump fat depth than PP heifers in the prepartum and postpartum periods, but rib fat depth was greater ($P = 0.03$) for MP than PP cows during only the postpartum period. Calf preweaning weights were not affected ($P = 0.89$) by the prepartum ME supplementation and neither ($P = 0.88$) was calf weaning weight. Treatment did not affect calf ADG ($P = 0.74$). These results suggest that providing ME above estimated requirements during late gestation corrects energy deficits experienced by the cow. This reduces the rate of prepartum body reserve mobilization in the cow but does not benefit calf growth.

Key words: beef, metabolizable energy, gestation

Using Synchrotron-Based on X-ray Fluorescence to Study Oat Varieties Impact on Minerals Distribution

Ganqi Deng[1], Miranda Lavier[2], Gosia Malgorzata Korbas[2], Viorica F. Bondici[2], Chithra Karunakaran[2], David Christensen[1], Bart Lardner[1], and Peiqiang Yu[1]*

[1]Department of Animal and Poultry Science, University of Saskatchewan, Saskatoon, SK

[2] Canadian Light Source, University of Saskatchewan, Saskatoon, SK

Peiqiang.yu@usask.ca

Abstract

The objective of this study was to use the synchrotron based on X-ray fluorescence (XRF) to investigate the micro-elements (Mn, Se, Cu, Zn, Fe, Mo) and macro-elements (Ca, P, S, K) distribution in different cool-season oat varieties. All the samples are from the same location and soil conditions. Four varieties of oats including Arborg, Nasser, Haymaker and Summit provided by the Crop Development Center (CDC, Aaron Beattie) of University of Saskatchewan were soaked in ultra-pure water overnight. These samples were flash frozen using liquid nitrogen and then sectioned at 80 μm with the Leica CM1950 cryostat. The samples were scanned by synchrotron-based X-ray fluorescence with 5 μm × 5 μm solution and 15 keV energy beam at the BioXAS-Imaging beamline of Canadian Light Source (University of Saskatchewan). The results show that Ca, P and K were mainly distributed in aleurone layer, crease region and germ but S was also found in the starch endosperm in any varieties. The concentration of macro-elements was higher increase region for all varieties. Compared with other oat varieties, the concentration of macro-elements was relatively lower in germ for Summit variety. As for micro-elements, Mn, Zn, and Fe were mainly distributed in aleurone layer, crease region and germ except starch endosperm. However, Se, Cu and Mo could be found in the starch endosperm, especially for Selenium. The concentration of Fe was relatively lower compared to other micro-elements in any oat variety. Finally, Cu appeared very low for CDC Nasser oat. In conclusions, synchrotron-based X-ray fluorescence is a very useful tool and has a great potential to reveal the macro and micro-elements distribution in oat cereal varieties.

Key words: synchrotron; X-ray fluorescence; micro-elements; macro-elements; cereal oats

Effect of Dry or Temper Rolling of High or Low Protein Wheat and its Impact on Rumen Parameters, Liver Abscesses, and Growth Performance of Feedlot Cattle

A.C. Meadows [a,b], *S.A. Terry* [a], *G.B. Penner* [b], *T.A. McAllister* [a], *and G.O. Ribeiro* [b]
[a] *Lethbridge Research and Development Centre, Agriculture and Agri-Food Canada, Lethbridge, AB*
[b] *Department of Animal and Poultry Science, University of Saskatchewan, Saskatoon, SK*
acm942@usask.ca

Abstract

The objective of this study was to assess the impact of dry (DR) vs. temper rolled (TR) and low (13%, LP) vs. high crude protein (18%, HP) wheat grain on rumen parameters, liver abscess incidence, and growth performance of feedlot cattle. One hundred and sixty Angus-cross steers (302 ± 34 kg; 24 ruminally cannulated) were used in a backgrounding (BG) to finishing (FN) feedlot trial. Steers were blocked by weight and randomly assigned to 16 pens. The cannulated steers (3 per pen) were housed within 8 of the pens that were equipped with a feed intake monitoring system. The BG diet consisted of 60% barley silage, 35% wheat grain, and 5% supplement, and the FN diet contained 10% barley silage, 85% wheat grain, and 5% supplement (DM basis). Transition (TN) diets included sequential increases to the proportion of wheat grain in the total mixed ration. Ruminal samples were collected in each phase and ruminal pH was measured using indwelling loggers. Steers experienced lower ($P \leq 0.03$) ruminal pH with HP-DR and LP-TR than HP-TR wheat during the TN phase. Tempering HP wheat reduced ($P \leq 0.03$) the generation of fine particles and bouts of low ruminal pH. Steers fed HP wheat had higher ($P \leq 0.01$) ruminal NH_3 concentrations throughout the BG and TN phases. Ruminal NH_3 concentrations were higher ($P = 0.01$) during the FN phase for steers fed LP-DR compared to LP-TR wheat. The C2:C3 ratio for BG and FN steers was greater ($P \leq 0.01$) with HP wheat; with a reduction ($P < 0.01$) in the C2:C3 ratio by TR wheat during the FN phase. Steers fed LP-DR wheat during the FN phase had the shortest ($P < 0.01$) meal durations, and DMI was reduced ($P = 0.04$) with HP wheat. Steers fed TR diets had slower ($P = 0.01$) eating rates and consumed more ($P = 0.02$) meals with shorter ($P = 0.03$) intervals between meals. G:F and NEg were improved ($P \leq 0.04$) with HP wheat during the BG phase. NEg values were higher ($P \leq 0.03$) with HP wheat in the FN phase. Liver abscesses were notably more ($P < 0.01$) severe in steers fed HP wheat. Results suggest that HP wheat may offer advantages for BG cattle, but at the cost of an increase in the severity of liver abscesses. Processing method did not impact growth performance, liver abscesses, or carcass characteristics.

Key words: wheat, beef cattle, protein, processing, liver abscesses

Complete Replacement of Inorganic Sources of Supplementary Trace Minerals by Organic Sources Alters Systemic Trace Mineral Status, Rumen Fermentation, and Energy Balance in Dairy Cows

Bruna Mion[1], Bryn Van Winters[1], Kristina King[1], Jose F. W. Spricigo[1], Lori Ogilvie[1], Leluo Guan[2], Trevor J. DeVries[1], Michael A. Steele[1], Stephen J. LeBlanc[3], and Eduardo S. Ribeiro[1]
[1]Department of Animal Biosciences, University of Guelph, Guelph, ON
[2]Department of Agricultural, Food and Nutritional Science, University of Alberta, Edmonton, AB
[3]Department of Population Medicine, University of Guelph, Guelph, ON
bmion@uoguelph.ca

Abstract

Trace minerals (TM) are required for optimal health and performance of dairy cows and the source of supplementation can affect their availability to rumen microbes and absorption in the digestive tract. Our objectives were to evaluate the effects of replacement of supplementary inorganic TM (ITM) by organic TM (OTM) in pre- and postpartum diets on feeding behavior, rumen physiology, energy metabolism, TM status, and performance. Pregnant cows and heifers (n=273) were randomly assigned to ITM or OTM diets at 45 d before expected calving (study d 0). Groups received the same diet, except by the source of TM fed at 100% recommended levels. The ITM group received Co, Cu, Mn and Zn sulfates and Na selenite, and the OTM group received Co, Cu, Mn and Zn proteinates and selenized yeast (Bioplex® and Sel-Plex®, Alltech). Automated feed bins were used to assign treatments and to measure ingestive behavior. Rumination was monitored by sensors and ruminal fluid was collected on d -21, 23, and 65. Metabolites and TM were evaluated on d -45, -21, -14, -10, -7, -3, 0, 3, 7, 10, 14, 23, 65 and 105. OTM tended to have longer feeding time (197 vs 188 min/d) and greater dry matter intake (DMI) (13.3 vs 12.9 kg/d), had a more positive energy balance (4.2 vs 3.6 Mcal/d) and shorter rumination time (37.5 vs 40.1 min/kg of DM) than ITM in the prepartum period. In the postpartum period, OTM increased DMI in multiparous (24.7 vs 24.1 kg/d) but not in primiparous cows (18.9 kg/d). Milk yield was not affected in multiparous cows (44 kg/d), however, OTM primiparous cows had lesser yield than ITM primiparous cows (29.8 vs 31.9 kg/d). OTM had a greater milk protein percentage (3.17 vs 3.11%), reduced NEFA in serum (0.40 vs 0.45 mmol/L) and rumination activity (27.8 vs 30.1 min/kg DM). On d 23, OTM had reduced molar proportion of acetate and pH, and tended to have greater concentration of total VFA in ruminal fluid. OTM had greater Se in serum and milk, reduced Se in urine, and greater prepartum Co and reduced postpartum Co in serum. In conclusion, complete replacement of ITM by OTM caused moderate changes in rumen physiology, behavior, energy metabolism, TM status, and performance, and seemed to reduce the energy deficit during the transition period.

Key words: minerals, feed intake, rumen

Assessment of Rumen and Fecal Short Chain Fatty Acids Profiles and Microbial Populations in Rumen Fluid and Feces of Beef Cows under Two Different Feeding Regimes

Sang Weon Na[1], Mi Zhou[1], Edward Bork[1], Carolyn Fitzsimmons[1,2], and Leluo Guan[1]
[1] Department of Agricultural, Food & Nutritional Science, University of Alberta, Edmonton, AB
[2] Agriculture and Agri-Food Canada, Edmonton, AB
sangweon@ualberta.ca

Abstract

For cow-calf production of beef industry in Canada, grazing on grasslands and seeded pastures provide a chance for farmers to reduce feeding costs compared to feeding grain-based diets. Beef cows usually graze on seeded pastures or native grasslands with highly diverse vegetation and plant species from spring to fall and feed silage or hay in winter. However, the knowledge on the relationship of grazing pattern, gut microbial activity, and feed efficiency of beef cows is limited. In this study, rumen fluid and fecal samples were collected from beef cows (n = 40) to identify the changes of short chain fatty acids (SCFAs) and microbial profiles between two different feeding regimes (fall deferred native grass grazing and winter silage feeding) and feed efficiency (efficient and inefficient) groups. The SCFA concentration was analyzed with gas chromatography and microbial populations were identified by using qPCR with specific primers for bacteria, archaea, protozoa, and fungi, respectively. The concentrations of acetate, propionate, and butyrate in the rumen were higher ($P < 0.01$) in fall grazing compared to winter feeding and rumen butyrate molar proportion was higher ($P < 0.05$) in low-residual feed intake (RFI) cows (efficient group) than high-RFI cows (inefficient group). However, except for the rumen butyrate proportion, there was no difference in both rumen and fecal SCFAs profiles between RFI groups. Rumen microbial profiles were not significantly different between fall grazing and winter feeding, but the populations of bacteria and archaea in feces were higher ($P < 0.001$) in winter feeding compared to fall grazing. Rumen protozoal population was higher ($P < 0.05$) in high-RFI group, while no difference between different RFI groups was observed for bacteria, archaea, and fungi. Furthermore, feeding management x RFI ($P < 0.05$) interactions only affected rumen and fecal protozoal population but not the other three microbial groups. These results suggest that different feeding managements with different diets and environments may affect rumen microbial population and fermentation which are closely associated with feed efficiency of beef cows. Further study investigating microbial changes during the whole grazing patterns throughout the year might contribute to comprehensive understanding on grazing cattle feed efficiency and the development of specific grazing strategies.

Key words: beef cows, grazing, feed efficiency, short chain fatty acid, microbial population

Predicting Composition, Intake, and Digestibility of Nutrients in Grazing Cattle Using Near Infrared Spectroscopy (NIRS) of Feces

J. F. Peters[1, 2], M. L. Swift[1,2], G. B. Penner[1], H. A. Lardner[1], T. A. McAllister[3] and G. O. Ribeiro[1]
[1]Department of Animal and Poultry Science, University of Saskatchewan, Saskatoon, SK
[2]Trouw Nutrition, Okotoks, AB
[3]Lethbridge Research and Development Centre, Agriculture and Agri-Food Canada, Lethbridge, AB
jenilee.peters@usask.ca

Abstract

The objective of this study was to determine the potential of near infrared-spectroscopy (NIRS) scanning of feces to predict fecal composition, diet digestibility, and intake in beef cattle fed high forage diets. Beef heifers were fed 12 different forage-based diets (> 95% forage dry matter basis) in 3 total collection digestibility studies, resulting in individual fecal samples and related spectra (n = 135) corresponding to nutrient, apparent total tract digestibility (aTTD), and intake data. Dried and ground fecal samples collected from each animal were scanned using a FOSS DS2500 scanning monochromator (FOSS, Eden Prairie, MN). Spectra were mathematically treated for scatter correction and modified partial least squares (MPLS) regression was performed to develop equations with acceptable accuracy and low error. The standard error of cross validation (SECV), and coefficient of determination for cross validation (R^2_{cv}) were used to evaluate the accuracy of calibration equations. Prediction equations were developed for fecal composition [organic matter (OM), nitrogen (N), neutral detergent fiber (NDF), acid detergent fiber (ADF), acid detergent lignin (ADL), undigestible NDF (uNDF), calcium (Ca), and phosphorus (P)], digestibility [DM, OM, NDFD, N], and intake [DM, OM, NDF, N, uNDF]. The calibrations for fecal OM, N, NDF, ADF, ADL, uNDF, Ca, P resulted in R^2_{CV} between 0.81 and 0.97 and SECV of 0.84, 0.60, 1.55, 0.47, 1.68, 0.15, and 0.06, respectively. Accuracy of equations predicting intake of DM, OM, N, NDF, ADL and uNDF resulted in R^2_{CV} values between 0.59 and 0.91, SECV for intake (kg/d) of 1.12, 1.10, 0.02, 0.69, 0.06, 0.24, respectively, SECV for intake % body weight (BW) between 0.00 and 0.16; and SECV for g/kg $BW^{0.75}$ between 0.60 and 0.86. Digestibility calibrations for DM, OM, NDF, and N resulted in R^2_{CV} ranging from 0.65 to 0.74 and SECV from 2.20 to 2.82. This study confirms the potential of NIRS to predict fecal chemical composition, as well as digestibility and intake of cattle fed high forage diets. Future steps include expansion and further validation of the calibration equations with spectra from varying grazing systems.

Key words: near-infrared spectroscopy, feces, fecal composition, apparent total tract digestibility, intake, beef, forage

In-vitro Dry Matter Digestibility, an Alternative Method for Silage Quality Assessment

C. Plett[1], N. Mclean[2], C. Lafrenière[3], S. Bittman[4], K. Ominski[1], and J.C. Plaizier[1]
[1]Department of Animal Science, University of Manitoba, Winnipeg, MB
[2]Dalhousie University, Truro, NS
[3]Université du Québec en Abitibi-Témiscamingue, Rouyn-Noranda, QC
[4]Agriculture and Agri-Food Canada, Agasiz, BC
kees.plaizier@umanitoba.ca

Abstract

Canadian dairy farmers rely heavily on silage as the primary feed ingredient and strive for optimal quality. Relative feed value (RFV) is a common index used within industry to compare forage quality and prices however it has been suggested that RFV may be a better reflection of plant maturity rather than digestibility. To improve the accuracy of silage quality assessment, an analytical technique which reflects the silages biological value is required, such as *in-vitro* dry matter digestibility (IVDMD). Between 2018-2020, samples of first cut legume, grass, and legume-grass mixed silages were collected from bunkers, piles, wrapped bales, and vertical silos across Canada (n=274). Subsamples were collected and sent for wet chemistry analysis of neutral detergent fiber (NDF) and acid detergent fiber (ADF) and values were used to calculate digestible dry matter (DDM) and dry matter intake (DMI) with the following equations, DDM= 88.9 – (0.779 x %ADF), DMI = 120/(%NDF). DDM and DMI were then used to calculate RFV with the following equation RFV= (DDM x DMI) / 1.29. IVDMD was conducted via the ANKOM DAISYII method with 48-hour incubation and NDF ending. RFV for grass, legume, and mixed silages at mid maturity were 96, 136, and 110 respectively while IVDMD was 55%, 66%, and 61% respectively. These results indicate RFV favours legume forages as grass and legume-mixed forages had considerably lower RFV while IVDMD values showed similar digestibility among the three forage types, indicating IVDMD is a more accurate comparison of silage quality than RFV as it represents the amount of dry matter available for digestibility. Incorporating IVDMD into the livestock and feed industries would allow for a more accurate comparison of forage types across regions not only for economic purposes but efficiency in allocating forages to specific group of animals.

Key words: silage quality, digestibility, relative feed value

Feeding Trial and Dairy Performance Evaluation Using Faba Bean Seeds: Effects of Raw and Steamed Pressured Faba Bean Seeds in Diets of High Producing Dairy Cows

María E. Rodríguez Espinosa[1], Dave Christensen[1], Rex Newkirk[1], Yongfeng Ai[2], Victor H. Guevara Oquendo[1], and Peiqiang Yu[1]

[1] Department of Animal and Poultry Science, University of Saskatchewan, Saskatoon, SK

[2] Department of Food and Bioproduct Sciences, University of Saskatchewan, Saskatoon, SK

peiqiang.yu@usask.ca

Abstract

The existence of alternative feed ingredients of good nutritional value is important for the livestock industry. Introducing new feeding options requires reliable information to prove beneficial or detrimental impacts on animal productivity. Hence, this study aimed to evaluate the use of raw and steamed Faba bean seeds (FBS) as an alternative for traditional feeding ingredients used in lactating dairy rations such as soybean and barley grain. A pilot study, using a batch reactor (Saskatoon Boiler Mfg.) for thermal hydrolysis, was performed to process Snowbird whole Faba bean seeds at 120 °C for 0, 7.5, 15, and 30 min. Eight high producing Holstein cows (2nd and 3rd lactation, 69 ± 15 days in milk, and 720 kg mean body weight) were fed in a total mixed ration (TMR) with 10% FBS inclusion for 120 days. Data were analyzed using the MIXED procedure of SAS 9.4. A Latin Square (4x4) design was used with fix effect (treatment), and random effect models (cows and periods). Polynomial contrasts were used to evaluate the effects of treatments with significance declared at $P<0.05$. Results showed a response to the TMR starch intake (linear $P=0.01$) increasing from 5.46 kg/d with raw seeds (TMR_0) up to 5.97 kg/d with seeds processed for 30 min (TMR_30). The average milk yield and fat content for all the diets were 39.4 kg/cow/day and 3.86 %, respectively ($P>0.05$). Feed efficiency reflected by the relation between energy corrected milk and dry matter intake (ECM/DMI) decreased as processing times increased (linear $P=0.02$) from 1.63 with TMR_0 to 1.52 with TMR_30. Milk urea nitrogen (MUN) decreased from 12.18 mg/dl with TMR_0 to 11.10 mg/dl with TMR_30 (cubic $P=0.04$). Based on the current findings, a potential use for FBS in ruminant diets is presumed, as no negative effects were observed on the production performance of high producing dairy cows fed total mix rations (TMR) with 10% inclusion. Thus, we concluded that raw or heated FBS could be a promising feed alternative for the dairy industry as partial replacement for soybean meal and barley grain. Future studies are still needed to sustain the reported results. Optimal processing method, processing time, and level of inclusion of Faba bean seeds remain broad fields of research in ruminant systems.

Key words: faba bean; steam-pressured toasting; dairy cows.

MARK YOUR CALENDARS

The Animal Nutrition Conference of Canada (**ANCC**) 2023 will take place **May 9-11, 2023** at Le Centre Sheraton Montreal Hotel.

MARQUEZ VOS CALENDRIERS

Le Colloque de nutrition animale du Canada (**CNAC**) 2023 aura lieu les **9 au 11 mai 2023** à l'hôtel Le Centre Sheraton Montréal.

ANCC

ANIMAL NUTRITION CONFERENCE OF CANADA

CNAC

COLLOQUE DE NUTRITION ANIMALE DU CANADA

Thank you to our sponsors / Merci à tous les commanditaires

Platinum / Platine

KEMIN

Pre-conference / Pré-colloque

ADM
Unlocking Nature. Enriching Life.

JM Bell Memorial Lecture / Conférence honneur de JM Bell

University of Saskatchewan

Gold / Or

Jefo

O&T Farms – Naturally Better

Pestell Nutrition

SGS – One Stop Shop for Your Testing Needs

Silver / Argent

Cbs Bio Platforms

Alltech

DSM – Bright Science. Brighter Living.

Bronze / Bronze

Agribution

Canola Mazing – Discover the amazing value of canola meal

Probiotech

Timab magnesium

Lallemand – Lallemand Animal Nutrition

Robinson Bioproducts Inc

O&T Farms – Naturally Better

Halchemix – research proven nutrition

Protekta

More Than Just Feed

WLT Distributors Inc

dcl

PMT Inc.

Supporting / Parrainage

NSC Minerals

Adisseo – A Bluestar Company

new-life mills – Feeding Your Future – A division of Parrish & Heimbecker, Limited

Romer – Romer Labs

Agribution

Huvepharma

APC

Floradale Feed Mill Limited

Co-op

ANS – Agri-Nutrient Solutions

Masterfeeds – an Alltech company

ABVista

ANCC – 2022 Animal Nutrition Conference of Canada

CNAC – 2022 Colloque de Nutrition Animale du Canada

animalnutritionconference.ca

Animal Nutrition Society of Canada
150 Metcalfe Street, Suite 1301
Ottawa, Ontario K2P 1P1
Canada

ISBN 978-1-7138-5548-4

CURRAN ASSOCIATES INC.
proceedings.com

Majan International Conference (MIC 2018)

Promoting Entrepreneurship and Technological Skills: National Needs, Global Trends

Muscat, Oman
19 March 2018

Majan College
University College